Multiple Roles for Landscape Ecology in Future Farming Systems

Multiple Roles for Landscape Ecology in Future Farming Systems

Editors

Diane Pearson
Richard Aspinall
Julian Gorman

MDPI • Basel • Beijing • Wuhan • Barcelona • Belgrade • Manchester • Tokyo • Cluj • Tianjin

Editors

Diane Pearson	Richard Aspinall	Julian Gorman
Massey University	James Hutton Institute	Massey University
New Zealand	UK	New Zealand

Editorial Office
MDPI
St. Alban-Anlage 66
4052 Basel, Switzerland

This is a reprint of articles from the Special Issue published online in the open access journal *Land* (ISSN 2073-445X) (available at: https://www.mdpi.com/journal/land/special_issues/landscape_farmingsystems).

For citation purposes, cite each article independently as indicated on the article page online and as indicated below:

LastName, A.A.; LastName, B.B.; LastName, C.C. Article Title. *Journal Name* **Year**, *Volume Number*, Page Range.

ISBN 978-3-0365-4049-8 (Hbk)
ISBN 978-3-0365-4050-4 (PDF)

Contents

About the Editors vii

Preface to "Multiple Roles for Landscape Ecology in Future Farming Systems" ix

Diane Pearson, Julian Gorman and Richard Aspinall
Multiple Roles for Landscape Ecology in Future Farming Systems: An Editorial Overview
Reprinted from: *Land* **2022**, *11*, 288, doi:10.3390/land11020288 1

Diane Pearson
Key Roles for Landscape Ecology in Transformative Agriculture Using Aotearoa—New Zealand as a Case Example
Reprinted from: *Land* **2020**, *9*, 146, doi:10.3390/land9050146 7

Duy X. Tran, Diane Pearson, Alan Palmer and David Gray
Developing a Landscape Design Approach for the Sustainable Land Management of Hill Country Farms in New Zealand
Reprinted from: *Land* **2020**, *9*, 185, doi:10.3390/land9060185 33

Uta Schirpke, Erich Tasser, Georg Leitinger and Ulrike Tappeiner
Using the Ecosystem Services Concept to Assess Transformation of Agricultural Landscapes in the European Alps
Reprinted from: *Land* **2022**, *11*, 49, doi:10.3390/land11010049 63

Richard Aspinall, Michele Staiano and Diane Pearson
Emergent Properties of Land Systems: Nonlinear Dynamics of Scottish Farming Systems from 1867 to 2020
Reprinted from: *Land* **2021**, *10*, 1172, doi:10.3390/land10111172 77

Julian Gorman, Diane Pearson and Penelope Wurm
Old Ways, New Ways—Scaling Up from Customary Use of Plant Products to Commercial Harvest Taking a Multifunctional, Landscape Approach
Reprinted from: *Land* **2020**, *9*, 171, doi:10.3390/land9050171 105

André Eduardo Biscaia Lacerda, Ana Lúcia Hanisch and Evelyn Roberta Nimmo
Leveraging Traditional Agroforestry Practices to Support Sustainable and Agrobiodiverse Landscapes in Southern Brazil
Reprinted from: *Land* **2020**, *9*, 176, doi:10.3390/land9060176 125

Caroline Wentling, Felipe S. Campos, João David and Pedro Cabral
Pollination Potential in Portugal: Leveraging an Ecosystem Service for Sustainable Agricultural Productivity
Reprinted from: *Land* **2021**, *10*, 431, doi:10.3390/land10040431 145

Nándor Csikós and Péter Szilassi
Modelling the Impacts of Habitat Changes on the Population Density of Eurasian Skylark (*Alauda arvensis*) Based on Its Landscape Preferences
Reprinted from: *Land* **2021**, *10*, 306, doi:10.3390/land10030306 159

Vasilii Erokhin, Tianming Gao and Anna Ivolga
Structural Variations in the Composition of Land Funds at Regional Scales across Russia
Reprinted from: *Land* **2020**, *9*, 201, doi:10.3390/land9060201 177

About the Editors

Diane Pearson is Professor of Environmental Management at the Farmed Landscape Research Centre within the School of Agriculture and Environment at Massey University. Diane describes herself as a landscape ecologist who is passionate about sustainable landscape management. She has over 20 years' experience undertaking studies that investigate the implications of land system change for landscapes and livelihoods in England, Scotland, Australia, and New Zealand, and the focus of her recent research looks at how landscape ecology can be used to plan and design future sustainable landscapes.

Richard Aspinall is an Honorary Research Fellow at the James Hutton Institute, Aberdeen, Scotland. Previously, he was Chief Executive of the Macaulay Land Use Research Institute (now the James Hutton Institute). He is founding editor of the Journal of Land Use Science. His research addresses land as a coupled human–natural system.

Julian Gorman works as a Senior Lecturer in the Planning Program of the School for Environmental Planning at Massey University, Palmerston North, New Zealand. Prior to this, between 1999 and 2019, he worked as a lecturer, researcher, and practitioner at the Research Institute of Environment and Livelihoods at Charles Darwin University, Darwin, Australia, and with the Caring for Country section of the Northern Land Council (who assist in Aboriginal land and sea management). His research focuses on wildlife-based enterprise development, and he has worked closely with Aboriginal people and their communities across the Northern Territory. His professional interests relate to the maintenance of intact and functional cultural landscapes and ecosystems, resilient communities, sustainable development, and valued Indigenous knowledge and practice.

Preface to "Multiple Roles for Landscape Ecology in Future Farming Systems"

As the world's population continues to grow and we face increasing challenges to meet the basic food requirements of this population against the backdrop of global environmental issues such as climate change, the future of farming systems is of utmost concern.

Farming systems must be intensified to be able to provide for the growing global population whilst continuing to generate livelihoods for the farmers that manage them in addition to supporting other important ecosystem services that can help to mitigate the impacts of climate change, reduce biodiversity loss, and maintain water quality.

To achieve this, future farming systems will need to be informed by a wide range of scientific and technological developments to rise to the challenges that lie ahead. Landscape ecology is well positioned to utilize its transdisciplinary focus to present some guidance on how to plan, design, and modify landscapes to achieve sustainable landscapes and livelihoods. It can help to provide appropriate tools, approaches, and frameworks that can be applied to better understand the dynamics of the landscape whilst also facilitating the development of new knowledge and improved advice that can bring about change.

Landscape ecology has the potential to provide scientific evidence and theory that can assist farmers in the creation of future farming systems that meet societal needs whilst also responding to environmental challenges. This book is a result of the Special Issue of Land entitled "Multiple Roles for Landscape Ecology in Future Farming Systems", dedicated to the symposium of the same name held at the 10th IALE World Congress in Milan in 2019, at which some of the many areas in which landscape ecology can assist in the development of future sustainable farming systems were discussed. Roles for landscape ecology range from creating a better understanding of how habitat change impacts on individual species, utilizing concepts of ecosystem services to quantify the impacts of transforming agricultural landscapes, incorporating indigenous and cultural knowledge into agricultural production, through to helping to design multifunctional landscapes. This book also throws down the gauntlet to landscape ecologists across the world to apply their knowledge and experience of this useful meta-discipline to help farmers across all areas of the globe—be they small or large in scale—to play their vital part in ensuring both food security and environmental sustainability.

Diane Pearson, Richard Aspinall, Julian Gorman
Editors

land

MDPI

Editorial

Multiple Roles for Landscape Ecology in Future Farming Systems: An Editorial Overview

Diane Pearson [1,*], Julian Gorman [2] and Richard Aspinall [3]

1 Farmed Landscape Research Centre, School of Agriculture and Environment, College of Sciences, Massey University, Palmerston North 4442, New Zealand
2 School of People, Environment and Planning, College of Social Sciences, Massey University, Palmerston North 4442, New Zealand; j.gorman@massey.ac.nz
3 James Hutton Institute, Craigiebuckler, Aberdeen AB15 8QH, UK; rjaspinall10@gmail.com
* Correspondence: d.pearson@massey.ac.nz

Citation: Pearson, D.; Gorman, J.; Aspinall, R. Multiple Roles for Landscape Ecology in Future Farming Systems: An Editorial Overview. *Land* **2022**, *11*, 288. https://doi.org/10.3390/land11020288

Received: 30 January 2022
Accepted: 8 February 2022
Published: 14 February 2022

1. Introduction—Challenges Facing Future Farming Systems

Farming faces new and urgent pressures, with an array of mounting social, environmental and economic challenges, and growing public and political expectations for improved stewardship of natural resources. Responses demanded of farming include changes that reduce greenhouse gas emissions, improve environment quality, restore and increase biodiversity, feed a growing global population, and support national economies, all while providing livelihoods for farmers themselves. Further, there is an immediacy and urgency to respond to an array of challenges as multiple planetary boundaries are exceeded or approached [1] while tackling important sustainable development goals which largely rely on sustainable future food production and livelihoods at local, regional and global scales.

Future farming systems need to respond to a recognition that a changing climate is impacting the capacity of farming and forestry across nations and regions [2]. This has implications for production and supply of food and fibre as the century progresses, with both flooding and drought events increasing in frequency [3], and water quality and quantity becoming increasingly problematic. We are also in an era where biodiversity and ecosystem services provided by the natural environment are declining, as land clearing continues to open up new land for cultivation and land cover change occurs associated with intensification of land use. Threats to biosecurity are proliferated with the movement of people and products, and are exacerbated by the implications of climate change. Market and consumer influences and preferences are changing as people become more aware of animal welfare issues, concerns about biodiversity loss and the need for sustainable production. Compounding this is a growing public awareness and dissatisfaction with the environmental impacts that result from high input and intensified agricultural production, and an increasing preference for products that identify as sustainable that are produced by businesses with environmental credentials. Agriculture also needs to respond to heightened concerns around the relationships between animal protein and human health issues, whilst recognising trends towards increased plant-based protein and flexitarian diets.

These pressures and demands challenge current patterns and practices of land use worldwide, and require development of sustainable agriculture and land-use practices that can address the climate, biodiversity, population, and water, energy, and food security issues [4,5]. Deliberate and directed change in land systems and practice requires clear and careful thought and guidance based on the best available evidence. Although farming continuously responds to market and other signals, it is perhaps less responsive to signals from factors typically considered as externalities, such as costs to natural capital. A growing body of information from a diverse set of disciplines and perspectives is being generated and has the potential to inform choices and decisions for future farming. Knowledge of the greenhouse gases associated with different farming types and activities have been established, the impacts of land clearing and intensification on biodiversity are well-known, although the specific details are contested, and the impacts of land management on erosion and water quality and quantity are becoming increasingly recognised. The task of assembling, synthesising and integrating these multiple evidence bases, objectives and priorities to make informed decisions about specific land, land uses and practice change, requires support.

Although this support will come from multiple sources, landscape ecology can play a particularly significant role. The interconnected variety of issues currently faced requires understanding based on "whole system" approaches. A number of authors have discussed the relationships, overlap and complementary perspectives of land systems science, landscape ecology, and political ecology, as well as their links to sustainability science [6–9]. All are interdisciplinary in scope and approach, recognise and address land as a coupled human–environment system, and all have a focus on land system dynamics. Landscape ecology also has a well-developed set of tools and methodologies for analysis across multiple spatial, temporal, and organisational scales. The holistic and interdisciplinary nature of landscape ecology positions it not only to address the specific human and environmental challenges facing agriculture, but also to offer advice on how to plan, design, modify and develop understanding for new land-use patterns and farming systems in specific geographic landscapes that can function with the best environmental, economic and social outcomes in mind [6]. Landscape ecology can provide appropriate tools, approaches and frameworks that can facilitate the action, knowledge and advice required to help work towards the creation of future farming systems that meet societal needs, respond to the environmental challenges and that can sit within sustainable landscapes and societies. However, it is important that these can be practically applied and are seen to be relevant for policy makers and farmers to be able to implement.

2. Contributions of these Special Issue Papers

The papers in this Special Issue explore the potential for these contributions and discuss the evolving roles for landscape ecology in future agricultural systems. Papers individually focus on specific parts of the challenges facing farming for the future. Biodiversity and its management and role in agricultural systems is examined in studies on the potential for improving land management to support pollination across Portugal, using approaches based on land cover data and modelling tools [10] and modelling impacts of habitat changes on Skylarks in Hungary [11]. Changes in land funds at regional scales in Russia [12] and exploration of the benefits of agroforestry systems in southern Brazil [13] address habitats and land covers associated with land uses, while other papers report on the roles and potentials for a new method of scaling from customary harvests at local scales to wider markets [14]. The concept of ecosystem services is used to assess the impacts associated with the transformation of agricultural landscapes in the European Alps [15], and the use of geodesign and the theoretical strengths of landscape ecology are explored to help to design multifunctional landscapes in New Zealand [16]. The ability to monitor and determine dynamics in farming systems is investigated for the Scottish pastoral and arable industries between 1867 and 2020 [17], illustrating both endogenous and exogenous drivers of change which provide important knowledge for managing future farming systems and understanding how they will respond to future stimuli. Although each of these case studies use a specific geographic location, they also establish principles that are of general application and provide useful foundational knowledge to input into the management of future farming systems that offer more sustainable solutions to food production. Pearson [18] builds on the theoretical strengths of landscape ecology in transformative agriculture, using the environments and agricultural activities in Aotearoa, New Zealand as a case study in linking theory and practice within farming, to cocreate future farming systems. The ability for farmers to have some control over the destiny of the systems in which they operate will be key to maintaining resilient, functioning farm systems that respond to global environmental and population challenges.

Taken together, the papers address a variety of questions about the nature of future farming systems, and the changes necessary to achieve those future systems, as well as the utility and capacity of landscape ecology as an approach to integrate and synthesize scientific information for effective regional and global landscape management. Since it is important that this integration and synthesis aims towards practical outcomes that create sustainable landscapes and futures for environments and people, landscape ecology needs to demonstrate its relevance and develop in basic, strategic and applied directions, and in participatory codesign of land management practices at various relevant scales.

Issues raised and discussed in these papers include: How can landscape ecology concepts be more practically applied to assist farmers and policy makers in facilitating sustainable land management decisions and planning and designing future farm landscapes? How can landscape ecology assist in the establishment of effective transdisciplinary projects that focus on the codevelopment of strategies to identify and address problems? How can knowledge and cultural connections and values that indigenous people associate with landscapes be incorporated into more western production systems for more sustainable outcomes? This includes exploring and better understanding how we determine the potential for diversification of agricultural production systems towards alternative practices, which integrate with customary knowledge and practice towards the growth and harvest of novel bushfoods and capitalize on organic practices which are nondestructive. Other questions raised in the Special Issue include: What are current potentials for geodesign and geospatial technology to propose and evaluate alternative patterns of farming land uses and create multifunctional landscapes? What lessons does a long view of land-use change in agriculture provide for understanding future change management? What are the implications of changes to the intensity associated with rural land use? What role do pollinators play in production and how can we incorporate the ecosystem services pollinators play into farm production? How can we improve agrobiodiversity by incorporating local ecological knowledge? What are the implications of losing productive land to urban and industrial use? How do we maintain important ecosystem services within future farming systems?

Key roles for landscape ecology that emerge from the papers in this Special Issue include the importance of landscape ecology in assisting in the design and creation of multifunctional landscapes that preserve important biodiversity and ecosystem services, and the ways in which this can help with the maintenance and preservation of vital landscape functioning and processes to ensure sustainable production into the future. The papers also demonstrate that landscape ecology can help with the development and application of relevant monitoring and evaluation tools to assist in quantifying the status and condition of farmland and the species that reside within it. Landscape ecology is also seen to have an important role in the creation of bottom-up approaches that consider farmers' and other stakeholders' world views and perspectives, and facilitate stewardship of the landscape by embracing cultural connections to landscapes and utilising indigenous and other forms of traditional knowledge. In doing this, landscape ecology has an important role to play in linking science and practice and in the coproduction of knowledge for sustainable futures. As a metadiscipline, landscape ecology can also have an important part to play in the integration of knowledge and approaches from a variety of disciplines and in solving agriculture-related environmental problems, thus facilitating transdisciplinary approaches for transformative outcomes. Further, by considering both socioeconomic as well as ecological considerations, landscape ecology can help to secure not only sustainable environmental outcomes but also, very importantly, sustainable business ones too.

3. Conclusions

Overall, the papers in this Special Issue highlight the challenges that farming systems and rural communities face whilst throwing down the gauntlet for a future landscape ecological research agenda that can support farming and farmers through important transformative change, which will help to create future sustainable farming systems that can feed a growing population. It is hoped that this Special Issue will: inspire landscape ecologists to explore theory and practical tools that can assist in the planning, design, modification and development of new farming landscapes with the best environmental, economic and social outcomes in mind; contribute towards developing land systems and land management practices for specific landscapes that meet the goals of increased nutritious food production in the face of market and climatic variability whilst reducing environmental impacts and enhancing natural capital; and assist in driving and supporting the transformative changes required to the socioeconomic and environmental systems of rural areas and food production for the future.

By exploring these issues through developing research agendas, landscape ecologists can demonstrate their importance in providing the scientific guidance to ensure farm systems of the future meet environmental and production targets. By considering the farm within the broader landscape mosaic in which it sits, and by treating the farm as an important, coupled human–environment system, recognising important drivers of change and acknowledging the farmer/landowner as an important participant in future design making, it is possible to help farmers and policy makers to be able to address economic requirements whilst preserving important ecosystem services that ensure sustainable landscapes and livelihoods into the future.

Author Contributions: Conceptualization, D.P., J.G. and R.A.; writing—original draft preparation, D.P., J.G. and R.A.; writing—review and editing, D.P., J.G. and R.A. All authors have read and agreed to the published version of the manuscript.

Funding: This research received no external funding.

Acknowledgments: We are grateful to the academic editors and reviewers for the collaboration and support necessary to ensure the academic excellence of this Special Issue. We would also like to acknowledge the 10th IALE World Congress, Milan 2019 from which the authors drew inspiration for this Special Issue from their symposium of the same name. Thanks also go to Massey University for supporting the ongoing collaboration of the authors.

Conflicts of Interest: The authors declare no conflict of interest.

References

1. Rockström, J.; Steffen, W.; Noone, K.; Persson, A.; Chapin, F.S., III; Lambin, E.F.; Lenton, T.M.; Scheffer, M.; Folke, C.; Schellnhuber, H.J.; et al. A safe operating space for humanity. *Nature* **2009**, *461*, 472–475. [CrossRef]
2. IPCC. *Climate Change and Land. Summary for Policymakers*; WMO: Geneva, Switzerland; UNEP: Nairobi, Kenya, 2019; p. 43.
3. EASAC. *Extreme Weather Events in Europe*; EASAC: Halley, Germany, 2018.
4. Beddington, J. Food security: Contributions from science to a new and greener revolution. *Philos. Trans. R. Soc. B-Biol. Sci.* **2010**, *365*, 61–71. [CrossRef] [PubMed]
5. UN General Assembly. *Transforming our World: The 2030 Agenda for Sustainable Development, 21 October 2015, A/RES/70/1*; UN General Assembly: New York, NY, USA, 2015.
6. Mayer, A.L.; Buma, B.; Davis, A.; Gagne, S.A.; Loudermilk, E.L.; Scheller, R.M.; Schmiegelow, F.K.A.; Wiersma, Y.F.; Franklin, J. How Landscape Ecology informs Global Land-Change Science and Policy. *Bioscience* **2016**, *66*, 458–469. [CrossRef]
7. Roy Chowdhury, R.; Turner, B.L. The parallel trajectories and increasing integration of landscape ecology and land system science. *J. Land Use Sci.* **2019**, *14*, 135–154. [CrossRef]
8. Turner, B.L., II; Robbins, P. Land-change science and Political Ecology: Similarities, differences, and implications for Sustainability Science. *Annu. Rev. Environ. Resour.* **2008**, *33*, 295–316. [CrossRef]
9. Wu, J. Linking landscape, land system and design approaches to achieve sustainability. *J. Land Use Sci.* **2019**, *14*, 173–189. [CrossRef]
10. Wentling, C.; Campos, F.S.; David, J.; Cabral, P. Pollination Potential in Portugal: Leveraging an Ecosystem Service for Sustainable Agricultural Productivity. *Land* **2021**, *10*, 431. [CrossRef]
11. Csikós, N.; Szilassi, P. Modelling the Impacts of Habitat Changes on the Population Density of Eurasian Skylark (*Alauda arvensis*) Based on Its Landscape Preferences. *Land* **2021**, *10*, 306. [CrossRef]
12. Erokhin, V.; Gao, T.; Ivolga, A. Structural Variations in the Composition of Land Funds at Regional Scales across Russia. *Land* **2020**, *9*, 201. [CrossRef]
13. Lacerda, A.E.B.; Hanisch, A.L.; Nimmo, E.R. Leveraging Traditional Agroforestry Practices to Support Sustainable and Agrobiodiverse Landscapes in Southern Brazil. *Land* **2020**, *9*, 176. [CrossRef]

14. Gorman, J.; Pearson, D.; Wurm, P. Old Ways, New Ways—Scaling Up from Customary Use of Plant Products to Commercial Harvest Taking a Multifunctional, Landscape Approach. *Land* **2020**, *9*, 171. [CrossRef]
15. Schirpke, U.; Tasser, E.; Leitinger, G.; Tappeiner, U. Using the Ecosystem Services Concept to Assess Transformation of Agricultural Landscapes in the European Alps. *Land* **2022**, *11*, 49. [CrossRef]
16. Tran, D.X.; Pearson, D.; Palmer, A.; Gray, D. Developing a Landscape Design Approach for the Sustainable Land Management of Hill Country Farms in New Zealand. *Land* **2020**, *9*, 185. [CrossRef]
17. Aspinall, R.; Staiano, M.; Pearson, D. Emergent Properties of Land Systems: Nonlinear Dynamics of Scottish Farming Systems from 1867 to 2020. *Land* **2021**, *10*, 1172. [CrossRef]
18. Pearson, D. Key Roles for Landscape Ecology in Transformative Agriculture Using Aotearoa—New Zealand as a Case Example. *Land* **2020**, *9*, 146. [CrossRef]

land

MDPI

Review

Key Roles for Landscape Ecology in Transformative Agriculture Using Aotearoa—New Zealand as a Case Example

Diane Pearson

Farmed Landscapes Research Centre, School of Agriculture and Environment, College of Sciences, Massey University, Palmerston North 4474, New Zealand; d.pearson@massey.ac.nz; Tel.: +64-6-951-7837

Received: 5 April 2020; Accepted: 8 May 2020; Published: 11 May 2020

Abstract: Aotearoa—New Zealand (NZ) is internationally renowned for picturesque landscapes and agricultural products. Agricultural intensification has been economically beneficial to NZ but has implications for its clean green image. Contaminated waterways, high carbon emissions, and extensive soil erosion demonstrate the downside of high stocking rates and land clearing. Transformative farming systems are required to address the challenge of balancing production with the environment. Whilst navigating through the process of change, farmers need to be supported to make informed decisions at the farm and landscape scale. Landscape ecology (LE) is ideally positioned to inform the development of future farming landscapes and provide a scientific context to the criteria against which land-related information can be evaluated. However, to do this effectively, LE needs to demonstrate that it can link theory with practice. Using NZ as a case example, this paper discusses the key roles for LE in future farming systems. It looks at the way LE can help quantify the state of the landscape, provide support towards the co-creation of alternative futures, and assist with the inclusion of land-related information into design and planning to ensure mitigation and adaption responses assist in the transformation of farming systems for sustainable outcomes.

Keywords: future farming systems; sustainable landscapes; landscape planning; environmental challenges; transdisciplinary

1. Introduction

Globally, agriculture finds itself on the brink of transformation. Agriculture needs to respond with immediate action to the current global environmental challenges like climate change and issues of contamination from intensively managed land. It also needs to be capable of feeding an increasing world population [1]. Reformation in terms of agricultural practice is also on the cards due to the increasing availability of new technological advancements applicable to farming systems and the changes in consumer preferences towards food products that have greater ethical and environmental credentials [2]. These challenges and pressures will likely result in a contemporary revolution for agriculture in which farming practices and systems move to a different paradigm. The imminent implications and consequences of which will mean landscapes, and the communities residing in them, will face outcomes and situations that have never been experienced before.

Aotearoa-New Zealand (NZ) is a nation that relies heavily on its tourism as well as its agricultural exports (beef, lamb, and dairy products). To balance the demands from both sectors, NZ is becoming increasingly aware of its need to change some farming practices and demonstrate its sustainable farming systems on a global platform. NZ prides itself on its picturesque landscapes, which are a drawcard for many of its international visitors. During 2019, international visitor arrivals to NZ were in the order of approximately 3.9 million [3]. With a population of only approximately 4.9 million [4], tourist numbers considerably bolster the spending capacity of residents and therefore have a significant economic impact in NZ. To attract tourists, NZ is very proficient at marketing the connection to nature that is part of its production cycle, especially that associated with its creation of meat products (e.g., the current "Taste Pure Nature" Campaign) [5]. This 'natural' image has been shown to have considerable market value overseas [6]. With heightened environmental concerns coming to the forefront of global consumer behaviour, NZ is grappling with the challenges of maintaining its status as a great primary producing nation and one of landscapes that are environmentally healthy. As a result, NZ wants to position itself as a leader in the global agricultural sector by demonstrating 'gold standard' environmental practice. However, knowing exactly what to do to have the best outcomes is currently difficult for farmers who are presented with lots of reasons to change their practices but not much information on the best way to change to have the biggest environmental impact whilst maintaining a viable livelihood.

Landscape ecology (LE) has a long association with agriculture and farming systems through its early focus on landscape pattern, habitat fragmentation, and the impact on biodiversity caused by productive and human-modified landscapes [7,8] However, the focus of LE research to date, especially in NZ, has largely been one of considering LE in terms of biodiversity loss, habitat fragmentation, restoration of natural systems, and trying to improve degraded landscapes and enhancing conservation outcomes at the landscape scale [9–11]. There is little evidence to date to demonstrate the potential LE has in NZ to work alongside farmers towards future farm system design. There appears to be untapped potential for a role in creating both sustainable production systems and sustainable local landscapes. This means appropriately advising farmers on how to improve production outcomes for their farms as well as environmental ones.

The holistic and transdisciplinary nature of landscape ecological science positions it well to not only address the environmental challenges facing the broader landscape but also to contribute to production issues by providing advice on how to plan, design, modify and develop new landscapes (both at the farm and wider landscape or catchment scale) with best environmental, and livelihood outcomes in mind. However, despite LE theory having been developed and extensively discussed over the last 20–30 years, it has not had the on-ground impact that it perhaps would have been expected to have [12].

This paper builds on the work of Opdam et al. [12] that makes recommendations for how LE can become 'a science for action' at the local landscape scale. Their narrative intended to "reinvigorate and refocus the aim of landscape ecology towards cooperative knowledge production in order to better integrate landscape science into local practice and to adjust scientific methods to better support actions at the local level" [12] (pg. 1444). Given that individual farm systems operate at a local level but collectively have significant broad landscape impacts, this call for action is particularly important to guide the research agenda for LE in the context of farm systems. This paper also expands on ideas developed 20 years ago by Meurk and Swaffield [13] around sustainable agriculture in New Zealand. In doing this, the paper reflects on the position of LE in NZ in terms of being able to effectively demonstrate its practicality as an applied science with local impact and direct relevance to NZ farmers and the challenges that that farmers face around landscape sustainability in the 21st century.

This paper uses NZ as a case example, to explore how the science of LE can capitalise on its potential to work closely with, and appropriately inform, an agricultural sector facing difficult challenges and radical change. The most significant challenge for agriculture being an ability to increase food production in the face of market and climatic variability whilst reducing environmental impact. The paper first provides the context for agricultural change by describing the main issues facing agriculture globally and in NZ. It then goes on to identify some key roles for LE, and recommends some priorities that should be addressed to help make LE science more useful for farmers. Finally, whilst reflecting on progress to date, the paper presents advice that can contribute to a LE research agenda that can support farming systems and rural communities through the transformative change that is required to address the challenges they face.

2. The Need for Agricultural Change

2.1. The Global Situation

McCalla [14] stated that agriculture globally would face three challenges in the 21st century. These challenges are: i) food security, ii) poverty reduction, and iii) sustainable natural resource management. However, facing challenges is not something new for the agriculture sector. The challenge of increasing production has proved to be something that agriculture has been able to respond to. In the past, the ability to keep pace with a growing population has included expansion of production onto 'new' land, technological advancements, and policy incentives [14]. Agricultural production more than tripled between 1960 and 2015 as a result of these responses [15]. However, 21st-century challenges for agriculture look likely to be more difficult to address than the ones that came before because individual issues are compounded. Challenges now include addressing wicked problems like being able to cope with an expected global population increase to almost 10 billion people by 2050 [15,16]; the impacts of climate change; and economic growth in low and middle-income countries that is likely to facilitate the transition in these countries to a diet consisting of more meat, fruit, and vegetables [2,15]. All these challenges will put increasing pressure on the world's natural resource base [15]. The ability to adequately address the issues associated with food security and environmental concern will require the creation of sustainable farming systems that can double their production whilst maintaining environmental integrity [14]. Farming systems will also need to have the ability to adapt to and mitigate impacts of climatic variability of unknown quantity and severity. For this to happen, there is a lot of pressure on farmers to make the right choices for their businesses, in terms of both production and the environment [2]. As custodians of the land, the role of farmers and their land use will become increasingly more important as they attempt to increase production whilst curtailing their own environmental footprint [2]. Farmers will also have to work for the greater good by putting in place strategies around carbon sequestration that can help to mitigate carbon emissions at both the global and national level [2]. Ensuring that farmers get the best advice possible to assist them to make the optimum decisions will be crucial for the sustainability of their livelihoods and the environment.

2.2. The Situation in Aotearoa—New Zealand

Aotearoa - New Zealand (NZ) covers a landmass of 268, 021 km^2 (26.8 million hectares) [17]. It comprises of 3 main islands – North Island, South Island, and Stewart Islands plus some smaller islands including the Chatham Islands. The main islands are located geographically between a latitude of 34 and 47S and a longitude of 166 to 179E and from the very top to bottom NZ covers approximately 1600 km and is 450 km wide at its widest point [18]. Most of the population of NZ is focused around major population centres (e.g., Auckland, the biggest city in NZ, had a population of 1.657 million in 2017, and the next biggest and capital city of Wellington had just under 500,000), leaving approximately 78% of the total area of NZ described as having "no inhabitants recorded per square kilometer" [19], demonstrating the rural and remote nature of much of the country.

The landscapes of NZ have many famous tourist destinations. Tourism currently contributes approximately 20% of foreign exchange earnings [20], making it NZ's biggest export industry. The financial injection tourism provides to the economy consists of a direct annual contribution to GDP of $16.2 billion, or 5.8% of GDP [20], and a further indirect contribution of $11.2 billion, another 4% of GDP [19]. Despite being worth less than tourism, agriculture is also hugely important to the NZ economy. The benefit of agriculture to the NZ economy consists of an annual contribution of around $11 billion, approximately 4% of gross domestic product (GDP) [21], with the dairy sector contributing 3.5% or $7.8 billion to NZ's total GDP [22].

Farming dominates the land use of NZ, with 45.3% of the total land area of NZ (i.e., 12.1 million ha) being utilised for agriculture or horticulture and 7.8 million ha of this farmland being grassland [23] (see Table 1 summarising the proportions of the major land uses in NZ). Livestock farming dominates agricultural land use, with sheep and beef farming occupying approximately 32% of land and dairying nearly 10% [20]. As of June 2019, NZ farmed 10.3 million cattle (3.9 million beef cattle and 6.4 million dairy cattle) and NZ farmed 26.7 million sheep [24]. In comparison, less than 0.5% of NZ's total land area is under fruit and berry production and 0.3% was used for growing vegetables [23].

Table 1. The proportion of the New Zealand (NZ) total land area occupied by major land use types (This work is based on/includes Stats NZ's data which are licensed by Stats NZ for reuse under the Creative Commons Attribution 4.0 International licence [23]).

Major Land Use Type	Proportion of NZ Total Land Area
Agriculture	45%
Conservation	32%
Forestry	8%
Urban	0.8%

Compared with neighbouring Australia, NZ has a relatively recent history of human settlement. The Polynesians who developed the Māori culture first settled NZ about 750 years ago, with European settlers arriving nearly 200 years ago. Impacts on biodiversity and land cover change started with early Māori deforestation but the expansion of the European settler population had the most significant impact on land use and land cover change, as they cleared the landscape to make way for more intensive agricultural production. By 2016, 55,473 farm holdings were recorded in NZ with an average size of 252 ha [25]. Since European settlement, the rural landscape has largely been dominated by the 'family farm' [26].

Despite NZ farming mostly being undertaken by farmers of European settler origin, the contribution of Māori agriculture to the farming economy is significant. About 5% of the total land area of NZ is Māori land. Of this, a total of 49.5% is administered by *ahu whenua* trusts (a land trust that promotes the use and administration of one or more Māori land blocks) [27] and 13.7% by Māori incorporations [28]. Most of the incorporations and a big share of the *ahu whenua* trusts have interests in agriculture. In 2016, 450,600 ha of land was categorised as being Māori farms used for primary production and of this, nearly half the total was under grassland or pasture (218,000 ha), followed by forest plantation (at 110,400 ha), bush and scrub (at 75,400 ha), and horticulture (at 2700 ha) [29]. Agriculture is estimated to account for about 20% of Māori authority enterprises [29]. In early 2000 more than 15% of NZ sheep and beef exports came from Māori farming interests [28]. The Treaty of Waitangi *Te Tiriti o Waitangi* signed in 1840 between the British Crown and Māori chiefs also plays a significant role in influencing land use and land rights in NZ [30].

The policy context for agriculture in NZ is presented in Swaffield [31]. In terms of land use change, significant events to drive change over the last 50 years include Britain entering the European Union in 1973, which affected the market for meat, wool, and dairy produce, and agricultural subsidies being removed in 1984, which has meant that agriculture has since operated under a free market system where consumer demand and markets started to drive agriculture change. The population of NZ has been steadily rising, with the resident population increasing from 3.5 million in 1991 to its near 5 million today [4]. During this time on-ground, observed changes to the land system have been recorded with a 10% increase in urban areas between 1996 and 2012 corresponding to a growing population and a 7% decrease in land used in agriculture production between the same time [21]. Between 2002 and 2016, there was also an approximate 20% decrease in land used for beef and sheep farming and during the same period there was a 40% increase in land used for dairy production [21] (see Figure 1 for change in land use during this time). However, despite less land being used for agriculture, there has been a continued intensification of farming with higher stocking rates on dairy farms over the last 15 years (the total number of dairy cattle increased 68.6%, from 3.84 million in 1994 to 6.47 million in 2017) [21]. The amount of nitrogen applied in fertiliser has also seen a more than a six-fold increase, with figures going from 59,000 tonnes in 1990, to 429,000 tonnes in 2015 [21].

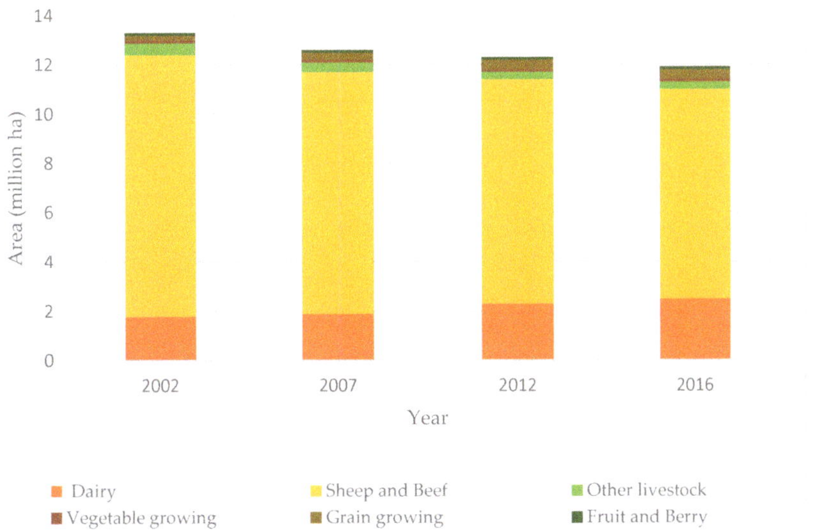

Figure 1. Area covered by main agricultural and horticultural land use types in NZ during 2002–2016 (This work is based on/includes Stats NZ's data which are licensed by Stats NZ for reuse under the Creative Commons Attribution 4.0 International licence [23].

When marketing NZ overseas, Tourism New Zealand has used the "100% Pure New Zealand" campaign since 1999 to deliver the 'story of NZ' to overseas consumers, the travel industry, and the media. This is a story of landscapes and a range of visitor activities and experiences that are unique to NZ [32]. This story is similarly replicated in the marketing campaign for Beef and Lamb NZ, which utilises the phrase "Taste Pure Nature" attached to its grass-fed meat products [33]. Although not intentionally focusing on the environment, by creating an image of "100% Pure" and "Pure Nature", NZ inadvertently lays claim to considerable environmental credentials, which may be in question given the country is currently facing many environmental challenges. Of note are issues associated with water quality (nitrate leaching from agricultural soils was estimated to have increased 29% percent between 1990 and 2012) [34] and erosion (the modelled rate of soil erosion is 720 tonnes per square kilometre per year) [35]. Increased erosion and soil loss (44% of soil lost per year from exotic grasslands) also increases the concentration of sediment in rivers, lakes, and coastal environments [21].

Another key environmental challenge faced in NZ includes the large-scale land clearance that has occurred to make way for agriculture. NZ is now in a situation where only 26% of its native forest remain, mostly in hilly and mountainous areas and less than half of the land area of NZ today is covered in some form of indigenous vegetation [36]. Whilst approximately 32% of the total land area is classed as having protected area status [23,36], just 10% of wetlands remain [35]. Added to this, 40% of the land cover of NZ is now under exotic pasture and exotic forest covers 8% of the land area [35] (see Figure 2 for a breakdown of the share of NZ total land by land cover [23]). With degraded ecosystem services provided by the fragmented native vegetation that remains, it is no surprise that nearly 80% of land vertebrates are now classified as threatened or at risk of extinction [37].

An intensive agriculture industry also has implications for emissions. The per-person rate of greenhouse gas (GHG) emissions is one of the highest for an industrialised country. This is in a country where 85% of electricity production comes from renewable sources, primarily from hydroelectric schemes [38]. Most emissions in 2017 were reported to have come from livestock and road transport with agriculture and livestock producing about half of NZ's total GHG emissions (see Figure 3) and of the 80,873 kilo-tonnes of greenhouse gases produced, most were methane and nitrous oxide (see Figure 4) [39].

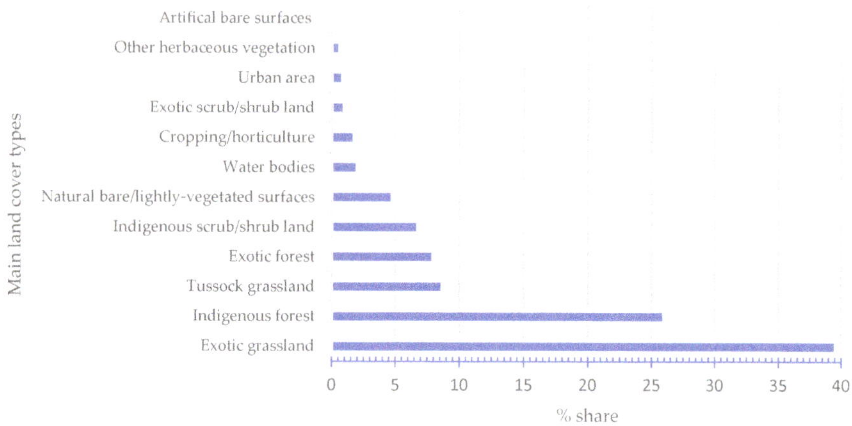

Figure 2. The percentage share of NZ total land area by main land cover types in 2012. (This work is based on/includes Stats NZ's data which are licensed by Stats NZ for reuse under the Creative Commons Attribution 4.0 International licence [21]).

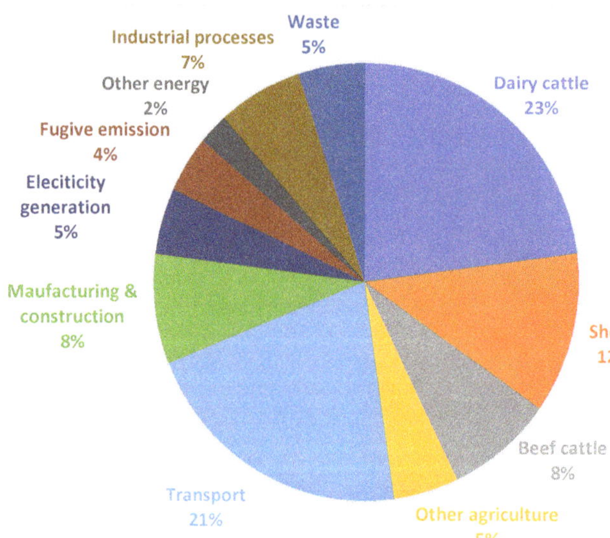

Figure 3. Breakdown of contributing sources of greenhouse gas (GHG) emissions in NZ (This work uses material sourced from the Ministry for the Environment, Stats NZ, and data providers, which is licensed by the Ministry for the Environment and Stats NZ for re-use under the Creative Commons Attribution 3.0 New Zealand licence. [39]).

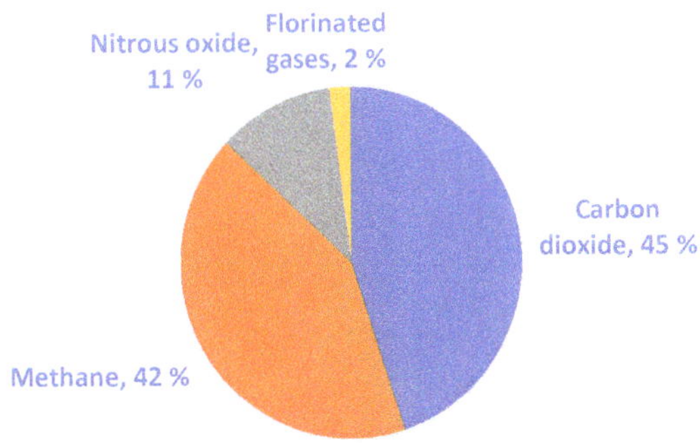

Figure 4. Percentage gas contribution to gross GHG emissions in 2017 (This work uses material sourced from the Ministry for the Environment, Stats NZ, and data providers, which is licensed by the Ministry for the Environment and Stats NZ for re-use under the Creative Commons Attribution 3.0 New Zealand licence [39]).

In 2016, livestock digestion was responsible for 82% of all methane emissions, whilst 94% of all nitrous oxide emissions were from agricultural soils, mainly from the urine and dung of grazing animals. In comparison, carbon dioxide emissions in 2017 were mainly attributed to having come from using fossil fuels in road transport and manufacturing [39]. Even though total agricultural emissions have increased by nearly 14% since 1990, efforts have been made by NZ farmers to reduce emissions and absolute emissions from the beef and sheep sector have been declining over the last 20 years and in 2017 were 30% below 1990 levels [39]. However, the 90% increase in the dairy herd, and a 650% increase in fertiliser use since 1990, has had a significant impact on GHG emissions [39], so considerable work still needs to be done to meet GHG targets. Under the Paris Agreement, NZ agreed to reduce its greenhouse gas emissions by 30% below 2005 levels by 2030 [40]. Recently (November 2019) the NZ government passed climate change legislation with bipartisan support to create a framework for NZ to develop climate change policies to meet targets agreed under the Paris Agreement. The resultant *Climate Change Response (Zero Carbon) Amendment Act 2019* aims to "reduce all GHGs (except methane) to net zero by 2050" and to "reduce emission of biogenic methane within the range of 24–47% below 2017 levels by 2050 including to 10% below 2017 by 2030" [41] and to enable NZ to prepare for and adapt to the effects of climate change [42,43]. This poses a big challenge to farming in NZ as each farm will need to be able to determine current emissions and come up with farm-specific plans that can demonstrate a reduction in emissions via a change in practice and management regimes.

Given the related nature of many of the environmental challenges facing agriculture, a holistic approach to farm management is required which addresses all issues (water quality, erosion, loss of biodiversity and ecosystem services, and GHG emissions). To addresses these challenges, farmers will need to be better at capturing and quantifying data relating to on-farm emissions and understanding approaches that can help to mitigate emissions. To date, accurate baseline monitoring data is limited with some modelling and farm management systems such as Overseer FM [44] being heavily relied upon to quantify emissions but their accuracy is dependent on what data are available. Overseer FM is currently used to generate models based on stock numbers and soil data that are based on broad-scale landscape monitoring. The outputs from which are being used to advise the process of creating farm environment plans with regulation determined under the *Resource Management Act 1991* (RMA 1991) [45] and jurisdictional control of Regional Councils. The creation of farm environmental is increasingly being used as a management approach for environmental sustainability in NZ but with limited fine-scale data, the ability of these to give accurate paddock-scale information is constrained.

More accurate and frequent data capture and modelling approaches will be required to improve modelling capability and to assist with better farm decision making into the future.

Another important challenge facing agriculture in NZ is that due to its free-market economy, it is under a great deal of pressure to respond to trends both domestically and from its global markets without buffering from government intervention. The most recent of these pressures that is currently recognised to be influencing the future development of NZ agriculture is an enhanced environmental consciousness as well as changing consumer views and preferences towards environmental sustainability and measurement of environmental effects and impacts from farming. This means that there are increasing consumer demands around the consideration of ethical production, environmental effects, and provenance/traceability. There are also faster mechanisms by which to spread and communicate these views e.g., via social media and internet-based news stories, meaning they can have a powerful influence both regionally and globally. The recent increase in urban land use and the rising urban versus rural population is also having a significant impact on expectations around farming. It has been noted that as the urban population grows, there is an increasing urban–rural divide when it comes to understanding agricultural production and the ethics around the use of animals as a food source as more people lose their connection to the land and become disconnected with where food comes from [2]. This is contributing to a demand for higher ethical standards around animal welfare and some change in consumer preferences. To maintain its market position, NZ agriculture must be suitably equipped to respond to these pressures and changing consumer trends. The result of a survey of key export consumers revealed that the NZ clean green image has "a significant export value" and that the environmental image of NZ is important to its goods and services in the international marketplace [6] (pg. 4). This means that the environmental credentials of NZ are vital not only for the sustainability of landscapes but also from a marketing perspective.

Other pressures being placed on NZ agriculture come from recent trends in transformational science associated with developments in genomics, and alternative proteins [2]. The changes that these are likely to make to farming and to the meat industry will no doubt impact further on consumer demand into the future. NZ meat consumption has already decreased by 57% in the last 10 years [2] and the rise of the "flexitarian" diet is becoming more evident in NZ and overseas [46]. A recent report commissioned by Beef and Lamb NZ showed that alternative proteins were likely to increasingly compete with meat, which has implications for NZ agriculture especially in the next five years, by which time alternative meat burgers are expected to be mainstream [46].

Other trends set to impact on agriculture include a rise in precision and digital agriculture. Developments and expansion in the uptake of on-farm technology to improve production are likely to lead agriculture into a new revolution [2]. Precision agriculture offers the ability to irrigate and apply chemicals with far greater accuracy and where they are needed most. Increasingly, farm machinery is supplied with onboard sensors that can provide spatial and temporal data useful for determining production [47]. Precision agriculture also offers opportunities for more sustainable management of inputs and thus has potential implications for more effective environmental management [2,47].

There is also a growing need to adapt and respond to a changing global climate. Climate change has the potential to bring about change that is to the detriment of some forms of agriculture in NZ whilst others might be provided new opportunities, with climate influencing where and what things will be able to be grown and produced into the future [41]. It is important to note that these trends and challenges are interlinked and interdependent and therefore are likely to compound each other and have multiple and far-reaching impacts. Farmers, therefore, need to be equipped with knowledge, tools, and approaches that are able to respond to these.

In summary, the challenges that NZ farming systems face are around i) ensuring sustainable production which will involve reducing nitrate emissions, reducing the loss of sediment into river systems and reducing GHG emissions especially methane; ii) dealing with the impacts of climate change which could prove to have both positive and negative consequence for NZ agriculture; iii) responding to an increasing disconnect between a growing urban population and the rural one and changes to consumer requirements along with the increasing need to demonstrate a social licence to operate; iv) increasing agricultural production to feed a growing global and domestic population and v) responding to a digital farming revolution. The implications for farmers associated with these challenges are that they will need to make different farm management decisions, with changes to land use being inevitable. Business decisions for the future will need to not only be based on profit and livelihood considerations but also on environmental outcomes. Overall, the new end goal for a sustainable future agriculture industry needs to be "maximising productivity within environmental constraints" [2] (page 26). As NZ farmers grapple with this, they are caught up in the dilemma of responding to open market agendas that push them to intensify to meet consumer demand and the need to meet sustainability agendas and preserve NZ's unique environmental integrity. Farmers will need access to the best information and relevant science to assist them to make informed decisions and to help them to navigate through the necessary transformation.

3. Landscape Ecology and Farming Systems

3.1. An Overview of Landscape Ecology in An Agricultural Context

Landscape ecology (LE) can be seen to be a science that is still relatively in its adolescence. It is characterised as a science that since its relatively recent inception in the mid-20th century has primarily focused on landscape structure, function and change [7]. Its main emphasis has been on the "spatial heterogeneity and pattern", "how to characterise pattern", "where pattern comes from", "why pattern matters", and "how pattern change over time" [48] (pg. 1170). However, the scientific theory behind LE has been evolving. Recent years have seen a change to the paradigm it represents with a move towards a stronger emphasis on socio-economic elements of landscapes. There is now a greater understanding of human nature relationships and a recognised need to quantify landscape/ecosystem service provision from an anthropogenic as well as environmental perspective [49].

Early roles for LE were largely concerned with focusing on conservation, restoration, and trying to improve degraded landscapes with biodiversity in mind [6]. Within this the links between LE and farm biodiversity are long-established. The relationship between habitat fragmentation and agriculture has been studied and reported [50], and literature can be found calling for an increased provision of habitat and connectivity in amongst a matrix of agricultural landscapes [51]. LE has also been discussed in terms of the role it can play in agroecosystem management [52]. In this context, LE can be seen as having a role to play in recognising and supporting regulatory landscape services like soil fertility and pollination that will ultimately impact on the biophysical capacity of the farm system. It is also seen to be considering the spatial arrangement and makeup of landscape features that support ecological systems separate to the farm. The credentials of LE towards informing the preservation and restoration of ecological integrity and biodiversity in the farm setting are widely acknowledged. An example of this type of application is demonstrated in recent research by Ekroos et al. [53]. They illustrate a role for LE in establishing an ecological focus over parts of the farm through the application of "greening" measures, where the specific placement of these measures across the landscape can enhance potential benefits to birds.

In recent times, LE has seen a shift in its predominant focus. The focus has moved from one of improved management strategies to facilitate the preservation of important landscape pattern and process [54–57] towards a stronger emphasis on the role of LE in landscape sustainability. This has resulted in LE increasingly incorporating planning and design as important considerations in its theory [48,58]. Within this is a stronger focus on the socio-economic aspects of landscapes and on the importance of livelihoods, value, and culture in framing future landscape systems. Associated with this is a recognition of the crucial role of community participation in problem-solving [48,59]. Applying this new holistic paradigm for LE [48] has great applicability to assist farmers who require socio-economic considerations in their decision making in order to develop sustainable landscapes and farming systems of the future. There is increasing debate on how LE can be integrated with current thinking related to on-farm system management and the need to consider landscape-scale functioning to assist in the development of more sustainable agriculture [60,61], presenting more opportunity for greater expansion into this area.

Despite having great potential to be applied in agricultural planning, LE has not had much opportunity to work in a more constructive sense and play a role in creating future farm landscapes. It has also had sparse practical flow through to on-ground farm planning and management and links between landscape science and farm system design are limited [62,63]. However, taking a landscape ecological approach to the integrated analysis of landscape function has the potential to help provide farmers and land managers with approaches, information, and important knowledge that can guide future management strategies and assist in land use decision making [64]. In this paper, I argue that landscape ecologists should be exploiting this potential and striving to assist farmers in places like NZ as they plan their future farm systems.

3.2. Landscape Ecology and Farming Systems in NZ

The idea that there is a role for LE in assisting farming systems to work towards more sustainable agriculture in NZ is not new. Twenty years ago, Meurk and Swaffield [13] made recommendations around a landscape ecological framework for the regeneration of indigenous vegetation in rural NZ, aspects of which were also discussed at the International Association of Landscape Ecology (IALE) World Congress in 1999. Meurk and Swaffield [13] stated that nature conservation and agri-business should not be seen as separate to each other, as they had been become viewed in association with the agricultural landscape in NZ. They suggested an alternative future vision for NZ agricultural landscapes where ecological integrity was reinstated via regeneration of indigenous vegetation within an exotic matrix. This is in keeping with the idea that the more a landscape resembles its natural state and maintains some key landscape features of importance, the more likely it is to have resilience and greater functionality, which ultimately assists with its sustainability [65]. Their vision promoted the establishment of landscape elements which enabled the integration of production and conservation. In this vision, they suggested creating a matrix of indigenous species along edge features that create functional elements like riparian systems and road verges. They also suggested having areas made up of predominantly indigenous species that could be useful for both production and functional purposes [13]. Theirs was an integrated vision in which natural processes and communities provide the underlying framework for the landscape rather than intensive agriculture dominating the landscape [13]. These ideas were expanded upon by Swaffield [26]. He argued for a strategy to create a long-term matrix of sustainable and self-regenerating ecosystems within which productive land uses can be undertaken. He suggested an 80:20 target i.e., for every 80 ha of production at least 20 ha should be allocated to regenerative functions such as biodiversity, riparian margins for water quality, shelter, and carbon sequestration. These sorts of ideas are becoming much more widely debated in the realm of regenerative and restorative farming [66–68].

Fundamental to a landscape ecological approach and the ideas put forward by Meurk and Swaffield [13] is a recognition of the interrelationships between landscape elements and how these interrelationships affect landscape function. The relationship between landscape structure and process/functioning is one that has received considerable attention in LE worldwide with the notion that patchiness in the landscapes is a way to help mitigate dysfunctional effects of intensive production [69]. Ideas put forward by Meurk and Swaffield [13], which focus on creating integrated landscapes, with a patchwork of landscape elements consisting of natural features and native species co-existing in unison with production in the landscape, is one of encouraging multifunctionality. The idea of creating multifunctional landscapes by applying LE principles has been widely discussed [70–73] and there is great potential to consider this more strongly in the planning and design of future farming systems especially in response to some of the challenges facing NZ [63].

Concerning the challenges faced by NZ agriculture, Swaffield [26] felt that a new vision for rural NZ was needed. That is one that could respond to some of the 21st century's greatest global challenges around the environment and sustainability and one that was developed with bottom-up considerations. He stated that being able to take this vision to reality required more integrated research, greater enhanced modelling of ecological processes within multifunctional landscapes, and the development of guidelines that could help to determine optimum locations where landscape reconstruction would be beneficial for the overall functioning of the landscape [26]. Thus, greater 'whole of landscape' thinking is necessary, and the redesign of rural landscapes is required to be able to transition from a divided landscape (production and environment as separate considerations) to an integrated one (where production and environment are equal and considered in unison) [26].

3.3. Landscape Ecology and Agroecology

During the late 20th and early 21st centuries, agroecological theory has developed in parallel to LE. Agroecology (AE) has been a term and a consideration that has increasingly been applied over the last 30 years [74]. It arose within the discipline of agriculture out of a need to address some of the growing concerns around intensification and the environmental impacts that result from a heavy production focused land management system. Recognising the importance of the whole agroecosystem, AE has been described as an integrative study that takes a holistic approach to look at the ecology of food systems [75]. It is an approach that works towards ecological, economic, and social sustainability considering the processes involved in food systems from farm gate to the plate [76]. Like LE, AE has evolved, in more recent years, to focus more heavily on transdisciplinary approaches, participation, and change-oriented research [74,76]. Differences between LE and AE have been described as relating to the scale of investigation associated with the management unit i.e., AE has tended to focus at the finer resolution of the farm or field scale (the ecosystems in which farm services are delivered) whilst LE has largely looked at the broader landscape or catchment scale [77] of which farm systems are just a part.

Since farm systems have been described as integrated components of rural landscapes [78], both AE and LE have been evolving to consider integrated models and approaches that help to understand and manage human-dominated landscapes [79]. Fundamental to agroecological thinking is the importance of a certain structure on-farm which can maintain key functionality that can play a part in overall sustainability. Gliessman [65] (page 300), in line with some of the arguments of Meurk and Swaffield [13], stated that "the greater the structural and functional similarity of an agroecosystem to the natural ecosystems in its biogeographic region, the greater the likelihood that the agroecosystem will be sustainable".

Dalgaard [78] (page 8) when discussing landscape agroecology stated "The ecological, economic, wildlife and visual functions of landscapes within modern society are determined by processes that operate over a range of scales in space and time. Integrating knowledge of these processes into tools that can be used by people who have stewardship over the land, such as farmers and regulators, requires an interdisciplinary approach. Such an approach demands significant effort as it must work against the trend of specialisation and fragmentation of knowledge that has occurred over the recent centuries. It also requires, substantial technical developments, relating to data collection from disparate sources, data manipulation, and integration of information about the multiple landscape functions". This statement recognises the need to preserve important landscape functioning in the farm context. It also emphasises the lack of collaboration and tools required for the collection of useful information that is important to progress the application of AE approaches for more sustainable agriculture. Dalgaard [78] also acknowledged the need for greater work across disciplines to pull together knowledge and approaches to develop new strategies for farm and landscape management. Even though as far back as 1992, Flora [80] recognised that sustainable agriculture required transdisciplinary teams which included farmers working together to address research priorities, there is still a need for more cross-disciplinary collaboration to address knowledge gaps. The call for greater farmer-researcher partnerships, as they strive to balance environmental outcomes, agricultural production farm profit, and community wellbeing, is still very valid today. Similarly, the fact that limiting uptake of practice change was put down to there being a lack of good indicators of sustainability [78].

Whilst more sustainable on-farm management is important for individual farms, just preserving or improving within farm ecosystem processes and functioning will have limited benefit for the wider landscape unless other farms adopt similar strategies. Collaboration on strategies and group action to improve functioning across a broader landscape will be more effective in having much more far-reaching and beneficial environmental outcomes that individual activity. Given the similarity of interest and focus between LE and AE and the requirement for landscape sustainability to incorporate across-scale activity, it would seem sensible for greater collaboration and integration of ideas not just between farmers but also between disciplines as they work towards addressing the gaps in research and application required for more sustainable future farming systems. An ideal platform for this being through concepts of designing and managing multifunctional landscapes [77] where landscape consideration is given at both the farm and catchment scale.

The merger of the disciplines of AE and LE has been described by the term agro-landscape ecology which has been mooted for some time [81–83] but with growing farm-based challenges, the integration of agro-landscape ecology into addressing problems through farm environment planning appears to be coming increasingly more relevant and important. A stronger integration of concepts and theories and scales of operation presents an excellent opportunity to expand the application of LE and to bring a more scientific context to both farm and catchment scale planning processes currently being undertaken in places like NZ. This is becoming increasingly more relevant in NZ as farmers are joining forces to create community catchment management groups and are reaching out to scientists to provide both relevant farm and catchment scale advice. Greater integration of AE and LE science make both disciplines better placed to aid farm decision making. Bringing together the combined thinking from both AE and LE has the potential to ensure greater on-ground benefits across scales. Working with colleagues in AE can help landscape ecologists to determine what good indicators of agricultural sustainability might look like and how to measure and quantify effectively on-ground farm-scale change that has beneficial environmental outcomes at both the farm and landscape scale. Greater collaboration between AE and LE also has the potential to enable scientists within the field of LE to obtain a clearer understanding of a farm system issues and considerations that can help to make them become a more effective advisory and supportive science towards transformative agricultural change.

LE is increasingly being recognised as an important framework capable of assisting in the planning and design of future landscapes [72,84–87]. In this context, LE has an important role to play in providing appropriate tools, approaches and frameworks that can facilitate the planning, design, and advice required to help work towards the creation of future farming systems that meet societal needs, respond to the environmental challenges and that can sit within sustainable landscapes. Despite its obvious potential in this area, LE has only recently started to be practically applied in this context to a NZ setting [62,63].

4. Identifying Future Key Roles for Landscape Ecology that Can Help Progress Sustainable Agriculture

4.1. Key Roles

4.1.1. An Informing and Decision-Making Science

Recognising some unrealised potential for LE to date, one of its key roles for future farming systems is that of placing itself at the forefront of being an informing science around landscape sustainability. It needs to provide the scientific context for informed farm and catchment decision making, whilst acknowledging the increasing recognition in the farming sector of the importance of AE and regenerative agriculture. This means that LE needs to demonstrate its ability to 'value add' to agroecological studies through a collaboration that clearly illustrates the importance of landscape and catchment scale studies for farm system sustainability.

In helping to work towards sustainability of farm systems, LE has an important role to play in helping with the definition of land use optimisation. By informing the pattern of land use and making suggestions around the allocation and reallocation of land, it is possible for LE to help to create multifunctional agricultural landscapes that have sustainability in mind. To do this effectively, LE needs to demonstrate its ability to quantify information relating to process and function. LE also needs to provide advice around the management of landscapes to preserve important landscape services and functioning on-farm and across the broader landscape. This means helping to create models and strategies that define important landscape characteristics that need to be maintained and preserved, as well as being able to provide advice around the positioning of introduced landscape features that could assist with sustainability. A good example of this for NZ could be around identifying key locations for reconstructed wetlands (as mentioned previously, NZ currently has a situation where 90% of its wetlands have been drained with significant impact on nutrient flows into river systems [35]). Reconstructed wetlands can put back into the landscape important functionality that can improve environmental outcomes by reducing the loss of nitrates from intensive farming practices like dairy farming that cause considerable pollution of rivers [2,21].

LE also has a key role to play in being able to easily incorporate the scientific evaluation of land-related information into design, planning, and decision making through helping to provide a scientific context to the criteria against which land-related information can be evaluated for sustainability decision making. This will be vital for informed planning and design of future agricultural landscapes capable of providing appropriate mitigation and adaptation responses to some of the biggest environmental challenges. LE should be ideally positioned to guide process informed and value influenced planning and design within a theoretic framework guided by LE principles to assist with more holistic farm environmental planning. Linked to this is the potential for LE to assist with the creation of spatial decision support tools. Suggested approaches include the application of a spatial decision support framework using approaches that can build on important LE concepts and design agendas in a spatial planning context to work towards the creation of multifunctional landscapes [63,88].

Geodesign [89–91] offers a framework for smart planning in a spatial environment that enables the analysis of multiple data layers so that planning proposals can be evaluated, and new scenarios tested. This type of approach paves the way for an integrated holistic approach to farm and landscape planning which is vitally needed in NZ [62,63]. Such a framework can provide a basis under which it is possible to analyse the impacts of on-farm land use change prior to a change being implemented. This type of modelling approach can help to integrate concepts of ecosystem/landscape services into farm and landscape-scale planning by incorporating information on functioning as an important data layer used in the model. The model created in a geodesign framework can help to spatially visualise patterns of land use and land cover that enable important landscape functions (economic and environmental) to operate side by side. The generation of future alternative scenarios for farm system management enables farmers to see and contribute towards the development of visualisations of what sustainable utilisation of the landscape might look like both on their farm and in the surrounding landscape. It can also allow them to subsequently evaluate the economic feasibility of different scenarios. Generating scenarios that suggest optimal sustainable land use could contribute towards designing an appropriate future landscape pattern that is based on process and economic analysis and the idea of creating multifunctional landscapes at the farm and landscape/catchment scale.

4.1.2. Tools and Techniques for Farmers

Change to land use and farming systems is inevitable but knowing what practice change is best, with environmental and business sustainability in mind, and therefore how to change, is crucial if farmers are to have optimum (livelihood and environmental) outcomes. This is where LE has a potentially pivotal role to play. It can position itself to provide valuable advice to farmers and land management policymakers in terms of the 'what' and 'how' when it comes to creating healthy and sustainable landscapes. Fundamental to fulfilling this role, especially in countries like NZ that face challenges of demonstrating their environmental sustainability, will be the ability to assist farmers through their transformation by providing the necessary tools and techniques. These will be needed to quantify the current state of landscape and audit environmental sustainability (at both the farm and landscape scale) so baseline data can be established, and on-going monitoring achieved. Often there is little baseline data from which to start monitoring. A lack of good baseline data makes it difficult to quantify changes that might occur or have occurred due to improved practice. This is a problem facing on-farm climate change mitigation responses in NZ as good current baseline data on GHG emissions at the farm-scale are currently not available, yet farmers have only a few years to demonstrate beneficial change. The provision of baseline data provides an opportunity for LE to offer better mechanisms to capture the landscape situation for baseline assessments and approaches to monitor change as 'best practice' activity is undertaken.

In NZ tools like Overseer FM [44] have become standard to assist in recommendations around nitrate emissions. This is because tools like this give a measure and a figure that farmers can work within an ongoing capacity to address their land management challenges. Overseer FM is currently being expanded to have more GHG emission monitoring capability which will widen its application to assist farm decision-making LE has the potential to provide similar quantification and measurement capability through the many metrics that have been applied to quantify pattern and process. Aspinall and Pearson [92] outline a suite of indicators and metrics based on the assessment of landscape characteristics that have been used in LE research and can be applied to quantify catchment condition and how it changes over time, many of which could equally be applied at the farm scale. However, despite the extensive application of LE metrics, there has been criticism about their ability to adequately quantify the links between pattern and process and the sensitivity of some processes to structural changes to the landscape [93–95]. The scale of the data used and the scale at which analysis is conducted are also fundamentally important to the reliability of the results. This is a challenge for places like NZ where mostly only broad-scale data exists, with only a small number of farms having fine-scale soil data available.

LE research agendas need to take note that work is still required to develop innovative mechanisms to help to capture the current landscape situation at an appropriate scale and to quantify what is happening on the ground in a relevant form by farmers and also their regulators (e.g., Regional Councils in NZ). This is important if LE is going to be able to proactively provide the necessary tools and techniques for application at the farm scale. It also important to allow farmers and regulators to appropriately monitor change as land use and agricultural practice changes occur. Measures and tools that are only capable of being used by scientists or are at too broad a scale will have limited application in on-ground farm system management situations. Ease of use and capability that is tailored to farm system applications and solutions is of utmost importance in demonstrating relevance to both regulators and farmers. Greater engagement and collaboration around tool and indicator development will help to develop approaches to make LE a more actionable science [12].

LE also needs to consider further work on determining and quantifying landscape/ecosystem services present in agricultural landscapes [96] and how these change over time. Recent research by Dominati et al. states that "the supply of multiple ecosystem services from farmland and agroecosystems and trade-offs between service remains under-researched" [61] (pg. 704). They feel that this impacts on the ability to determine how well farm systems can achieve environmental sustainability outcomes whilst maintaining economic profitability. This means that more work is needed to determine how multifunctionality aspirations within farm systems can work alongside farm profitability goals for mutually beneficial outcomes [61]. As a result, there appears to be a role for LE to work more closely with AE to contribute towards developing better solutions towards more effective monitoring and evaluation of functioning and processes on agricultural land using appropriate indicators and measures.

Techniques need to be developed that can monitor the condition of processes and how this relates to function [64]. In line with recognised LE principles, it is important that these techniques should be able to adequately and easily capture information relating to state (condition), trend (changes across space and time), and function (stability, resilience, and sensitivity) at appropriate scales. Where possible, these need to be simple measures that represent key components of the system and have meaning beyond the attributes that are directly measured. For ease of capture, many of these indicators need to be capable of measurement from on-ground and remote sensing data sources and be spatially and temporally explicit offering opportunities to link to data capture for precision agriculture. It is also important that scenarios be developed for increasing functionality and that these can also be related to measures associated with raising profit. This is important to adequately understand the impacts of increasing landscape services and functioning on a farm and determining the relevance of actions that increase landscape functioning for long term farm and environmental sustainability.

4.1.3. Integrating Value and Cultural Perspectives into Landscape Analysis and Decision Making

Recognising values and cultural perspectives as well as landscape functioning is important for landscape management and therefore sustainable farming systems [59,97–100] so as to ensure the creation of future landscapes that can adequately support a diversity of values. This means that it is crucial that LE has mechanisms that consider the value of production from an economic and livelihood context and can incorporate key stakeholder involvement in processes to advise on the make-up of future farming systems. In doing this, LE needs to draw on its strength as a transdisciplinary science and put greater emphasis on developing the support mechanisms required for transdisciplinary approaches to problem-solving in an on-ground capacity. This means that there is an essential role for LE to be able to guide the effective on-ground application of real transdisciplinary projects that bring together a variety of stakeholders and recognise different values, beliefs, and perspectives. However, LE still needs to rise to the challenge laid down by Wu and Hobbs, Opdam et al., and Swaffield [12,98,101] around integrating science and practical application and outreach. This means that landscape ecologists should also be striving towards making LE a facilitating discipline i.e., one that takes the initiative to bring together other science disciplines with key stakeholders to work to generate solutions to some of the complex issues facing farming systems and is not afraid to promote bottom-up stakeholder-led research.

LE also has a key role to play in recognising culture as an important aspect of landscape management [100]. In places like NZ not enough attention to date has been paid to Māori perspectives and relationships between people and landscapes and what this can bring to future farmsystem design. Māori relations with landscapes are expressed through *kaitiakitanga*. *Kaitiakitanga* is a Māori term that encompasses guardianship and stewardship for the environment. This includes the responsibility to ensure sustainable harvest and fair distribution of resources. Incorporating *kaitiakitanga* into cultural functional assessment and an appreciation of value will help to derive a more broadly integrated stewardship focused approach to land management that is likely to provide for generational sustainability. Other important environmental concepts in Māori culture include *Ki uta ki tai*, which is a Māori whole-of-landscape approach similar to the concept of integrated catchment management that looks at resources and ecosystems from the mountains to the sea; *Te Ao Turoa* which is an intergenerational concept of resource sustainability; and *Whakapapa* which is concerned with the connection, lineage, or genealogy between humans and ecosystems and all flora and fauna [102]. Māori seek to understand the total environment or whole system and its connections through *whakapapa*, meaning that this perspective towards human and environment interactions is holistic and integrated [102]. An example of where Māori cultural perspectives are being integrated into NZ primary production has come from the recent (December 2019) launch of the vision for future NZ Primary Industries 'Fit for a better world' by the Primary Sector Council which centres on the Māori concept of *Taiao* which emphasises "respect for and harmony with the natural world" [103]. These concepts and perspectives are very much in line with LE thinking, so NZ is in the unique and exciting position of being able to capitalise on its Māori cultural heritage and the values and perspectives it can encourage in order to direct future sustainability. Landscape ecologists in NZ have the potential to help to bring these to the forefront to be implemented in helping to determine farm system change.

4.2. Past Impediments to LE Uptake and What Can Be Done to Improve This?

Given that LE has such an evident role to play in assisting in the environmental management of agricultural landscapes, it seems obvious to ask why we have not up to now seen a far greater integration of LE principles and practices into the management of current farming systems? And why in particular, we still have work to do on sustainable farm systems in NZ given that landscape ecologists like Meurk and Swaffield [13] were giving advice 20 years ago on how to create more multifunctional and sustainable farm landscapes. To answer this, several potential reasons can be identified. Acknowledging and recognising limitations can be useful for the future implementation of LE in the agricultural context. The reasons appear to be largely related to an ongoing disconnect between science, policy, and practice [12,98,104].

LE has shown its strength in providing good quality research on the theory of landscape management but regrettably, there has not been much flow through to practical on-ground application relevant to farm management. In 1997, Hobbs [104] concurred with this statement when he reported that LE did not at the time truly meet the criterion of being an applied science. More recently, in 2013, Opdam et al. [12] questioned LE's on-ground applicability. They highlighted the need for LE to deal more effectively with the science-policy gaps. Opdam et al. [12] also pointed out that many papers published in the journal *Landscape Ecology* are analytical in approach. The shortfalls associated with a lack of public involvement in the co-production of science-based solutions were also demonstrated in Seppelt et al. [49] in regard to research on ecosystem services. In 2002, Wu and Hobbs [98] highlighted the reasons that limit LE's applicability as being issues of integration between basic research, applications, and outreach and communication with the public and decision-makers. LE appears to be still struggling to address key issues around integration and outreach which would make it more applied and landscape ecologists still do not seem to have adequately addressed these challenges. Work is still required to make LE more applicable to on-ground action [105]. This means that despite prompting by leading landscape ecologists LE has stayed very in a much more theoretical and scientific realm rather than a practical context. To address some of the shortfalls, Opdam et al. [12] highlighted the need for additional topics of investigation in LE. The topics for further research looked at 'the local landscape as a boundary object that builds communication among disciplines and between science and local communities', 'iterative and collaborative methods for generating transdisciplinary approaches to sustainable change' and 'the effect of scientific knowledge and tools on local landscape policy and landscape change' [12]. Given the lack of follow up commentary in the LE literature that adequately addresses these topics of investigation, it appears that LE is still struggling to effectively address these topics in a research capacity. The result being that these challenges for LE could be equally listed on a research agenda today.

Some of the disconnect between science and practice is evident through the fact that to date the theoretical concepts and methods have not transitioned well into useful and practical tools and techniques for non-scientists. These are approaches that can help farmers and policymakers to compose better-informed decisions through capturing good monitoring data and translating it into useful information that can guide practice change and thus assist with environmental management at the farm and catchment scale [98]. Another potential problem has been the lack of relevant information made available in a manner that can be digested by farmers and policymakers. Most LE research is presented in journal format restricting its readability by land managers. There is also an inability to be able to adequately relate 'the cause to effect' when it comes to issues on the ground [2], which is crucial to demonstrate the overall worth of approaches for any farmer-based change. An inability to do this restricts a farmer's ability to make good decisions and practice change based on a knowledge that certain practices will cause problems whilst others will help. Also, the advances in science have helped us know more about the problem but are failing to suggest appropriate mitigation responses that can clearly demonstrate the cause and effect relationships to farmers. So suggested change to practice cannot always be definitively related to better environmental outcomes.

The receptiveness of farmers will also influence the uptake of science-based solutions. The factors that influence the decision-making processes of farmers are complex. Failure to adequately contemplate worldviews of farmers and livelihood considerations are also significant in terms of uptake of scientifically based approaches to landscape management. Suggesting landscape options that are not financially viable for farmers is not going to facilitate uptake as fundamentally farmers must maintain livelihoods and lifestyles for themselves and their families. A lack of political will to drive a pro-environmental agenda in many first world countries also has implications for implementation of on-farm landscape sustainability responses especially when the emphasis is placed on maintaining production without risking economic impact.

Both Hobbs [104] and Opdam et al. [12] said that if landscape ecology was to be more widely applied in a practical sense it would require increased engagement with policy and management when it comes to future landscape design strategies and approaches particularly at the local scale. It is evident that LE has come a long way in the last 30 years and progress has been made but more work is needed if LE is to have a bigger impact in helping to design future farming landscapes. More directed effort is needed to make a coherent link between science, policy, and practice [58] and to help to remove some of the barriers which have prevented the adoption of LE to real-world problems [105].

4.3. *Understanding Farmer Behaviour and Motivations and Their Role in Changing Farming Systems*

As noted above, a key problem for LE is linking the science to practice and recognising the role of the farmer in the implementation of science-based solutions. With increasing pressure on farmers to change practices for greater environmental outcomes, it is vital that the science that advocates for change can understand the constraints on farmers as well as their motivators and drivers that will encourage behaviour change. A report that was undertaken by AgFirst (Independent Agriculture and Horticulture Consultant Network) on farming decision making in NZ about climate change action discusses farmer behaviour and response to change in detail [106]. This report states that being able to understand farmer change and their ability to change practice and uptake new innovations and technologies as part of new farming systems requires recognition of i) "farmer awareness of the innovation"; ii) "ease of trialling an innovation on-farm"; iii) the "perception that an innovation is worth trialling"; and iv) "the value of the innovation in achieving the farmers' objectives" [106]. These recommendations have been substantiated by research findings from Pannell et al. [107], who looked at the adoption of conservation practices by farmers in Australia. He found that innovations were most likely to be adopted when they could demonstrate "high relative advantage" i.e., benefit returned from innovation over current practice, and when they were readily "triable" on-farm. The AgFirst report also stated that other important aspects of an innovation that interact with farmer decision making include i) "compatibility" i.e., degree to which the innovation is comparable with current farm system; ii) "complexity" i.e., the perceived difficulty to understand or implement the innovation; and iii) "observability" i.e., the visibility of the results of the change [106]. This means that action to bring about change needs to focus on increasing awareness and understanding, highlighting the advantages of mitigating impacts, emphasising the compatibility, encouraging and making trialability easier, and making the significance of change more obvious.

Other social factors that need to be considered when it comes to farmer behaviour change and uptake of new ideas include i) at the individual level considerations like time availability, level of education, approach to risk, advice sought and from whom, as well as personal and family circumstances [106–108]; and ii) at the wider social level, they include considering that farming is a socio-cultural practice i.e., it is a way of life and that the notion of sustainability for most farmers means staying on-farm rather than having environmental outcomes [106–108]. It is also important to consider that not all farmers are the same i.e., that the farming community is very socially diverse [108]. Also, that farmers might not distinguish environmental issues from other farm management issues, and that farmers often create their own knowledge through experience and that their key source of information is other farmers [108]. This means that effective extension requires an understanding of the worldview of the farmer. Further complicating this for NZ is the challenge of bringing together worldviews of Pakhea farmers (of European descent) and Māori farmers, plus the worldview of other migrant farmers.

The review that was undertaken by Journeaux et al. [106] also revealed that extension promoting change needs to emphasise the need to provide whole farm system-level solutions to the farmer so any change that is to be promoted needs to be explained within the context of the system as a whole. Making things difficult for farmers is the shift in farming focus i.e., from a system concerned with development and production to one of production within environmental constraints. Whilst the public perception of farm sustainability tends to focus on environmental sustainability, farmers who grapple with the concept of sustainability are required to consider both economic and environmental implications of their actions, and these can often conflict with each other.

If LE is to be useful to farmers, it is important that it recognises farmer's needs and aspirations and recognises that farmers need to be better supported to make informed farm and landscape-scale management decisions. To better support change in agricultural practice, LE research agendas need to focus on transdisciplinary and collaborative approaches that have a greater emphasis on understanding markets and consumer preferences and demand. Also important to individual farmer adoption of changes to their practices is having access to useful and relevant information and advice that is easy to collect, understandable, and implementable by the farmer himself. Overall, it is paramount to recognise that when considering or instigating farmer change there are lots of legitimate reasons for non-adoption of change and generally farmers need to feel valued as they embark on their transformation journey [106].

4.4. The Way Forward

As we face increasing environmental challenges around intensive agricultural production, population rises and the potential for natural disasters to occur more frequently and with unprecedented consequences, the science to support agricultural sustainability needs to deal with unpredictable situations and be able to move quickly [2]. To address these challenges, it is becoming increasingly more important to find solutions that look not only at the farm but also beyond the farm i.e., to consider the broader landscape mosaic of which the farm is a part. By working more closely with AE, LE can prove itself as a science to guide and inform farm and landscape decision making for more sustainable landscape management both in NZ and elsewhere at scales that are relevant. However, to do this effectively, it must be able to link to policy and practice and understand farmer needs and aspirations. To ensure this occurs, it is important that stakeholders become more involved in landscape research and that they participate from the inception of future farm systems and landscape planning and design processes. With wide-ranging participation, a transdisciplinary approach that can link the latest analysis and monitoring tools and techniques with LE theory and spatially informed design principles, will have the ability to suggest a recommended spatial structure and function for farm systems that can help with problem-solving [64]. Working with, and embracing, cultural values like *kaitaikitanga* presents an excellent opportunity for NZ agriculture to capitalise on Māori values, using this important sense of stewardship towards the land when developing strategies for future farming systems in NZ to ensure long term sustainability.

A good example of an application that effectively makes the link between science and practice in an agricultural context and addressed the issue of how to bring about farm-scale change is reported in Bohnet et al. [109]. This is a useful illustrative example of a transdisciplinary LE-focused collaborative project which was undertaken between scientists and graziers in North Queensland, Australia. The study looked at grazier change in relation to sediment and nutrient flow into the Great Barrier Reef and identified broad types of farmers within the study. These types were identified as "traditionalists", "diversifiers", and "innovators" [109]. Understanding the values and motivations underlying the types of farmer was shown to have an impact in terms of on-ground action. This study demonstrates that land management strategies are the result of an interaction between how farmers view their land, their experiences, their knowledge, their values, their motivations and their socio-economic situation [109]. Understanding what provides the context for farmer decision making is important when attempting to effectively link science and practice. As Leopold [110] (page 263) said in 1939 – "*the landscape of any farm is the owner's portrait of himself*". Therefore, it is crucial to understand farmers and their individual and collective worldviews when trying to change land management practice. Simply applying 'a one rule fits all' approach will not work [109]. This means that LE needs to be better at recognising and dealing with stakeholder differences.

It is crucial to understand that a sustainable future for farming systems will consist of considering the range of values that agricultural landscapes have to offer and managing them in a different way. This means treating landscapes as being 'healthy', 'attractive', and productive' and places were people and nature are capable of thriving and prospering alongside each other [111]. It is also important to recognise that agricultural landscapes need to support 'profitable enterprises', and 'vibrant communities', as well as being in good physical condition [112]. In order to achieve this, a changing approach to agribusiness is required and a shared vision needs to be built collaboratively with farmers whilst developing the important mechanisms to achieve the vision. LE with its ability to be 'more than just another discipline' has the opportunity to rise to the challenges laid down. It can cross the boundaries of other more traditional disciplines to view land management issues from a more holistic context that considers social and natural sciences [113]. This means that LE can work to stimulate the integration of a variety of disciplines to solve problems. To do this well will require capitalising on its transdisciplinary nature and being a facilitator of science to help drive the sustainability agenda. It also means guiding the effective on-ground application of projects and approaches that bring together a variety of stakeholders. It also means recognising different values, beliefs, and perspectives, as well as working more readily with colleagues in AE to jointly advance the knowledge, science, and application of sustainable solutions for future farming systems.

From a practical sense, LE science also needs to back up the facilitation role it can provide by offering better mechanisms to practically capture and quantify the current landscape situation and monitor change as land use and practice changes occur. This, as discussed above, will require the development of tools and indicators that can help with understanding landscape processes especially tools that evaluate ecosystem and landscape services and monitoring if a situation is deteriorating or improving. This will also require supporting mechanisms that enable the incorporation of relevant science into design, planning, and decision making. This is vital for informed planning and design of future landscapes to provide appropriate mitigation and adaptation responses. If LE can do this, it has great potential to make a strong contribution to the development of future farming systems.

5. Conclusions

Landscape ecologists to date have been responsible for developing some excellent science and theories which have the potential to help agriculture respond to future challenges and play some key roles in the management and design of future sustainable farming systems. However, as Flora [80] (page 38) said "*sustainable agriculture is as much a process as an endpoint*". So despite the fact that LE has always had on-ground relevance and strong applicability to heavily modified and agricultural landscapes, more work needs to be done on the "process"' of helping farmers to transition to more sustainable enterprises. This will require ensuring that the current science and methods are more frequently applied to practical farm management situations so what we already know in LE is taken out of the 'ivory tower' and becomes more widely recognised as a practical tool for farm sustainability rather than a theoretical construct. This means increasingly focusing on bridging the divide between science and practice and working more closely with the farming related community.

It is also important that landscape ecologists rise to the challenge of directing future research agendas towards developing LE-informed practices and techniques that are relevant and easy to apply for those engaging in on-ground landscape management. Above all, LE needs to demonstrate that it can be more practically applied to farmers and farming systems for both business and environmental sustainability. This will potentially mean engaging with the relevant stakeholders to undertake more 'bottom-up' research. In doing this, an area of focus for LE should be to provide easy to use tools and techniques for quantifying the state of the environment at the farm and landscape scale that can be integrated into business management systems to advise around holistic farm sustainability. Capturing good baseline data is fundamental to determining what is happening on the ground and what happens in response to practice change. So, LE also needs to focus on providing measures that relate landscape pattern/cover to ecosystem services/landscape functioning as well as providing ongoing monitoring tools to evaluate on-ground activity. Monitoring approaches that reflect pattern, process, and change relevant to farm systems need to be further developed and innovative ways to analyse monitoring data captured through farming mechanisms such as precision agriculture practices need to be explored for their use towards determining environmental outcomes as well as production outcomes. After data acquisition, LE needs to demonstrate an ability to turn relevant spatial and temporal data into useful information for decision support and be able to actively contribute towards generating adaptive management advice and planning techniques to assist in the creation of multifunctional landscapes. LE also has an important role to play in contributing towards the generation of spatial decision support tools to assist planning (scenario and visualisation) at the farm and landscape scale that combine concepts and theory of LE framed within the context of business support systems. This is particularly important for places like NZ, which has an intensive agriculture industry backed up by a strong rural-based economy and population and an important tourism industry.

With a strong desire to maintain their place in the global tourism and food markets, NZ needs to demonstrate its environmental credentials and the idea that "Pure NZ" and "Taste Nature" are a reality and reflected in landscape management, not just marking campaigns. This means that NZ farmers need to increasingly factor more integrated ways to manage the land into their business management approaches. These are approaches that work in association with nature to reduce emissions and maintain healthy landscapes and waterways. For farmers searching for answers on how best to do this, LE is ideally positioned to work with them to develop future farming systems. These systems can be not only multifunctional and environmentally healthy, but also work towards creating sustainable business enterprises that are capable of increasing productivity outcomes to the point required to bolster farming livelihoods and to feed a growing global population. There is plenty of good science out there within the disciplines of both LE and AE and extensive knowledge amongst farmers about their farm systems. Collaboration and the co-production of integrated interventions and solutions that plan and design future farm systems with people and the environment in mind appears to be the best way forward for achieving sustainable agricultural production, both in NZ and globally.

Funding: This research received no external funding.

Acknowledgments: Acknowledgement is given to Massey University for supporting this research. I would also like to thank the valuable advice given by 3 anonymous reviewers that helped to improve this manuscript and Julian Gorman for comments on earlier versions of the paper.

Conflicts of Interest: The author declares no conflict of interest.

References

1. Hertel, T.W. The challenges of sustainably feeding a growing planet. *Food Secur.* **2015**, *7*, 185–198. [CrossRef]
2. Davies, P.; Moore, D.; Yarrall, D. *Current Land-Based Farming Systems Research and Future Challenges*; Report for the Ministry of Business, Innovation and Employment; Sapere Research Group: Wellington, New Zealand, 31 October 2018.
3. Stats NZ-Tatauranga Aotearoa. International Travel. October 2019. Available online: https://www.stats.govt.nz/information-releases/international-travel-october-2019 (accessed on 10 December 2019).
4. Stats NZ. Summary Figures for the NZ Population 1991–2018. Available online: https://www.stats.govt.nz/topics/population (accessed on 10 December 2019).

5. Beef and Lamb NZ. Taste Pure Nature–Fact Sheet. Beef and Lamb Annual General Meeting 2019. Available online: https://beeflambnz.com/your-levies-at-work/taste-pure-nature-origin-brand (accessed on 10 December 2019).
6. MFE (Ministry for the Environment). Our Clean Green Image: What's it Worth? Available online: https://www.mfe.govt.nz/publications/sustainability/our-clean-green-image-whats-it-worth (accessed on 10 December 2019).
7. Forman, R.T.T.; Godron, M. *Landscape Ecology*; John Wiley and Sons. Inc.: New York, NY, USA, 1986.
8. Bissonette, J.A.; Storch, I. Landscape Ecology and Resource Management. In *Linking Theory with Practice*; Island Press: Washington, DC, USA, 2003.
9. Richard, Y.; Armstrong, D.P. The importance of integrating landscape ecology in habitat models: Isolation-driven occurrence of north island robins in a fragmented landscape. *Landsc. Ecol.* **2010**, *9*, 1363–1374. [CrossRef]
10. Laforezza, R.; Coomes, D.A.; Kapos, V.; Ewers, R.M. Assessing the impacts of fragmentation on plant communities in New Zealand: Scaling from survey plots to landscapes. *Glob. Ecol. Biogeogr.* **2010**, *19*, 741–754. [CrossRef]
11. Viles, R.L.; Rosier, D.J. How to use roads in the creation of greenways: Case studies in three New Zealand landscapes. *Landsc. Urban Plan.* **2001**, *55*, 15–27. [CrossRef]
12. Opdam, P.; Iverson Nassauer, J.; Wang, Z.; Albert, C.; Bentrup, G.; Castella, J.C.; McAlpine, C.; Liu, J.; Sheppard, S.; Swaffield, S.R. Science for action at the local landscape scale. *Landsc. Ecol.* **2013**, *28*, 1439–1445. [CrossRef]
13. Meurk, C.D.; Swaffield, S.R. A landscape ecological framework for indigenous regeneration in rural New Zealand-Aotearoa. *Landsc. Urban Plan.* **2000**, *50*, 129–144. [CrossRef]
14. McCalla, A.F. Challenges to world agriculture in the 21st Century. *Agric. Resour. Econ.* **2001**, *4*, 1–11.
15. FAO (Food and Agriculture Organization of the United Nations). *The Future of Food and Agriculture–Trends and Challenges*; FAO: Rome, Italy, 2017; ISBN 978-92-5-109551-5.
16. United Nations Department of Economic and Social Affairs and Population Division. *World Population Prospects: The 2015 Revision*; Key Findings and Advance Tables; In Working Paper Number ESA/P/WP.241; United Nations Department of Economic and Social Affairs and Population Division: New York, NY, USA, 2015.
17. MfE (Ministry for the Environment). *Environment New Zealand*; MfE: Wellington, New Zealand, 2007; ISBN 978-0-478-30191-5.
18. McKenzie, D.W. *Heinemann New Zealand Atlas*; Heinemann Publishers: Portsmouth, NH, USA, 1987; ISBN 0-7900-0187-X.
19. Douglas-Clifford, A. Nobody Lives Here: Uninhabited Areas of New Zealand. The Map Kiwi. Available online: https://www.andrewdc.co.nz/project/nobody-lives-here-uninhabited-areas-of-new-zealand/ (accessed on 10 December 2019).
20. Stats NZ. Tourism Satellite Account: 2019. Available online: https://www.stats.govt.nz/information-releases/tourism-satellite-account-2019 (accessed on 10 December 2019).
21. MfE (Ministry for the Environment) & Stats NZ. *New Zealand's Environmental Reporting Series: Our Land 2018*; MfE (Ministry for the Environment) & Stats NZ: Wellington, New Zealand, 2019. Available online: https://www.mfe.govt.nz/publications/environmental-reporting/our-land-2018 (accessed on 10 December 2019).
22. NZIER (NZ Institute of Economic Research). *Dairy Trade's Economic Contribution to New Zealand*; NZIER: Wellington, New Zealand, 2017.
23. MfE (Ministry for the Environment) and Stats NZ. Agriculture and Horticulture Land Use. Available online: http://archive.stats.govt.nz/browse_for_stats/environment/environmental-reporting-series/environmental-indicators/Home/Land/land-use.aspx (accessed on 10 December 2019).
24. MfE (Ministry for the Environment) and Stats NZ. Agriculture Production Statistics–June 2019 (provisional). Available online: https://www.stats.govt.nz/information-releases/agricultural-production-statistics-june-2019-provisional (accessed on 10 December 2019).
25. Beef and Lamb NZ. *Compendium of New Zealand Farm Facts 2018*, 42nd ed.; Beef and Lamb NZ: Wellington, New Zealand, 2018.

26. Swaffield, S. Sustaining the Country: Ideals, Opportunities and Imperatives for Future Rural Landscapes. In *Presented at Environmental Defense Society 2008 Conference Conflict in paradise*; Auckland, New Zealand, 11–12 June 2008.
27. Ministry of Justice. Māori Land Court. Your Māori Land: Māori Land Trusts and Incorporations. Available online: https://Maorilandcourt.govt.nz/your-Maori-land/trusts-and-incorporations/#ahu-whenua-trust (accessed on 10 December 2019).
28. Kingi, T. Ahuwhenua. Māori land and agriculture. Te Ara–the Encyclopaedia of New Zealand. Available online: https://teara.govt.nz/en/ahuwhenua-maori-land-and-agriculture (accessed on 10 December 2019).
29. Stats NZ. Agricultural Production Statistics: June 2016 (final)–Māori Farms Tables (corrected). Available online: https://www.stats.govt.nz/information-releases/agricultural-production-statistics-june-2016-final (accessed on 10 December 2019).
30. NZ History. Treaty of Waitangi. NZ Government. Available online: https://nzhistory.govt.nz/politics/treaty-of-waitangi (accessed on 23 April 2020).
31. Swaffield, S. Sustainability Practices in New Zealand Agricultural Landscapes under an Open Market Policy Regime. *Landsc. Res.* **2014**, *39*, 190–204. [CrossRef]
32. Tourism, N.Z. What we do–100% Pure New Zealand. Available online: https://www.tourismnewzealand.com/about/what-we-do/campaign-and-activity/ (accessed on 10 December 2019).
33. Beef and Lamb NZ. Taste Pure Nature–Our Story. 2019. Available online: http://tastepurenaturenz.co.nz/ (accessed on 10 December 2019).
34. MfE (Ministry for the Environment) & Stats NZ. New Zealand's Environmental Reporting Series: Our Fresh Water 2017. Publication Number ME 1305. Available online: https://www.mfe.govt.nz/sites/default/files/media/Environmental%20reporting/our-fresh-water-2017_1.pdf (accessed on 10 December 2019).
35. MfE (Ministry for the Environment) & Stats NZ. New Zealand's Environmental Reporting Series: Environment Aotearoa 2019. Publication number: ME 1416. Ministry for the Environment and Stats NZ, 2019 April. Available online: https://www.mfe.govt.nz/environment-aotearoa-2019 (accessed on 10 December 2019).
36. Ciraad, E.; Walker, S.; Price, R.; Barringer, J. An updated assessment of indigenous cover remaining and legal protection of New Zealand's land environments. *New Zealand J. Ecol.* **2015**, *39*, 309–315.
37. Stats NZ. Conservation Status of Indigenous Land Species. Available online: https://www.stats.govt.nz/indicators/conservation-status-of-indigenous-land-species (accessed on 10 December 2019).
38. MBIE (Ministry of Business, Innovation and Employment). Energy in New Zealand 2018. Available online: https://www.mbie.govt.nz/assets/d7c93162b8/energy-in-nz-18.pdf (accessed on 10 December 2019).
39. MfE (Ministry for the Environment). New Zealand's Greenhouse Gas Inventory 1990–2017. Publication number: 1411. Available online: https://www.mfe.govt.nz/sites/default/files/media/Climate%20Change/nz-greenhouse-gas-inventory-2019.pdf (accessed on 10 December 2019).
40. MfE (Ministry for the Environment). About New Zealand's Emission Reduction Targets. Available online: https://www.mfe.govt.nz/climate-change/climate-change-and-government/emissions-reduction-targets/about-our-emissions (accessed on 10 December 2019).
41. MfE (Ministry for the Environment). Climate Change–Likely Climate Change Impacts in New Zealand. Available online: https://www.mfe.govt.nz/climate-change/likely-impacts-of-climate-change/likely-climate-change-impacts-nz (accessed on 12 January 2020).
42. MfE (Ministry for the Environment). Climate Change Response (Zero Carbon) Amendment Act 2019. Available online: https://www.mfe.govt.nz/climate-change/zero-carbon-amendment-act (accessed on 11 February 2020).
43. New Zealand Legislation. Climate Change Response (Zero Carbon) Amendment Act 2019. Available online: http://www.legislation.govt.nz/act/public/2019/0061/latest/LMS183736.html (accessed on 11 February 2020).
44. Overseer Ltd. Overseer FM. Available online: https://www.overseer.org.nz/ (accessed on 11 February 2020).
45. New Zealand Legislation. Resource Management Act (RMA) 1991. Available online: http://www.legislation.govt.nz/act/public/1991/0069/latest/DLM230265.html (accessed on 11 February 2020).
46. Beef and Lamb NZ. The Future of Meat. How should New Zealand's Red Meat Sector Respond to Alternative Protein Advancements? Available online: https://beeflambnz.com/sites/default/files/levies/files/Future%20of%20Meat%20Final%20Report.pdf (accessed on 10 December 2019).
47. Jiang, G.; Grafton, M.; Pearson, D.; Bretherton, M.; Holmes, A. Integration of precision farming data and spatial statistical modelling to interpret field-scale maize productivity. *Agric* **2019**, *9*, 237. [CrossRef]

48. Pearson, D.; Gorman, J. Exploring the relevance of a landscape ecological paradigm for sustainable landscapes and livelihoods: A case-application from the Northern Territory Australia. *Landsc. Ecol.* **2010**, *25*, 1169–1183. [CrossRef]
49. Seppelt, R.; Dormann, C.F.; Eppink, F.V.; Lautenbach, S.; Schmidt, S. A quantitative review of ecosystem services studies: Approaches, shortcomings and the road ahead. *J. Appl. Ecol.* **2011**, *48*, 630–636. [CrossRef]
50. Fischer, J.; Lindenmayer, D. Landscape modification and habitat fragmentation: A synthesis. *Glob. Ecol. Biogeogr.* **2007**, *16*, 265–280. [CrossRef]
51. Fahrig, L. Relative effects of habitat loss and fragmentation on population extinction. *J. Wildl. Manag.* **1997**, *61*, 603–610. [CrossRef]
52. Bezuidenhout, Y. The Role of Landscape Ecology in the Management of Agroecosystems. Available online: https://www.researchgate.net/publication/325119414_The_role_of_landscape_ecology_in_the_management_of_agroecosystems (accessed on 10 December 2019).
53. Ekroos, J.; Tiainen, J.; Seimola, T.; Herzon, I. Weak effects of farming practices corresponding to agricultural greening measures on farmland bird diversity in boreal landscapes. *Landsc. Ecol.* **2019**, *34*, 389–402. [CrossRef]
54. Forman, R.T.T.; Godron, M. Patches and structural components for a landscape ecology. *Bioscience* **1981**, *31*, 733–740.
55. Peterjohn, W.T.; Correll, D.L. Nutrient dynamics in an agricultural watershed: Observation on the role of a riparian forest. *Ecology* **1984**, *65*, 1466–1475. [CrossRef]
56. Opdam, P.; Rijsdijk, G.; Hustings, F. Bird communities in small woods in an agricultural landscape: Effects n area and isolation. *Biol. Conserv.* **1985**, *34*, 33–354. [CrossRef]
57. Turner, M.G. Landscape Ecology: The effect of pattern on process. *Annu. Rev. Pf Ecol. Syst.* **1989**, *20*, 171–197. [CrossRef]
58. Nassauer, J.I.; Opdam, P. Design in science: Extending the landscape ecology paradigm. *Landsc. Ecol.* **2008**, *23*, 633–644. [CrossRef]
59. Termorshuizen, J.W.; Opdam, P. Landscape services as a bridge between landscape ecology and sustainable development. *Landsc. Ecol.* **2009**, *24*, 1037. [CrossRef]
60. Landis, D.A. Designing agricultural landscapes for biodiversity-based ecosystem services. *Basic Appl. Ecol.* **2017**, *18*, 1–12. [CrossRef]
61. Dominati, E.J.; Maseyk, F.J.F.; Mackay, A.D.; Rendel, J.M. Farming in a changing environment: Increasing biodiversity on farm for the supply of multiple ecosystem services. *Sci. Total Environ.* **2019**, *662*, 703–713. [CrossRef] [PubMed]
62. Moore, A.; Johnson, M.; Gbolagun, J.; Miller, A.; Rombouts, A.; van der Ven, L.; Lord, J.; Coutts, S.; Pagan, M.; Hall, G.B. Integrating agroecology and sustainable tourism: Applying geodesign to farm management in Aotearoa New Zealand. *J. Sustain. Tour.* **2018**, *26*, 1543–1561. [CrossRef]
63. Tran, D.X.; Pearson, D.; Palmer, A.; Gray, D. Developing a planning and design approach for the sustainable management of hill country farms in New Zealand. *Land* **2020**. under review.
64. Pearson, D.; Nawaz, M.; Wasson, R. Creating sustainable future landscapes: A role for landscape ecology in the rangelands of Northern Australia. *Rangel. J.* **2019**, *41*, 13–21. [CrossRef]
65. Gliessman, S.R. *Agroecology: Ecological Processes in Sustainable Agriculture*; CRC Press: Boca Raton, FL, USA, 1998.
66. Rodale, R. Breaking New Ground: The Search for Sustainable Agriculture. *Futurist* **1983**, *17*, 15–20.
67. Rhodes, C.J. The imperative for Regenerative Agriculture. *Sci Prog.* **2017**, *100*, 80–129. [CrossRef]
68. LaCanne, C.E.; Lundgren, J.G. Regenerative agriculture: Merging farming and natural resource conservation profitably. *PeerJ* **2018**, *6*, e4428. [CrossRef]
69. Ludwig, J.; Tongway, D.; Freudenberger, D.; Noble, J.; Hodgkinson, K. *Landscape Ecology Function and Management*; CSIRO Publishing: Clayton, Australia, 1997.
70. Vejre, H.; Abildtrup, J.; Andersen, E.; Andersen, P.S.; Brandt, J.; Busck, A.; Dalgaard, T.; Hasler, B.; Huusom, H.; Kristensen, L.S.; et al. Multifunctional agriculture and multifunctional landscapes—land use as an interface. In *Multifunctional Land Use*; Mander, Ü., Wiggering, H., Helming, K., Eds.; Springer: Berlin, Heidelberg, 2007. [CrossRef]
71. Lovell, S.T.; Johnston, D.M. Creating multifunctional landscapes: How can the field of ecology inform the design of the landscape. *Front. Ecol. Environ.* **2009**, *7*, 212–220. [CrossRef]

72. Lovell, S.T.; Johnston, D.M. Designing landscapes for performance based on emerging principles in landscape ecology. *Ecol. Soc.* **2009**, *14*, 44. [CrossRef]
73. O'Farrell, P.J.; Anderson, P.M. Sustainable multifunctional landscapes: A review to implementation. *Curr. Opin. Environ. Sustain.* **2010**, *2*, 59–65. [CrossRef]
74. Mendez, V.E.; Bacon, C.M.; Cohen, R. Agroecology as a transdisciplinary, participatory and action-oriented approach. *Agroecol. Sustain. Food Syst.* **2013**, *37*, 3–18.
75. Francis, C.; Lieblein, G.; Gliessman, S.; Breland, T.A.; Creamer, N.; Harwood, R.; Salmonsson, L.; Helenius, J.; Rickerl, D.; Salvador, R.; et al. Agroecology: The ecology of food system. *J. Sustain. Agric.* **2003**, *22*, 99–118. [CrossRef]
76. Gliessman, S. Transforming food systems with agroecology. *Agroecol. Sustain. Food Syst.* **2016**, *40*, 187–189. [CrossRef]
77. McGranahan, D.A. Ecologies of Scale: Multifunctionality Connects Conservation and Agriculture across Fields, Farms, and Landscapes. *Land* **2014**, *3*, 739–769. [CrossRef]
78. Dalgaard, T. Landscape Agroecology: Managing Interactions between Agriculture, Nature and Socio-Economy. In Proceedings of the MACE Conference on 'Multi-level processes of integration and disintegration', Green Week Scientific Conference, International Congress Centre, Berlin, Germany, 14–15 January 2009.
79. Crooke, I.R.; Queenborough, S.A.; Mattison, E.H.A.; Bailey, A.P.; Sandars, D.L.; Graves, A.R.; Morris, J.; Atkinson, P.W.; Trawick, P.; Freckleton, R.P.; et al. Integrating socio-economics and ecology: A taxonomy of quantitative methods and a rereview of their use in agro-ecology. *J. Appl. Ecol.* **2009**, *46*, 269–277.
80. Flora, C.B. Building Sustainability Agriculture. *J. Sustain. Agric.* **1992**, *2*, 37–49. [CrossRef]
81. Barrett, G.W. Landscape Ecology. *J. Sustain. Agric.* **1992**, *2*, 83–103. [CrossRef]
82. Ryszkowki, L. Agriculture and Landscape Ecology. In *Landscape Ecology in Agroecosystems Management*; Ryszkowski, L., Ed.; CRC Press: Boca Raton, FL, USA, 2002; 8p, Chapter 16. [CrossRef]
83. Barrett, G.W.; Skelton, L.E. Agrolandscape ecology in the 21st century. In *Landscape Ecology in Agroecosystems Management*; Ryszkowski, L., Ed.; CRC Press: Boca Raton, FL, USA, 2002; 10p, Chapter 15. [CrossRef]
84. Fry, G.L.A. Multifunctional landscapes—Towards transdisciplinary research. *Landsc. Urban Plan.* **2001**, *57*, 159–168. [CrossRef]
85. Selman, P. Planning for landscape multifunctionality. *Sustain. Sci. Pract. Policy* **2009**, *5*, 45–52. [CrossRef]
86. Godfree, R.; Firn, J.; Johnson, S.; Knerr, N.; Stol, J.; Doerr, V. Why non-native grasses pose a critical emerging threat to biodiversity conservation, habitat connectivity and agricultural production in multifunctional rural landscapes. *Landsc. Ecol.* **2017**, *32*, 1219–1242. [CrossRef]
87. Stanek, E.; Lovell, S. Building multifunctionality into agricultural conservation programs: Lessons learned from designing agroforestry systems with central Illinois landowners. *Renew. Agric. Food Syst.* **2019**, 1–9. [CrossRef]
88. Mander, U.; Helming, K.; Wiggering, H. *Multifunctional Land Use: Meeting Future Demands for Landscape Goods and Services*; Springer: Berlin, Germany, 2007.
89. Campagna, M. Metaplanning: About designing the Geodesign process. *Landsc. Urban Plan.* **2016**, *156*, 118–128. [CrossRef]
90. Li, W.; Milburn, L.A. The evolution of geodesing as a design and planning tool. *Landsc. Urban Plan.* **2016**, *156*, 5–8. [CrossRef]
91. Steinitz, C. Design is a verb; design is a noun. *Landsc. J.* **1995**, *14*, 188–200. [CrossRef]
92. Aspinall, R.; Pearson, D. Integrated geographical assessment of environmental condition in water catchments: Linking landscape ecology, environmental modelling and GIS. *J. Environ. Manag.* **2000**, *59*, 299–319. [CrossRef]
93. Hulshoff, R.M. Landscape indices describing a Dutch landscape. *Landsc. Ecol.* **1995**, *10*, 101–111. [CrossRef]
94. Haines-Young, R.; Chopping, M. Quantifying landscape structure: A review of landscape indices and their application to forested landscapes. *Prog. Phys. Geogr.* **1996**, *20*, 418–445. [CrossRef]
95. Frohn, R.C. *Remote Sensing for Landscape Ecology: New Metric Indicators for Monitoring, Modeling and Assessment of Ecosystems*; Lewis Publications: Boca Raton, FL, USA, 1998.
96. Vialatte, A.; Barnaud, C.; Blanco, J.; Ouin, A.; Choisis, J.-P.; Andrieu, E.; Sheeren, D.; Deconchat, M.; Clement, F.; Esquerre, D.; et al. A conceptual framework for the governance of multiple ecosystem services in agricultural landscapes. *Landsc. Ecol.* **2019**, *34*, 1653–1673. [CrossRef]

97. Iverson Nassauer, J. Culture and changing landscape structure. *Landsc. Ecol.* **1995**, *10*, 229–237. [CrossRef]
98. Wu, J.; Hobbs, R. Key issues and research priorities in landscape ecology: An idiosyncratic synthesis. *Landsc. Ecol.* **2002**, *17*, 355–365. [CrossRef]
99. Antrop, M. Sustainable landscapes: Contradiction, fiction or utopia? *Landsc. Urban Plan.* **2005**, *75*, 187–197. [CrossRef]
100. Wu, J. Landscape of culture and culture of landscape: Does landscape ecology need culture. *Landsc. Ecol.* **2010**, *25*, 1147–1150. [CrossRef]
101. Swaffield, S. Empowering landscape ecology-connecting science to governance through design values. *Landsc. Ecol.* **2013**, *28*, 1193–1201. [CrossRef]
102. Harmsworth, G.R.; Awatere, S. Indigenous Māori knowledge and perspectives of ecosystems. In *Ecosystem Services in New Zealand—Conditions and Trends*; Dymond, J.R., Ed.; Manaaki Whenua Press: Lincoln, New Zealand, 2013; pp. 274–286.
103. Primary Sector Council of NZ. Fit for a Better World. A Strategic Approach. Available online: https://fitforabetterworld.org.nz/primary-sector-council/ (accessed on 12 February 2020).
104. Hobbs, R. Future landscapes and the future of Landscape Ecology. *Landsc. Urban Plan.* **1997**, *37*, 1–9. [CrossRef]
105. Pearson, D. Landscape Ecology: Its role as a transdisciplinary science for rangeland sustainability. *Rangel. J.* **2013**, *35*, 363–371. [CrossRef]
106. Journeaux, P.; van Reenen, E.; Pike, S.; Manjala, T.; Miller, D.; Austin, G. *Literature Review and Analysis of Farmer Decision Making with Regard to Climate Change and Biological Gas Emissions*; A report prepared for the Biological Emissions Reference Group by Independent Agriculture and Horticulture Consultant Network; AgFirst Waikato Ltd.: Hamilton, New Zealand, 2016.
107. Pannell, D.J.; Marshall, G.R.; Barr, N.; Curtis, A.; Vanclay, F.; Wilkinson, R. Understanding and promoting adoption of conservation practices by rural landholders. *Aust. J. Exp. Agric.* **2006**, *46*, 1407–1424. [CrossRef]
108. Vanclay, F. Social principles for agricultural extension to assist in the promotion of natural resource management. *Aust. J. Exp. Agric.* **2004**, *44*, 213. [CrossRef]
109. Bohnet, I.C.; Roberts, B.; Harding, E.; Haug, K.J. A typology of graziers to inform a more targeted approach for developing natural resource management policies and agricultural extension programs. *Land Use Policy* **2011**, *28*, 629–637. [CrossRef]
110. Leopold, A. The Farmer as a Conservationist, 1939. In *The River of the Mother of God and Other Essays by Aldo Leopold*; Flader, S.L., Callicott, J.B., Eds.; Univ of Wisconsin Press: Madison, WI, USA, 1991; p. 263.
111. Naveh, Z. Landscape ecology and sustainability. *Landsc. Ecol.* **2007**, *22*, 1437–1440. [CrossRef]
112. Ferguson, J. A sustainable future for the Australian rangelands. *Rangel. J.* **2012**, *34*, 27–32. [CrossRef]
113. Potschin, M. Landscape and Landscape Ecology Landscape ecology in different parts of the world. In *Development and Perspectives of Landscape Ecology*; Bastian, O., Steinhardt, U., Eds.; Kluwer Academic Publications: Amsterdam, The Netherlands, 2002; pp. 38–47.

land

Review

Developing a Landscape Design Approach for the Sustainable Land Management of Hill Country Farms in New Zealand

Duy X. Tran *, Diane Pearson, Alan Palmer and David Gray

School of Agriculture and Environment, College of Sciences, Massey University, Palmerston North 4414, New Zealand; D.Pearson@massey.ac.nz (D.P.); A.S.Palmer@massey.ac.nz (A.P.); D.I.Gray@massey.ac.nz (D.G.)

* Correspondence: D.Tran@massey.ac.nz; Tel.: +64-27-270-4426

Received: 11 May 2020; Accepted: 31 May 2020; Published: 3 June 2020

Abstract: Landscape modification associated with agricultural intensification has brought considerable challenges for the sustainable development of New Zealand hill country farms. Addressing these challenges requires an appropriate approach to support farmers and design a better landscape that can have beneficial environmental outcomes whilst ensuring continued profitability. In this paper we suggest using geodesign and theories drawn from landscape ecology to plan and design multifunctional landscapes that offer improved sustainability for hill country farm systems and landscapes in New Zealand. This approach suggests that better decisions can be made by considering the major landscape services that are, and could be, provided by the landscapes in which these farm systems are situated. These important services should be included in future landscape design of hill country by creating a patterning and configuration of landscape features that actively maintains or restores important landscape functioning. This will help to improve landscape health and promote landscape resilience in the face of climate change. Through illustrating the potential of this type of approach for wider adoption we believe that the proposed conceptual framework offers a valuable reference for sustainable farm system design that can make an important contribution to advancing environmental management globally as well as in New Zealand.

Keywords: multifunctional landscapes; landscape services; geodesign; landscape ecology; agricultural landscape planning

---

1. Introduction

The green revolution in agriculture that occurred during the second half of the 20th century has greatly contributed to increased global food and fiber production, which has enabled a rapidly growing world population to be fed [1]. In order to increase productivity, agricultural intensification has taken the form of an increase in single crop cultivation and chemical and mechanical inputs [2]. This has led to negative impacts on the environment, evident through a loss of biodiversity and a decline in soil and water quality [3]. In response to the resultant environmental issues and the need to feed a growing population, agriculture needs to evolve from a production paradigm that has focused primarily on productivity and profitability to a more sustainable paradigm that focuses on how to ensure productivity can support human needs whilst also preserving important land resources and environmental integrity [4]. Recently, society and the market have initiated a shift from a focus on agricultural productivity and intensification to a focus on sustainable farming (with an emphasis towards efficiency, sustainability and resilience) [5]. New Zealand (NZ) is a good example of an agriculturally-focused nation that faces sustainable production challenges. It has achieved great improvements in agricultural productivity and product quality over recent decades [6] but progress has come with significant environmental costs [7].

Although New Zealand is accredited as one of the more sustainable countries in the world and was ranked 11th globally in 2019 for sustainable development [8], its agricultural sector is facing a number of significant issues, such as soil degradation [9], water pollution [10], greenhouse gas (GHGs) emissions [11,12] and soil erosion [13]. Moreover, the possible impact of climate change (e.g., increased flood risk, storm damage and drought severity) is also a crucial threat to agricultural production [14]. To respond to these environmental issues farmers are now faced with a situation of having to operate farm systems that are productive and profitable as well as being sustainable with limited impacts on the environment [15]. This is a major challenge facing NZ farmers, as agricultural production could potentially become increasingly constrained by environmental regulations [16] as governments also respond to growing environmental concerns.

The environmental challenges facing future farming systems are likely to strongly impact upon NZ hill country farming. This is because environmental issues compound already high concerns for these farms, which are associated with the contemporary impacts of increasing production costs, market volatility, climate change, highly variable topography and climatic conditions, and more dispersed and isolated families and communities [17]. This means that future hill country farming systems will need to improve its profitability and build resilience in order to be able to adapt to a changing climate whilst reducing its impacts on the environment. To do this, farmers will need good support systems to help with land use decision-making. However, current land use planning and management approaches that support farm and landscape decision-making in NZ reveal several limitations, such as lack of data and model transparency, insufficient collaboration capability among researchers, policy-makers and other end users, and are limited in terms of the communication of modeling results to end-users [18]. Additionally, some land and environmental planning tools are not simple to implement, as farmers are overwhelmed with information and the process required to develop the land and environmental plans [19]. Consequently, these limitations will reduce the effectiveness of land and environmental planning strategies. Therefore, the development of an effective landscape design approach will be central to helping farmers develop profitable and sustainable farming systems in the future.

The multiple objectives of sustainable agriculture require a multifunctional agricultural landscape that promotes agricultural production whilst ensuring environmental standards are met [20], and landscape ecology can have an important role to play in this [21]. Developing a multi-functional agriculture landscape that provides multiple landscape services (i.e., ecosystem services) for society in addition to the service of food and fiber production [22] has become a key focus for sustainable agricultural research and policy-making, and this has been widely discussed internationally [23–26]. However, there is a gap between theory and practice [27], and transferring the concept of creating multifunctional landscapes into the practice of landscape planning and management has proved to be challenging [28]. The reason for this is that agricultural landscape planning needs to be implemented for a specific geographical region that is strongly associated with local knowledge [29]. This needs the planning process to involve the considerations of local people and therefore requires participation and collaboration of the main stakeholders [30]. Often this does not happen and as a result, local people (or "people of the place") may not agree or may not be able to afford the future landscape scenarios proposed by landscape planners [31], so it is critical that the relevant different stakeholder groups can actively contribute to designing the future landscape by bringing their knowledge and aspirations to the table [32]. It is important that effective landscape planning and scenario development involves an iterative collaborative process and that a design-driven perspective is taken [33].

Recently, geodesign has emerged as an efficient instrument for the implementation of sustainable landscape planning [34]. Geodesign integrates geospatial technologies and scientific methods (e.g., geospatial science, environmental science) to inform spatial decision-making based on the knowledge and information obtained from spatial data [35]. By integrating multiple layers of geographic information and spatial analysis models, geodesign enables the identification and development of a future landscape that has an appropriate spatial pattern or configuration of landscape features [36]. This also enables the rapid generation of future landscape scenarios for a study area, the ability to visualize change scenarios, and the assessment of the impacts of future landscape designs on multiple landscape functions and services [37]. In addition, visualization tools and iterative quantitative modeling used in geodesign can promote collaboration between participants, as they enable stakeholders to enter into the discussion and express their opinions and aspirations as part of the design procedure [38]. Among the geodesign frameworks that have been published, the operation framework developed by [31] has been disseminated to a wide range of landscape and environmental design situations [39]. This framework considers landscape design as an iterative process in which the collaboration among the group of people involved in the design process (which includes design professionals, the people of the place, information technologists and geographic scientists) is an integral part of the design procedure, and the relevant stakeholders play a central role in all of the design stages [40].

The adaptation of the framework outlined by Steinitz offers a potential solution to guide farm system decision-making for the creation of multifunctional landscapes. This paper develops these ideas by proposing a landscape design approach for the sustainable land use planning and management of hill country farms in NZ. The approach developed utilizes geodesign and the concepts of landscape function and services as informed by landscape ecology. The specific objectives of this paper are to: (i) define the major challenges facing current and future agriculture in the NZ hill country that need to be considered in future farm landscape planning; and (ii) design a framework that can assist in the creation of multifunctional landscapes for sustainable agricultural production. In doing this, the paper highlights the benefits of integrating geodesign into multifunctional landscape planning for the creation of multifunctional farm landscapes in NZ. This research offers a valuable reference for sustainable farm system design that can make an important contribution to advancing environmental management globally as well as in NZ.

2. Multifunctional Landscape and Geodesign

2.1. Multifunctional Landscapes and its Application in Agricultural Landscape Planning

A multifunctional landscape is seen as being one capable of providing a wide range of landscape services (i.e., ecosystem services) covering three main areas relevant to landscape management, i.e., ecological, cultural and production functions [41]. Natural and semi-natural landscapes are considered as multifunctional landscapes because they provide a variety of goods and services to people, such as food and fiber, climate regulation and water purification [42]. However, multifunctional landscapes of the past have been transformed into more simple landscapes (e.g., single-function landscapes), which have a dominant land use type (e.g., croplands). This is because land managers and decision-makers have focused on increasing agricultural productivity rather than considering the benefits that can be provided by a multifunctional landscape [43]. The transformation of a natural landscape into an agricultural landscape, especially one that is farmed intensively, leads to landscape simplification. This occurs as diverse stands of native vegetation are cleared and replaced with a monoculture, resulting in a loss of biodiversity and a reduction in landscape functions and services [44]. Many studies have demonstrated the negative effects of landscape simplification, such as an increase of insecticide use [45], loss of habitats [46] and a reduction in biological control [47]. As such, developing a multifunctional landscape is increasingly being recognized as offering an appropriate solution for solving the issues and challenges that have arisen from agricultural intensification (i.e., landscape simplification) [20].

A landscape ecological approach based on the concept of the multifunctional landscape has been widely applied in sustainable agricultural landscape planning [48–51]. In the European Union (EU), multifunctional agriculture is significantly encouraged, as it is a key concept of the Common Agricultural Policy for the EU countries [52]. This concept is also applied in many developed countries, like the United States of America, Canada and Australia (as cited in [48]). The overall goal of agricultural landscape planning that is based on the concept of the multifunctional landscape is to develop future or alternative landscapes that can enhance and increase the multifunctionality of the current landscape, in order to achieve a better balance between agricultural production and other landscape services [53].

A landscape services approach has been applied in order to examine a wide range of issues in NZ, such as biological control [54], biodiversity [55] and land use planning and management [56–58]. However, some limitations have been identified, such as the obstacles associated with incorporating the landscape services concept into agricultural land use decision-making and the lack of participation and contribution of farmers in the creation of a future multifunctional landscape [59]. Another important limitation is the inadequacy of the link between landscape service supply and demand. For instance, there is a lack of research that assesses the imbalance between landscape service supply and non-market demand in a spatially explicit manner (e.g., where and to what extent in the landscape are certain services generated by agro-ecosystems needed to maintain desirable environmental conditions) [59]. In addition, current research involves limited measurements of landscape services (e.g., biodiversity) other than production services (e.g., food and fiber) across small areas (e.g., farm scale) [60]. Therefore, the ability to fully integrate multiple landscape services into land use planning and the implementation of a collaborative planning process will provide a greater opportunity to address these gaps.

In this research we have used the terms landscape functions and landscape services instead of ecosystem functions and ecosystem services. Although these concepts are often used as synonyms, it is advocated that the use of the terms landscape functions and landscape services is more appropriate, as these terms are more attractive to people outside the ecological sciences and may be more related to local people [61]. In addition, landscape functions and services are more appropriate to landscape planning, which is strongly associated with human involvement, whereas ecosystems are often perceived as merely natural and semi-natural systems [62].

2.2. Geodesign

Geodesign is defined as "a design and planning method which tightly couples the creation of design proposals with impact simulations informed by geographic contexts, systems thinking, and digital technology" [37] (p. 29).

Geodesign often involves collaboration among essential groups (e.g., the design experts, geographical information system (GIS) scientists, information technologists and the stakeholders) to develop and decide sustainable scenarios for the future landscape of their area [31]. These groups comprise a geodesign team, and collaborate based on a set of questions and methods, typically within a framework that consists of six key questions [63]:

1. How should the landscape be described in content, space and time?
2. How does the landscape operate?
3. Is the current landscape working well?
4. How might the landscape be altered? By what policies and actions, where and when?
5. What differences might the changes cause?
6. How should the landscape be changed?

Six models are employed to answer each of the six questions, ranging from the description of the study area to the decision on a desired future landscape. The process presented in the framework is an integrated and continuous procedure, because the outcome of each phase serves as an input for the subsequent phase, and all the stages of the design (understand study area, specify methods and perform study) are incorporated into one unified system.

Recently, geodesign has emerged as an innovative design approach, developed to provide alternative scenarios for future landscapes, based on a rich knowledge base about the environment [35]. Geodesign has been extensively applied to different landscape planning and management case studies, such as urban development [64–66], environmental management [67,68], and sustainable agricultural land use [69–71]. This approach is also flexible in terms of the scale of application (e.g., a street, a farm, small town, catchment and regional scales) [72,73]. Various examples of geodesign applications were discussed at the Geodesign Summit in 2019 [74]. In the case of agricultural landscape planning, a typical example of the application of geodesign is illustrated through the use of the approach to increase food production and biofuel commodities and improve water quality and habitat performance in the Seven Mile Creek watershed, Minnesota, United States [75]. At the farm scale, another example is a geodesign project that utilizes 3D modeling and geospatial analysis to design strategies for climate change mitigation on a farm in Iowa, United States. This project applies geodesign for real-time scenario development and interactively evaluating alternative farm design [76].

In New Zealand, GIS tools and techniques have been widely applied to solve environmental problems [77–81], but the tools and approaches that link design and GIS have not been readily available [82] and there is a limited number of applications that follow the geodesign framework to solve problems in landscape planning, especially at the farm scale. For instance, only one previous paper was identified that applied geodesign to plan a route for visitor access across a farm in NZ [83]. Meanwhile, there is an absence of geodesign applications that focus on developing a multifunctional agricultural landscape. Hence, research that utilizes geodesign procedures in an agricultural landscape, especially at the farm scale, has the potential to contribute to environmental management studies in NZ but has not yet been fully explored.

2.3. The Benefits of Integrating Geodesign into Multifunctional Landscape Planning

Geodesign offers an efficient solution to implement the adaptive design of multifunctional landscape planning. It is an effective approach because it can (1) promote collaborative and adaptive landscape design among different stakeholders, (2) advance landscape multifunctionality in agricultural landscape planning and (3) enable the implementation of the landscape design problem on a large scale. One key advantage of geodesign compared to traditional landscape planning approaches is that it allows for collaboration among researchers, policy-makers, and other end users, because it divides the landscape planning into different processes (with six distinct phases) and allows the participants involved to provide feedback and suggestions at any step in the process [84]. With the latest geospatial technologies (e.g., WebGIS application, human–computer interaction tools), participants can directly interact with both the data and the analysis procedure. This is considered an efficient way to initiate discussion among different stakeholders about alternative futures or visions for the new landscape [85]. In addition, a geodesign framework includes a decision model [63] so this can make the application of landscape planning more adaptive and practical. It supposes that decision-makers may agree with or oppose the proposed change, so the decision model that includes a negotiation process (e.g., discussion) and method (e.g., Delphi method) will be able to effectively build consensus among decision-makers and other stakeholders, as well as able to suggest necessary modifications to the proposed changes or the development of new adapted plans [40]. Additionally, alternative landscape plans are not always going to provide a first and ultimate fix, so decision-makers can iteratively discover the trade-offs and synergies inherent in different design scenarios until a final decision is achieved [27]. In the case of NZ, where agricultural land is under private ownership and farmers are the final decision-makers, the inclusion of a decision model in landscape planning is critical because it increases the role of farmers in the landscape design process. This can potentially facilitate the approval by private landowners of proposed landscape change and therefore make the implementation of future landscape change more feasible [86].

Compared to other landscape design methods and techniques, geodesign has a great potential to break new grounds in the design industry, as it is based on advanced geospatial technologies [87]. State-of-the-art remote sensing, image processing, and GPS tools and techniques enable the collection and processing of large amounts of biophysical data in high spatial and temporal resolution. This means that geodesign can be implemented at various scales [88]. This is an asset in the case of NZ hill country, where geospatial data, and especially data for farm scale application, is poor. For instance, it is common that there is a lack of detailed land use land cover (LULC) data at the farm scale, so in this case high-resolution remotely sensed data can be used to produce necessary LULC information. In addition, a wide range of tools, techniques, and models that have arisen from GIS, geospatial information, spatial statistics and computer programming can be incorporated into one spatially informed planning platform so as to allow comprehensive landscape design issues to be resolved (as it is a multidisciplinary or transdisciplinary problem) and to provide a more efficient communication mechanism for the modeling processes and results [89]. Geodesign can also integrate different kinds of environmental and socio-economic models to quantitatively and spatially measure the cost and benefit of implementing alternative land use scenarios [90]. The outcome from each geodesign question, such as landscape structure and pattern, environmental sensitivity and risk, and future landscape scenarios, are presented in a meaningful and intuitive visualization (e.g., dynamic map, table and graph) so as to provide better assistance for decision-makers. Once the farmers can see the environmental issues on their farms and measure how much they must invest and can benefit from the future landscape, they will be more confident to make a decision.

In order to effectively co-design future multifunctional landscapes, non-technical people (i.e., farmers) may require an understanding of the basic landscape concepts, such as different socio-economic and ecological landscape functions and services [34]. Through collaboration with other participants, farmers can receive support from technical people (e.g., scientists) to acquire the necessary knowledge. More importantly, geodesign employs GIS models, tools and applications to incorporate numerous layers of geospatial information and transfer the key multifunctional landscape concepts into realistic visualization forms (e.g., map, graph) [91], as well as to develop future landscape scenarios, visualize them and analyze the impacts of the different proposed landscapes on multiple landscape services [37]. This may encourage farmers to pay attention to not only commerce and food production but also the role of the non-trade functions of agricultural landscapes. In addition, the adaptive design capability of geodesign enables farmers' priorities to be considered, as their preferences or requirements can be set in the land change model and this can subsequently increase the ability to reach a consensus between farmers and other stakeholders on future multifunctional landscape scenarios.

3. The Case Study

3.1. Introduction to New Zealand Hill Country and its Environmnetal Challenges

New Zealand hill country is defined as land with slopes above 15° and located below an altitude of 1000 m above sea level [92]. This landscape type covers a variety of land class types, climatic conditions, geology, and topography properties [93]. The hill country landscape is a mixture of steeplands, rolling land and flat land [94] (Figure 1).

Figure 1. Hill country landscape: (**a**) earth flow; (**b**) slump/earth flow; (**c**) steep slopes > 25°; (**d**) flat topped ridges; (**e**) hilly slopes 15–25°. Photographed by Duy X. Tran in 2019.

Most of the hill country is classified as land use capability classes (LUC[1]) 5–7, which are suitable for pastoral grazing, tree crops or production forestry [95]. Other LUC (e.g., classes 3, 4 and 8) often occupy a small proportion of hill country land. Overall, approximately 10 million hectares of NZ's total land area is classified as hill country (approximately 37.5% of the NZ land surface), with the majority located in the North Island (6.3 million hectares or 23.5% of NZ's total area) [96]. Approximately half of the hill country land (5 million hectares or 18% of NZ's total area) is allocated to pastoral farmland used for sheep and cattle farming [97]. It has been reported that sheep and cattle farms, the bulk of which are located on hill country, also own some 25% of the total native vegetation remaining in NZ [98]. This significant proportion of native vegetation plays an important role in carbon sequestration and biodiversity conservation [99].

In recent years, hill country farms have become increasingly concerned about environmental issues [100]. For instance, Beef and Lamb NZ, an industry organization representing NZ's sheep and beef farmers, has defined four pillars for an environment strategy (created in 2018) for sheep and cattle farms. These include working towards cleaner freshwater, healthy and productive soils, thriving biodiversity and reduced emissions in order to achieve the goal of being carbon neutral by 2050 [98]. However, several environmental problems and the negative effects of climate change are challenging the sustainable development of this type of farming [17,101].

Understanding the major environmental challenges facing hill country farming is vital to ensure that good planning for future landscape and farm systems is made for the future. In the following section, the five major issues that need to be considered prior to landscape planning in order to make progress towards a more sustainable future for hill country farming are examined in the discussion below. These are land use change and deforestation, soil erosion, climate change, agricultural intensification and change in consumers preferences.

Large areas of native forests and shrubland on the steep erodible terrain of NZ hill country were cleared for pastoral farming by the European settlers [102]. Although limited deforestation has occurred since the 1980s, the response to historic deforestation and land clearing is still affecting the current landscape and environment [103]. The negative impact of deforestation has been reflected in a significant increase in soil erosion [104]. Over the last three decades, reforestation and regenerating of native vegetation has been increasingly implemented on hill country [105] to reduce sediment loss from steep slopes into river channels [106] and to increase the capacity for climate change mitigation and adaptation [107]. Plantation forestry has a number of positive effects on the environment, such as a reduction in soil erosion and flooding, an increase in carbon sequestration and a reduction in the GHGs emissions, and it has also reduced pressure on native forests for timber [108]. For instance, a report on erosion-prone hill country (for the period of 1997 to 2002) reported that the area prone to soil erosion had been reduced by 36,000 hectares (3% of the total erosion-prone area) due to the planting of exotic forest or through reversion to native shrublands [104]. However, removal of forest cover at harvest on steeplands can result in significant environmental impacts, such as landslides, debris flows and significant impacts on water quality due to sediment loss into waterways [102].

Over the period of 1990–2015, the total area of hill country sheep and beef farms decreased by approximately 1.3 million ha [100]. This is because the more productive land was converted to dairy farming or higher-value horticultural crops [97] whilst the steeper, less productive land, which is more vulnerable to erosion and generates lower financial returns [109], was converted to an alternative land use, such as forestry, manuka[2] for honey production or retirement and a return to native vegetation [17]. Recently, carbon farming, which is a conversion from pasture to forest, is emerging as an alternative to sheep and beef farming in hill country due to the dramatic increase in the price of carbon credits, and this conversion can bring high economic profit if this occurs in eligible areas (the land areas where

---

1 LUC class 1 is flat highly productive land and LUC class 8 is very steep unproductive land.

2 Manuka honey is a monofloral honey produced from the nectar of the manuka, a native tree (Leptospermum scoparium) that grows in New Zealand and parts of Australia.

there has been a net land use conversion to new forests since 1 January 1990) [110]. Therefore, it is important that relevant scientific information (e.g., mapping of suitable areas for alternative land use options) is available so as to allow landowners to make appropriate decisions [111].

In the NZ hill country, soil erosion is a critical issue that contributes to land degradation [112]. The hill country has a high level of both natural and human-induced erosion [113] due to the amalgamation of coarse-textured soils, high slope terrain, high precipitation and agricultural intensification [114]. Soil erosion presents a significant problem to the practices of current pastoral land, and it is especially severe on hill country, which has substantial areas of steep slopes and erodible rocks (e.g., soft rock) [115], especially in combination with high rainfall and high-intensity rainstorms [104]. It is estimated that 192 million tons of soil are lost every year because of erosion and 44% of this takes place on grassland [116]. Soil erosion does not only represent a reduction in NZ's natural resources, but it also results in a decline in soil productivity and a reduction in water quality [113]. In relation to the economic cost, the effects of soil erosion on hill country can be on-site (e.g., a reduction in productivity) and off-site (e.g., an increase in flood damage in downstream regions) [95]. The cost of erosion control and mitigation has often surpassed the value of the production that can be obtained from that land [117], and an increase in vegetation cover (e.g., regenerating native trees, tree planting and reforestation) has been described as being the most efficient solution for this problem [118]. For instance, it is argued that the reforestation of unstable and degraded land can not only effectively control current erosion problems, but also preclude the formation of new forms of erosion [106]. For these reasons, soil erosion control is important in land use planning and management in hill country. Characterizing the detailed spatiotemporal pattern of soil erosion and the capability of landscape options to reduce this environmental problem are central to managing this issue.

Climate change is recognized as one of the significant challenges facing agricultural development in NZ hill country [109], as the country's land-based economy is profoundly reliant on climatic conditions for the growth of pasture and crops [119]. Increased frequency of intense rainfall events is a threat to soil erosion, predominantly on hill country steepland [113]. The expected increase in drought frequency and intensity in some drier regions may severely affect the water supply, agricultural production and magnitude of wildfire risk [120–123]. Climate change may also directly affect pastoral production, because the seasonal variation of pasture growth is influenced by rising temperatures, CO_2 fertilization and changes in rainfall patterns [119]. Thus, climate change may result in greater variation in sheep and cattle growth and productivity [124]. Adaptation solutions have been developed to reduce risks and build resilience to climate change impacts in NZ. Some of the major adaptation strategies put emphasis on a long-term perspective and suggest an integration of climate change adaptation into the decision-making process [125].

The impacts of climate change on hill country farming may also be off-site and long-term [126]. For example, climate-concerned international consumers or markets might result in an increased demand for the outputs from production that has low GHGs emissions [127], which will mean that NZ agricultural production will have to change accordingly to maintain their market share. Considerable effort has been made by both the public and private sector to determine climate change mitigation solutions in NZ, and central to this is to reduce the GHGs emissions caused by agricultural production [128]. For instance, in the agricultural sector it is suggested that changes to land use and pasture management will be key solutions for reducing GHGs emissions along with other strategies (e.g., innovation in animal genetics and breeding) [129]. It is therefore suggested that multiple land use options (e.g., pasture, forestry, horticulture) need to be considered in relevant areas of the hill country and the integration of climate change scenarios needs to be made into future land use plans for more comprehensive land use planning and management models capable of addressing issues related to climate change.

Intensive pastoral farming in hill country increased rapidly from the late 1940s to early 1980s. This was due to the increasing demand and rising prices for meat and wool products on the world market [130]. It was also supported by government subsidies for land development, as well as the emergence of new technological developments (e.g., aerial topdressing—application of aircraft for fertilizers spreading and pasture seeding) [131]. Intensive farming during this period was reflected in a re-clearance of a substantial area of native vegetation that was planted in pasture grass for meat and wool production, an extensive application of fertilizers and agrichemicals, and a high stocking rate [132]. Agricultural intensification and inappropriate agricultural practices in the hill country have resulted in negative impacts on the environment. This includes an increase in soil erosion on steeplands where native bush and shrubs were cleared for pasture, a decrease in biodiversity [133], an increase of nutrient leaching [134], a reduction in water quality [135] and a reduction of future carbon stocks [136].

Since 1984, hill country farming has undergone a dramatic reduction in sheep numbers, as more productive pastoral land was converted to other land use types, and farmers also reduced the stocking rate [130]. Recently, sustainable practices such as organic farming have also been increasingly implemented on some NZ hill country farms [137]. These sorts of changes have resulted in both productivity improvement and better environmental outcomes [138,139]. However, despite these successes, some hill country farms have been managed intensively to improve economic profitability and unsuitable agricultural practices are still happening [140,141]. For example, farmers tend to eliminate the reinvading bush, shrubs and exotic weeds in some high-altitude farms, or marginal land is not fenced off, and this limits the restoration of native forest, which can cause problems associated with soil erosion as well as reducing future carbon stocks [136].

With increasing concerns about the environmental impacts of agricultural intensification and the need to mitigate the impacts of climate change, it is necessary to promote a wider uptake of more sustainable agricultural practices in the hill country [131,142]. Several studies have shown that applying appropriate farming practices, such as developing shelterbelts and hedges, using native plants, or riparian plantings can significantly enhance the provision of landscape services (e.g., increase biodiversity, pest control, water purification) [143–146]. Moreover, by applying appropriate land management decisions it is possible to increase farm productivity whist reducing the impacts on the environment [147]. For instance, using soil data, topographic maps and spatial analysis can help to determine optimum fertilizer application to the appropriate areas and assist in the reduction of nitrate runoff [148]. Making informed decisions requires good land use planning and management tools, which can provide detailed land use and environmental information at the farm scale.

Meat and fiber from NZ hill country farms are well recognized on the world market because they are safe, nutritious and grass-fed [109]. However, international consumers are increasingly becoming aware of environmental issues that arise from intensive agricultural production and are requesting more eco-friendly agricultural products or products that respect environmental standards [149,150]. Therefore, the way food is produced (i.e., considering factors such as environmental impact, animal welfare and carbon footprint) is becoming an important focus of consumer preference that now needs to be considered alongside the more traditional values associated with high quality [109]. Subsequently, environmental and sustainability standards are being added to the traditional quality and health standard requirements for produce. As a result of changes in consumers' preferences, NZ hill country farmers are required to adopt more sustainable farming systems that take into account the impact of their practices on the environment [151,152]. Adopting more sustainable farming practices will not only improve the environmental health of NZ hill country; it also presents an opportunity for farmers to capitalize on the growing market for environmentally-friendly products. The utilization of effective tools for land use planning and appropriate resource allocation will contribute to solving many of the issues faced by NZ's hill country.

3.2. *Tools and Approaches for Supporting Sustainable Land Use Planning used in New Zealand*

Government organizations, research institutions and the private sector have developed a wide range of land use models and tools to help to address some of the impacts associated with land use issues and environmental concerns in NZ [153] as well as supporting farm and landscape decision-making in hill country [154]. Various types of models have enabled the user to deal with specific environmental concerns, such as carbon sequestration [155], greenhouse gas emissions [12], soil erosion [156], nutrient loss [157] or water use [158]. There are also various applications to help farmers deal with the issues of farm production: AgInform [159], BiomeBGC [160], MitAgator [161] and Farmax [162]. There are more complex land use models (e.g., Agent-Based Rural Land Use New Zealand (ARLUNZ) [163], New Zealand Forest and Agriculture Regional Model (NZ-FARM) [164], Waikato Integrated Scenario Explorer (WISE) [101]), which can take into account different factors, such as land use information, socio-economic conditions and environmental parameters (climate, water quality and biodiversity) to provide projected outcomes for land use and environmental, economic and demographic indicators.There are also Whole Farm Plans (WFP), which are a long-established land management tool that is being widely used across NZ to deal with both economic considerations and environmental constraints on farming systems [165]. Recently, the Land and Environment Plan (LEP) was developed by Beef and Lamb NZ to support sheep and beef farmers to have a better understanding about the land and environmental issues that exist on their farms so that they can develop a land use and environment plan to manage these issues [154].

Land use and environmental planning tools and models have contributed significantly to agricultural development as well as supporting farmers in decision-making to address sustainability issues in NZ [18]. However, several improvements are required to increase the effectiveness of the model outcomes. A review conducted by Motu Economic and Public Policy Research in 2018 [18] pointed out some gaps that NZ land use modeling needs to take into account in order to improve its usability. These include increasing the reliability of the data and increasing model transparency, improving collaboration capability among researchers, policy-makers and other end-users, enhancing the communication associated with the model results to stakeholders and enabling a climate change mitigation framework in the land use planning process [18].

Of upmost importance for improving model reliability is the use of data with a better spatial and temporal resolution. It was conceded that NZ lacks good GIS data when compared to many other developed countries [18]. Using data that are too generalized means that it is not possible to achieve accurate analysis, especially at the finer scales (e.g., farm and paddock) [166], as it will fail to capture the variability present at the a farm scale in relation to factors such as variations in slope, soil types, soil fertility and effective rainfall [97]. Therefore, it is important to consider acquiring better data at a high resolution so that land use optimization models can adequately represent the environmental and ecosystem services variability within small farm-scale areas [59]. It is also important to have an appropriate amount of time-series data to enable trends in environmental issues to be examined over time [167]. This is critical for predicting change to the future environment and is an important basis on which to develop long-term land use and environmental planning.

It is also important that land use planning takes into account the collaboration between different stakeholders so that they can be involved in the planning process [18]. Farm system research has evolved to recognize that there needs to be a shift towards more trans-disciplinary approaches to farm system management, which require collaboration and integration of knowledge and ideas between different people, disciplines and methods [168]. A framework that allows the collaboration among researchers, policy-makers and users will enable them to easily and actively be involved in the planning process and develop a comprehensive land use plan that satisfies multiple objectives (i.e., socio-cultural, economic and environment issues).

A land use planning framework needs to enable the integration of different models and tools to better solve different aspects of land use planning. Various tools and applications have been developed to deal with a wide range of the land use and environmental issues in NZ, and these continue to receive support and investment from the government, research institutions and the private sector [169]. However, the integration of different models into a single framework to solve interdisciplinary questions has been limited in NZ [169]. Hence, future land use models need to consider the synergies between different models and techniques so that they can be utilized to solve real world problems.

It is necessary to improve the communication of both the modeling processes and the outcomes from this process. Some land use and environmental planning tools are not simple to implement, as they require farmers to prepare and enter a large and complicated set of data into the model. Such models may also use several complex spatial analysis processes (e.g., map overlay, multicriteria analysis) to define environmental issues on a farm, which are often difficult to interpret [170]. In fact, land and environmental planning is a spatially complex problem, since it requires the integration of a wide range of geographic information (e.g., soil, land use types, climate variables) to define issues and allocate and plan resource use. Without an appropriate spatial support system, the process is intimidating for farmers, as they are overwhelmed with information [19]. An adaptive spatial-based decision support system incorporating spatial analysis tools and techniques would provide models with the capability to capture, store, manipulate, analyze, manage and visualize land resources and environmental data and information [171]. This would make the results more transparent to the various decision-makers through the use of different forms of visualization (such as interactive maps, graphs and reports).

3.3. *Why Have Multifunctional Landscapes on Hill Country Farms?*

Landscape simplification is significant in the hill country landscapes in NZ, as there has been extensive conversion of the natural vegetation to pastoral land associated with the expansion of agriculture since European settlement [172] (Figure 2). The area under pasture has increased rapidly from less than 70,000 hectares in 1861 to 1.4 million hectares in 1881, 4.5 million hectares in 1901, and 7.7 million hectares in 2016 [173,174]. The conversion of natural ecosystems (e.g., forest, shrubs) to pasture has led to a degradation of landscape functions in the sense that provisioning services (e.g., grazing production) are dominant and increasing, whereas regulating services are weak and declining. In other words, the human need to produce food has eroded the capacity of the ecosystems to produce other essential services (e.g., regulating services) [175]. The negative impacts of landscape intensification on hill country are well documented, such as the impacts on the provision of freshwater [176], soil and plant biodiversity [177,178] or soil biogeochemical cycling of nutrients [179].

Figure 2. Example of NZ hill country landscapes: high simplification with low regulating services (**left**); low simplification with high regulating services (**right**). Photographed by Duy X. Tran in 2019.

It is suggested that the issues that originate from landscape simplification due to agricultural intensification could only be solved by taking into account the redesign of agricultural landscapes [180]. The goal of the approach suggested in this paper is to redesign (or plan) the agricultural landscape to achieve a better balance between ecological, cultural and production functions [180]. The cultural and production functions reflect the capability of the landscape to produce goods and services that support human demand from a socio-economic perspective [181]. Whereas maintaining and improving the ecological functions of the landscape is thought to increase biodiversity and landscape connectivity, which has important conservation and landscape resilience implications, including the ability to adapt to climate change and disturbance [182–185]. The creation of this kind of landscape is expected to be an effective solution to solve the problems related to landscape simplification in NZ hill country farms.

The justification for this is that a multifunctional agricultural landscape that is made up of a mosaic of natural habitat areas and agricultural production areas could help to maximize the balance of ecological and socio-economic demands and minimize the conflicts between them [186]. This allows the landscape to provide multiple services and achieve multiple objectives (both agricultural production demand and environmental standards) [51,187]. By diversifying farming activities, farmers can secure various income sources whilst at the same time promoting the cultural and natural heritage [188]. For instance, a sustainable multifunctional agricultural landscape may provide the option to develop agritourism or environmental education. Consequently, this contributes to an added income for farmers and increases public interest in the social and environmental values that the farms bring to the community. However, the challenge comes in determining how to implement the multifunctional landscape approach as a practical application to develop a sustainable agricultural landscape where different land use and land cover types (e.g., wetland pasture, forest, and horticulture) co-exist and the land use pattern is appropriate to maintain and promote sufficient heterogeneity so that different landscape functions work properly [189,190].

4. A Conceptual Framework that Combines Multifunctional Landscapes and Geodesign Concepts for Sustainable Agricultural Landscape Planning

In this paper, we propose a conceptual framework for sustainable agricultural landscape planning (Figure 3) that integrates the concept of multifunctional landscapes with a geodesign approach. It also draws on several studies that have focused on developing a framework for landscape planning [87,187,191]. The geodesign processes in this framework follows the approach outlined by Steinitz [31], which comprises six phases. These phases are: (1) Landscape description, (2) Landscape process; (3) Landscape evaluation; (4) Future landscape scenarios development; (5) Impact assessment of alternative landscape scenarios; and (6) Decision-making. Within this framework, the basic concepts of a multifunctional landscape and a landscape services approach can be fully integrated.

4.1. Landscape Description

The landscape description phase is used to describe a general picture of the study area. The first task is to define an appropriate boundary for the study area. It is suggested to consider both the social and ecological boundaries (i.e., boundaries that cover both the ecological and socio-political/cultural functions of the landscape) when defining the boundary for the study area [192,193]. The ecological boundary of the study area may be determined based on ecological processes or biophysical constraints (e.g., land management unit, catchment or sub-catchment boundaries) [194]. The cultural functions of the landscape sometimes may not align with the boundary of the ecological functions, so it is recommended to work with the "people of the place" to properly define an appropriate boundary [192]. In NZ, a catchment group is a community network of farmers who operate in a particular catchment. They are increasingly committed to tackling environmental issues and responding to a long-term sustainable development plan for the catchment [195,196]. Working with such groups offers the potential to assist in developing a relevant cultural boundary.

Once the study area boundary is defined, the next step is collecting necessary physical and socio-economic data, especially data for characterizing landscape services and environmental issues (e.g., soils, topography, LULC, climate). In the case of NZ hill country, the lack of data is a limiting factor for analysis. To navigate around this requires an integration of multiple data sources that may come from the government, research institutions, remote sensing and field surveys. In fine-scale applications, such as those undertaken at the farm and paddock scales, information provided by farmers (e.g., stocking unit, grazing rotation) is an important source of data. The integration of local and global data to model landscape services is therefore a valuable option to address data deficiencies in remote and data-poor areas [197]. In addition, data are normally archived in different formats, standards and scales, so data standardization is an important step to make sure multiple data layers can be appropriately integrated and used.

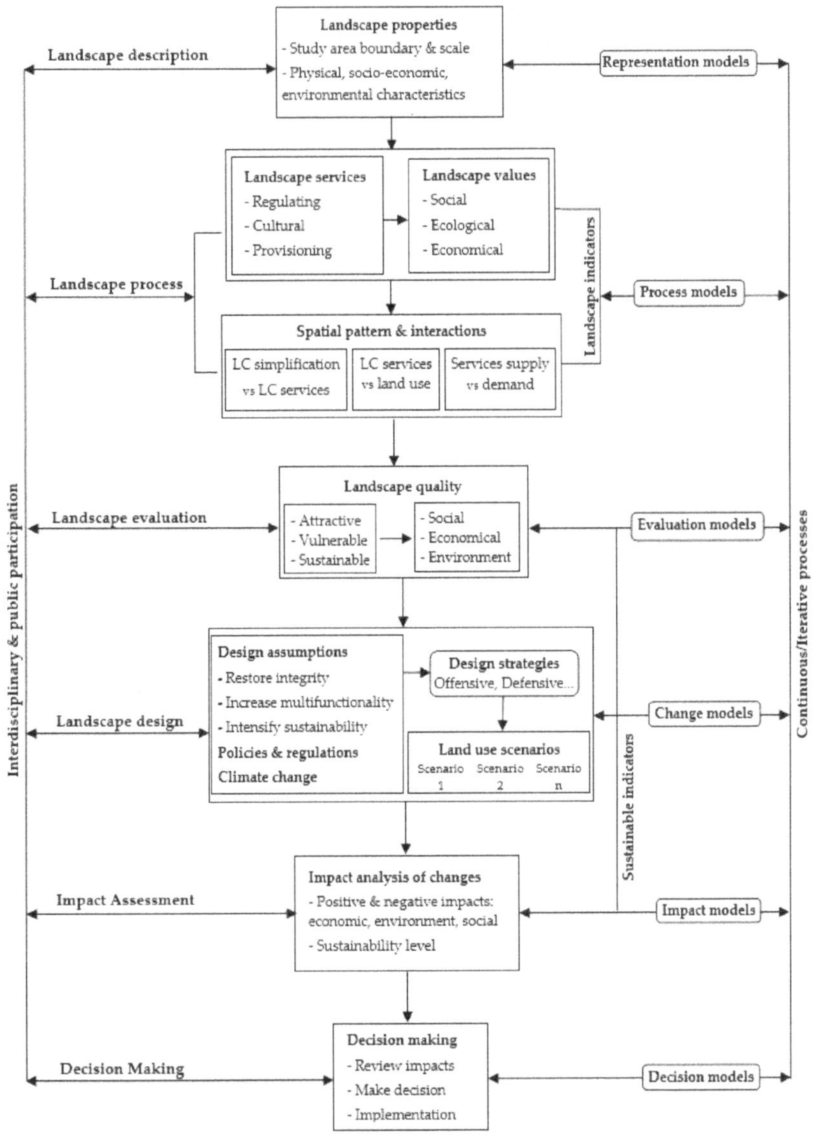

Figure 3. A conceptual framework for a multifunctional landscape-based geodesign for sustainable landscape planning, adapted from Steinitz [31].

A representation model (e.g., a raster-based 2D data model) is used to organize and visualize data collected for the study area through space and over time. For example, maps visualize LULC types of a farm or rainfall and temperature patterns in a catchment from 20–30 years ago to the present. This gives a general understanding of the landscape (from the past to the present) and provides necessary input for the other stages of the framework. Data resolution and availability will affect all other processes of landscape design, as the difference in the resolution and level of data accuracy in the input process could lead to completely different results. For instance, small landscape features (e.g., small plots of shrubs or ponds) play an important role in a farm, such as providing biodiversity, water resources and shade for stocks. However, these features are often eliminated in the low-resolution data (e.g., LULC at the catchment or smaller scale), so landscape services provided by these features may not be quantified when using such coarser data.

4.2. Landscape Process/Operation

The landscape process phase aims to define key processes in the study area that include both physical/ecological drivers and socio-economic drivers. The first step is spatially and quantitatively characterizing major landscape functions, services, and values. This provides insights into the landscape operation in which important landscape characteristics are examined. An example of major landscape functions or services supply and their indicators on a hill country landscape in NZ is presented in Table 1. It is important that landscape service supply is estimated in monetary units so that the overall benefit that a landscape provided can be easily measured. Various economic valuation methods have been used for estimating the value of landscape services, such as market prices, replacement cost and provision cost [57]. For instance, the market price method can be applied directly to convert several landscape services (e.g., pasture and timber production, carbon sequestration) to appropriate monetary units. Many indirect use services (e.g., drought mitigation, flood mitigation, nutrient retention) may require using provision cost or replacement cost methods to transfer their qualities to monetary value. Additionally, the economic value of landscape aesthetics in an area can be evaluated by estimating people's willingness to pay for visiting heritage or tourist sites distributed in the landscape.

Table 1. Example of landscape services in the hill country New Zealand.

	Landscape services	Indicators	Units/Measurements
Provisioning	Stock feed production	Pasture productivity	Pasture yield (kg Dry matter/ha/yr)
	Timber production	Timber productivity	Volume of harvest (tons/ha/yr)
	Provision of Manuka honey	Honey production	Honey yield (kg/ha/yr)
	Fresh water supply	Water availability for irrigation or drinking	Water supply (m^3/yr)
Regulating	Erosion control	Capacity of landscape for retaining sediments	Retained soil (ton/ha/yr)
	Flood regulation	Rainfall absorbed by soil	Runoff (mm/ha/yr)
	Drought mitigation	Capacity of landscape for retaining moisture	Drought severity (mm/ha/yr)
	Carbon sequestration	Landscape capacity to trap/absorb carbon	Sequestered carbon (ton/ha/yr)
	Nutrient retention	Part of nutrient retained by the soil	Retained N and P (kg/ha/yr)
	GHG emissions mitigation	Amount of GHG emissions regulated	CO_2, N_2O, CH_4 (tons/ha/yr)
Supporting	Forest biodiversity	Landscape capacity to support natural habitats	Native/natural forest (%/ha)
	Plant habitat	Rare, endemic, and indicator plant species	Conservation Value index
Information	Aesthetic and amenity values	Sites of beauty and heritage	Number of interest points/km^2
	Recreation and ecotourism	Attractive landscape for recreation activities	Recreation activities suitability

Sources: adapted and revised from [57–60,206,207].

After that, the spatial interaction between the provision of landscape services and landscape simplification and LULC dynamics are analyzed to determine how these processes are linked to each other. A substantial number of studies have stated that the provision of landscape services has been significantly affected by LULC dynamics [198–203]. Quantifying these relationships will be a key to transferring a multifunctional landscape design to a future land use plan. Landscape indicators that reflect the landscape simplification (i.e., agricultural intensification) well (e.g., the proportion of cropland and semi-natural land obtained from LULC data [45]), the variations in landscape services provision (e.g., landscape services change index [204] or multifunctionality index [205]) and spatial regression analysis will be used to characterize the spatial interactions between the change in LULC and variations in landscape services.

Quantifying and mapping landscape services can help farmers recognize and understand the multiple values of their farms. This is an advantage compared to using land cover information, as many landscape services may not be directly quantified by using land cover data alone [208]. Understanding major landscape processes and the interaction between them is the key basis for designing a sustainable multifunctional landscape.

4.3. Landscape Evaluation

The landscape evaluation phase seeks to assess whether the landscape is working well or not [28], in other words, assessing the overall quality of the landscape [209]. In a multifunctional landscape this can be understood as assessing the quality of goods and services that a landscape provides to humans and the environment. To determine landscape quality, an evaluation model that utilizes comprehensive indicators will be used to evaluate the attractiveness, vulnerability and sustainability of the study area. Attractiveness refers to the advantages that landscapes may have for a specific land use purpose or for socio-economic activities (e.g., suitable soil and climate conditions for fruit production). The vulnerability relates to characteristics that negatively contribute to socio-economic development or the environment (e.g., impacts of extreme climate and steep slopes on agricultural production, or negative effects of agricultural intensification on water quality and biodiversity). Sustainability reflects the landscape's capacity for steadily supplying long-term landscape services that are critical for maintaining human and environmental well-being (e.g., a landscape that has different functions and services that co-exist and balance) [210].

Landscape assessment indicators, which can be of various types, including single (e.g., GHG emissions mitigation index), multiple (e.g., a combined-index integrating several parameters, such as soil erosion control, carbon sequestration and drought mitigation), static (the sustainable threshold being classified into a fixed category) and dynamic (the sustainable threshold being subjected to the dynamic interaction between indicators) [87,211], and come from various sources (e.g., expert consultant, environmentalist, empirical analysis, law and regulation) [31], could be used to assess past and present situations of a study site, monitor the design process and compare design alternatives [87]. Hence, choosing appropriate indicators is important for the success of a landscape design project. Suitable landscape indicators should satisfy several requirements, such as the capability to reflect a wide range of landscape services to analyze the trade-offs between landscape service provision and land use change options [212], providing reliable, detailed, understandable, comparable and spatially explicit information to support decision-making [213], and providing cost-effective indicators by utilizing available data or employing low-cost generated data and models [214].

Landscape evaluation models also need to reside within the geographical context in the sense that assessment indicators should recognize and align with existing legitimized environmental strategy and policy and reflect major landscape processes in the study area. For example, in the case study of hill country in NZ, water quality, soil erosion control, drought mitigation, pasture productivity and GHGs emission mitigation could be used as some of the indicators for landscape sustainability assessment.

4.4. Future Landscape Scenarios Development

Based on the results achieved from the landscape evaluation process, change models will be used to define a series of alternative future scenarios for the proposed multifunctional landscapes. In this stage, stakeholders can follow the scenarios developed by scientists or propose their scenarios (a user-defined plan) for the future landscape. Alternative scenarios for future landscape design can be implemented by applying the following procedure:

First, the information on landscape process (characters, services and values) as well as major socio-economic drivers and environment issues are used to define how the landscape should be changed. Determining the expected future landscape is based on several assumptions, such as the preferences of local people, the landscape functions or services that the future landscape will be capable of providing, and the implications of policies and regulations [215]. In agricultural landscapes, the design goal for future landscapes is mainly based on the level of agricultural intensification (or landscape simplification) [216]. Landscapes that have been highly simplified may need to be redesigned in order to restore integrity between provisioning, supporting, regulating and cultural services, whereas the likely design goal for less simplified landscapes is to increase provisioning services while maintaining current levels of other services [180]. Climate change scenarios can be integrated in this step to measure how the changes in climate variability can affect the landscape operation through the interaction with landscape functions.

Afterwards, a design strategy that could take an offensive approach (where the design goal is utilizing the advantageous or attractive landscape characteristics to develop a future landscape), or a defensive approach (where the development of a future landscape is based on one that avoids vulnerability or risks), or a combination of these approaches, will be used to create a specific change model to simulate future change for the landscape [31]. There are different methods of designing for landscape change, such as rule-based, optimized, and agent-based approaches (see [31] (pp. 56–59) for further details). Among these, the use of multi-criteria decision-making (MCA) can be an efficient method to propose future landscape scenarios in the study area, as the creation of a future landscape can be regarded as a complex MCA process [217]. Each land use scenario or option often requires multiple objectives (e.g., erosion control, carbon sequestration, pasture productivity, GHGs emission) and the final decision will be a compromise between the interests of the different stakeholders involved in the design process. The results from these approaches are maps showing the future landscape with the distribution and pattern of different LULC types. Associated with each LULC map will be the provision of landscape services and landscape multifunctionality maps. For each scenario and stage, different alternatives can be created and reassessed iteratively until consensus is achieved.

4.5. Impact Assessment

In the impact assessment of alternative landscape options, the criteria and indicators used in landscape evaluation will be applied to assess the positive and negative impacts (benefits, risks and sustainability) of the future landscape. In a geodesign project, an environmental impact assessment is often implemented to characterize the consequences of the proposed change. In the context of developing a multifunctional agricultural landscape, the impact assessment is related to quantifying the costs and benefits (including both socio-economic and environmental costs) of recovering landscape functions or re-designing the landscape to increase landscape diversification (or landscape multifunctionality). The results of this stage include maps and statistical data showing the cost–benefit ratio of each alternative landscape option. For instance, associated with each land use scenario will be maps showing landscape services provision and value of carbon sequestration, GHGs emissions, erosion control, drought mitigation and pasture productivity, as well as the total benefit (value) of that scenario. This includes the cost to implement such a landscape (e.g., loss of pastoral area, fencing cost, tree planting cost). This will be critical for the decision-making stage.

4.6. Decision-Making

In the last phase, the scenario analysis and group discussion will be conducted with the public, experts and stakeholders. The results of the future landscape scenario development and impact analysis will be utilized for discussion, and this will from a basis for making the final decision. According to Steinitz [31], participants in the geodesign process might give different answers, including "Yes", "Maybe" and "No", in response to proposed scenarios. If decision-makers agree with one of the proposed plans, the next stage is to develop the implementation plan. In case stakeholders are not sure about their decision, further study or analysis is needed to provide more information to help them decide. Sometimes decision-makers may not approve the designed landscape. If this is the case it is necessary to get comments and feedback on why this is so. This will be valuable information to integrate into the landscape project in the future.

The proposed framework in this research inherits the major advantages exhibited by a geodesign approach. These include the fact that it can be a continuous procedure, a multidisciplinary or transdisciplinary approach, and a participatory collaborative planning technique. Moreover, this framework integrates concepts drawn from landscape ecological theory (such as incorporating information on landscape functions and services, landscape simplification and landscape pattern). This means that the theory provides the scientific context to informed and collaborative decision support processes for farm systems that are faced with the need to change in response to environmental pressures and market influences.

5. Conclusions

This paper reviewed the major challenges facing NZ hill country farms and proposed an approach for sustainable agricultural landscape planning. The significant issues facing hill country farming include land use changes and deforestation, soil erosion, agricultural intensification, climate change and the impacts of changes in consumers' preferences. These challenges are considerations for farmers striving towards the long-term sustainable development of NZ's hill country. Currently, landscape simplification associated with agricultural intensification is a significant feature of hill country farms. This may reduce the landscape's capacity to mitigate and adapt to the environmental challenges and climate change effects. Therefore, we have suggested that designing a more sustainable multifunctional landscape is a possible solution to tackle the issues facing NZ hill country. The development of multifunctional agricultural landscapes can contribute towards innovative future farming systems that can deal with emerging environmental issues [218]. In addition, the design of multifunctional landscapes can improve their resilience to change and disturbance [219], which will be crucial for ongoing sustainability in NZ hill country.

This is one of the first studies to propose a geodesign framework for sustainable multifunctional agricultural landscape planning in NZ. By integrating a multifunctional landscape approach in a geodesign context we offer a solution to address some of the implementation problems that have restricted uptake. Considering landscape planning in a design-driven perspective, geodesign embraces collaborative planning (among different stakeholders) as the key to landscape design. It also enables the incorporation of stakeholder values and aspirations as a central element to this process. By dividing the landscape design process into different phases and utilizing geospatial technologies (e.g., human–computer interaction), geodesign allows important stakeholders to be effectively involved and contribute to the planning process. In addition, geodesign enables the use of multiple sources of relevant spatial and temporal resolution data for landscape planning, especially in large-scale applications, as well as being better at dealing with different aspects of land use planning.

The proposed framework in this paper considers the major concepts associated with a multifunctional landscape approach, including landscape functions and services, landscape supply and demand, the value of landscape services, sustainable landscape indicators, spatial patterns and interactions. This facilitates a comprehensive implementation of the multifunctional landscape approach in land use planning and management. A landscape ecological approach has been talked about conceptually for landscape sustainability but has not been widely applied practically in NZ. Therefore, the comprehensive integration of a landscape services approach in landscape planning offers a solution to address some of the limitations faced by current land use planning and management practices in NZ [59,60]. The proposed approach and associated framework can provide a scientific basis towards the development of a future commercial land and environmental planning tool. This will hopefully give farmers and rural professionals more options to conduct useful land use planning at the farm scale.

We believe that the proposed conceptual framework of an integrated landscape ecological (the scientific theory behind a multifunctional landscape concept) and geodesign approach will be a valuable reference for future work about agricultural landscape planning. Ideas around creating multifunctional farm landscapes have been discussed [24,220,221], the role that geodesign can play in future planning has been explored [34,210] and frameworks for developing sustainable landscape based on an integration of geodesign and landscape ecology have been proposed [87,222]. However, there is a lack of a detailed framework that can demonstrate how concepts associated with the generation of multifunctional landscapes can be incorporated into a geodesign process to create a planning tool at the farm scale. Hence, the approach proposed in our paper, which covers a comprehensive description of a type of geodesign process applied to the management of a multifunctional agricultural landscape, will significantly contribute to environmental management studies and illustrate the potential of this type of approach for global application.

Although the framework proposed in this paper demonstrates a comprehensive approach for agricultural landscape planning that can be applied to NZ hill country farms, we acknowledge that future work needs to consider and investigate the issue regarding the financial resources required to support the farmers to overcome their economic concerns associated with changes in land use. Farmers may recognize and be motivated by the great value that extra landscape services can provide and agree with a proposed landscape design, but a barrier to implementation of this design might be the lack of the long-term support that is needed to enable them to be able to afford the cost of implementation and to follow the suggested revised land use and environmental plan. For instance, increasing native woody vegetation on a farm provides a great range of landscape services, but it may potentially affect economic profit in the short term due to the fact that it would decrease land available for grazing and has a low growth rate [223], and thus have less earning capacity in its early life stages. A solution to this is for policy-makers in NZ to consider payment for landscape services. In many countries a wide range of regulating and supporting services are estimated in terms of economic value, and the farmers (i.e., landowners) are able to get a payment for these services [224–226]. Currently, farmers in NZ can only receive payment for carbon sequestration services, so there are no strong incentives to encourage farmers to implement a land use plan that promotes multiple landscape services on their farm. An approach such as the one outlined in this paper can help to demonstrate a proof of concept to policy-makers so that they recognize the greater environmental value that farmers can provide by designing future landscapes for multifunctionality and landscape services and therefore build financial support into future policy-making.

Author Contributions: All authors contributed to the research. D.X.T. conceptualized the main idea of the study. D.X.T. and D.P. designed the structure of the paper. D.X.T., D.P., A.P. and D.G. contributed to the writing of the manuscript. D.P., A.P. and D.G. supervised the final paper content and edited the writing. All authors have read and agreed to the published version of the manuscript.

Funding: This research received no external funding.

Acknowledgments: Acknowledgement is given to Massey University for supporting this research. The authors would like to acknowledge the valuable comments of three anonymous reviewers that assisted with the finalization of this manuscript.

Conflicts of Interest: The authors declare no conflict of interest.

References

1. Welch, R.M.; Graham, R.D. A new paradigm for world agriculture: Meeting human needs: Productive, sustainable, nutritious. *Field Crop. Res.* **1999**, *60*, 1–10. [CrossRef]
2. Kenmore, P.E.; Stannard, C.; Thompson, P.B. *The Ethics of Sustainable Agricultural Intensification*; FAO: Rome, Italy, 2004; Volume 4.
3. Kanianska, R. *Agriculture and Its Impact on Land-Use, Environment, and Ecosystem Services*; IntechOpen: London, UK, 2016; pp. 1–26.

4. Food and Agriculture Organization of the United Nations (FAO). *Transforming Food and Agriculture to Achieve the SDGs: 20 Interconnected Actions to Guide Decision-Makers*; FAO: Rome, Italy, 2018.
5. Proudfoot, I. *KPMG Agribusiness Agenda: The Big Opportunities and Challenges Facing New Zealand Agriculture: Reflections of the Views of Industry Leaders*; KPMG: Auckland, New Zealand, 2010.
6. Robertson, M. Agricultural Productivity in Australia and New Zealand: Trends, Constraints and Opportunities. Available online: https://www.grassland.org.nz/publications/nzgrassland_publication_11.pdf (accessed on 20 May 2020).
7. Baskaran, R.; Cullen, R.; Colombo, S. Estimating values of environmental impacts of dairy farming in New Zealand. *N. Z. J. Agric. Res.* **2009**, *52*, 377–389. [CrossRef]
8. Sachs, J.; Schmidt-Traub, G.; Kroll, C.; Lafortune, G.; Fuller, G. Sustainable Development Report 2019. Available online: https://s3.amazonaws.com/sustainabledevelopment.report/2019/2019_sustainable_development_report.pdf (accessed on 25 July 2019).
9. Ministry for the Environment and Stats NZ. *Land Report Highlights Issues with Soil Degradation*; Ministry for the Environment and Stats NZ: Wellington, New Zealand, 2018.
10. Duncan, R. The challenges of regulating diffuse agricultural pollution to improve water quality A science policy perspective on approaches to setting enforceable catchment load limits in New Zealand. *Case Stud. Environ.* **2017**, *1*. [CrossRef]
11. Kelliher, F.M.; Cox, N.; Van der Weerden, T.J.; De Klein, C.A.M.; Luo, J.; Cameron, K.C.; Di, H.J.; Giltrap, D.; Rys, G. Statistical analysis of nitrous oxide emission factors from pastoral agriculture field trials conducted in New Zealand. *Environ. Pollut.* **2014**, *186*, 63–66. [CrossRef] [PubMed]
12. Safa, M.; Nuthall, P.; Nejat, M.; Greig, B. Predicting CO_2 Emissions from Farm Inputs in Wheat Production using Artificial Neural Networks and Linear Regression Models "Case study in Canterbury, New Zealand". *Int. J. Adv. Comput. Sci.* **2016**, *7*, 268–274.
13. Fernandez, M.A. Adoption of erosion management practices in New Zealand. *Land Use Policy* **2017**, *63*, 236–245. [CrossRef]
14. Reisinger, A.; Mullan, B.; Manning, M.; Wratt, D. *Global and Local Climate Change Scenarios to Support Adaptation in New Zealand: Future Scenarios and Some Sectoral Perspectives*; New Zealand Climate Change Centre: Wellington, New Zealand, 2010; pp. 26–43.
15. Dewes, A.M. Economic Resilience and Environmental Performance of Dairy Farms in the Upper Waikato Region. Master's Thesis, University of Waikato, Hamilton, New Zealand, 2014.
16. Ministry for the Environment. *National Policy Statement for Freshwater Management*; Ministry for the Environment: Wellington, New Zealand, 2019.
17. Scrimgeour, F. Pathways ahead for New Zealand hill country farming. *J. N. Z. Grassl* **2016**, *78*, 73–82.
18. Hendy, J.; Timar, L.; White, D. *Land-Use Modelling in New Zealand: Current Practice and Future Needs*; Motu Economic and Public Policy Research: Wellington, New Zealand, 2018.
19. Synge, K. An evaluation of the Land and Environment Planning Toolkit for Advancing Soil and Nutrient Management in Hill Country and Steepland Farm Systems. Master's Thesis, Massey University, Palmerston North, New Zealand, 2013.
20. McGranahan, D. Ecologies of scale: Multifunctionality connects conservation and agriculture across fields, farms, and landscapes. *Land* **2014**, *3*, 739–769. [CrossRef]
21. Pearson, D. Key Roles for Landscape Ecology in Transformative Agriculture Using Aotearoa—New Zealand as a Case Example. *Land* **2020**, *9*, 146. [CrossRef]
22. Lankoski, J. *Multifunctional Character of Agriculture*; Maatalouden taloudellinen tutkimuslaitos; Agricultural Economics Research Institute: Helsinki, Finland, 2000.
23. Estrada-Carmona, N.; Hart, A.K.; DeClerck, F.A.J.; Harvey, C.A.; Milder, J.C. Integrated landscape management for agriculture, rural livelihoods, and ecosystem conservation: An assessment of experience from Latin America and the Caribbean. *Landsc. Urban Plan.* **2014**, *129*, 1–11. [CrossRef]
24. Dalgaard, T.; Kjeldsen, C.; Hutchings, N.; Happe, K.; Osuch, A.; Damgaard, M.; Zander, P.; Piorr, A. Multifunctional farming, multifunctional landscapes and rural development. In *Multifunctional Land Use*; Springer: Berlin/Heidelberg, Germany, 2007; pp. 183–193.
25. Atwell, R.C.; Schulte, L.A.; Westphal, L.M. How to build multifunctional agricultural landscapes in the US Corn Belt: Add perennials and partnerships. *Land Use Policy* **2010**, *27*, 1082–1090. [CrossRef]

26. Huang, J.; Tichit, M.; Poulot, M.; Darly, S.; Li, S.; Petit, C.; Aubry, C. Comparative review of multifunctionality and ecosystem services in sustainable agriculture. *J. Environ. Manag.* **2015**, *149*, 138–147. [CrossRef] [PubMed]
27. Burgi, M.; Ali, P.; Chowdhury, A.; Heinimann, A.; Hett, C.; Kienast, F.; Mondal, M.K.; Upreti, B.R.; Verburg, P.H. Integrated Landscape Approach: Closing the Gap between Theory and Application. *Sustainability* **2017**, *9*, 1371. [CrossRef]
28. Di Lucia, L.; Usai, D.; Woods, J. Designing landscapes for sustainable outcomes—The case of advanced biofuels. *Land Use Policy* **2018**, *73*, 434–446. [CrossRef]
29. Zasada, I.; Häfner, K.; Schaller, L.; Van Zanten, B.T.; Lefebvre, M.; Malak-Rawlikowska, A.; Nikolov, D.; Rodríguez-Entrena, M.; Manrique, R.; Ungaro, F. A conceptual model to integrate the regional context in landscape policy, management and contribution to rural development: Literature review and European case study evidence. *Geoforum* **2017**, *82*, 1–12. [CrossRef]
30. Natarajan, L. Socio-spatial learning: A case study of community knowledge in participatory spatial planning. *Prog. Plan.* **2017**, *111*, 1–23. [CrossRef]
31. Steinitz, C. *A Framework for Geodesign: Changing Geography by Design*; Esri: Redlands, CA, USA, 2012.
32. Karrasch, L.; Maier, M.; Kleyer, M.; Klenke, T. Collaborative Landscape Planning: Co-Design of Ecosystem-Based Land Management Scenarios. *Sustainability* **2017**, *9*, 1668. [CrossRef]
33. Opdam, P.; Nassauer, J.I.; Wang, Z.F.; Albert, C.; Bentrup, G.; Castella, J.C.; McAlpine, C.; Liu, J.G.; Sheppard, S.; Swaffield, S. Science for action at the local landscape scale. *Landsc. Ecol.* **2013**, *28*, 1439–1445. [CrossRef]
34. Slotterback, C.S.; Runck, B.; Pitt, D.G.; Kne, L.; Jordan, N.R.; Mulla, D.J.; Zerger, C.; Reichenbach, M. Collaborative Geodesign to advance multifunctional landscapes. *Landsc. Urban Plan.* **2016**, *156*, 71–80. [CrossRef]
35. Campagna, M. Metaplanning: About designing the Geodesign process. *Landsc. Urban Plan.* **2016**, *156*, 118–128. [CrossRef]
36. Lee, D.J.; Dias, E.; Scholten, H.J. *Geodesign by Integrating Design and Geospatial Sciences*; Springer: Cham, Switzerland, 2014; Volume 111.
37. Flaxman, M. Fundamentals of geodesign. In *Proceedings of Digital Landscape Architecture*; Anhalt University of Applied Science: Bernburg, Germany, 2010; pp. 28–41.
38. Albert, C.; Vargas-Moreno, J.C. Testing geodesign in landscape planning–first results. In *Digital Landscape Architecture Conference*; Anhalt University of Applied Science: Bernburg, Germany, 2012.
39. Hollstein, L.M. Retrospective and reconsideration: The first 25 years of the Steinitz framework for landscape architecture education and environmental design. *Landsc. Urban Plan.* **2019**, *186*, 56–66. [CrossRef]
40. Campagna, M.; Carl, S.; Di Cesare, E.A.; Cocco, C.; Hrishikesh, B.; Tess, C. *Collaboration in Planning: The Geodesign Approach*; University of Cagliari: Cagliari, Italy, 2016.
41. Wiggering, H.; Müller, K.; Werner, A.; Helming, K. The concept of multifunctionality in sustainable land development. In *Sustainable Development of Multifunctional Landscapes*; Springer: Dordrecht, The Netherlands, 2003; pp. 3–18.
42. Assessment, M.E. *Ecosystems and Human Well-Being*; Island Press: Washington, DC, USA, 2005; Volume 5.
43. De Groot, R.; Hein, L. Concept and valuation of landscape functions at different scales. In *Multifunctional Land Use*; Springer: Berlin/Heidelber, Germany, 2007; pp. 15–36.
44. Emmerson, M.; Morales, M.B.; Onate, J.J.; Batry, P.; Berendse, F.; Liira, J.; Aavik, T.; Guerrero, I.; Bommarco, R.; Eggers, S.; et al. How Agricultural Intensification Affects Biodiversity and Ecosystem Services. *Adv. Ecol. Res.* **2016**, *55*, 43–97. [CrossRef]
45. Meehan, T.D.; Werling, B.P.; Landis, D.A.; Gratton, C. Agricultural landscape simplification and insecticide use in the Midwestern United States. *Proc. Natl. Acad. Sci. USA* **2011**, *108*, 11500–11505. [CrossRef] [PubMed]
46. Poveda, K.; Martinez, E.; Kersch-Becker, M.F.; Bonilla, M.A.; Tscharntke, T. Landscape simplification and altitude affect biodiversity, herbivory and Andean potato yield. *J. Appl. Ecol.* **2012**, *49*, 513–522. [CrossRef]
47. Grab, H.; Danforth, B.; Poveda, K.; Loeb, G. Landscape simplification reduces classical biological control and crop yield. *Ecol. Appl.* **2018**, *28*, 348–355. [CrossRef] [PubMed]
48. Otte, A.; Simmering, D.; Wolters, V. Biodiversity at the landscape level: Recent concepts and perspectives for multifunctional land use. *Landsc. Ecol.* **2007**, *22*, 639–642. [CrossRef]

49. Groot, J.C.J.; Jellema, A.; Rossing, W.A.H. Designing a hedgerow network in a multifunctional agricultural landscape: Balancing trade-offs among ecological quality, landscape character and implementation costs. *Eur. J. Agron.* **2010**, *32*, 112–119. [CrossRef]
50. Fagerholm, N.; Eilola, S.; Kisanga, D.; Arki, V.; Kayhko, N. Place-based landscape services and potential of participatory spatial planning in multifunctional rural landscapes in Southern highlands, Tanzania. *Landsc. Ecol.* **2019**, *34*, 1769–1787. [CrossRef]
51. Rallings, A.M.; Smukler, S.M.; Gergel, S.E.; Mullinix, K. Towards multifunctional land use in an agricultural landscape: A trade-off and synergy analysis in the Lower Fraser Valley, Canada. *Landsc. Urban Plan.* **2019**, *184*, 88–100. [CrossRef]
52. Van Zanten, B.T.; Verburg, P.H.; Espinosa, M.; Gomez-y-Paloma, S.; Galimberti, G.; Kantelhardt, J.; Kapfer, M.; Lefebvre, M.; Manrique, R.; Piorr, A.; et al. European agricultural landscapes, common agricultural policy and ecosystem services: A review. *Agron. Sustain. Dev.* **2014**, *34*, 309–325. [CrossRef]
53. Chopin, P.; Blazy, J.M.; Guinde, L.; Wery, J.; Dore, T. A framework for designing multi-functional agricultural landscapes: Application to Guadeloupe Island. *Agric. Syst.* **2017**, *157*, 316–329. [CrossRef]
54. Fiedler, A.K.; Landis, D.A.; Wratten, S.D. Maximizing ecosystem services from conservation biological control: The role of habitat management. *Biol. Control* **2008**, *45*, 254–271. [CrossRef]
55. Gillespie, M.; Wratten, S.D. The importance of viticultural landscape features and ecosystem service enhancement for native butterflies in New Zealand vineyards. *J. Insect Conserv.* **2012**, *16*, 13–23. [CrossRef]
56. Van den Belt, M.; Schiele, H.; Forgie, V. Integrated Freshwater Solutions—A New Zealand Application of Mediated Modeling. *J. Am. Water Resour.* **2013**, *49*, 669–680. [CrossRef]
57. Dominati, E.; Mackay, A.; Green, S.; Patterson, M. A soil change-based methodology for the quantification and valuation of ecosystem services from agro-ecosystems: A case study of pastoral agriculture in New Zealand. *Ecol. Econ.* **2014**, *100*, 119–129. [CrossRef]
58. Dominati, E.J.; Mackay, A.D.; Bouma, J.; Green, S. An Ecosystems Approach to Quantify Soil Performance for Multiple Outcomes: The Future of Land Evaluation? *Soil Sci. Soc. Am. J.* **2016**, *80*, 438–449. [CrossRef]
59. Van den Belt, M.; Blake, D. Ecosystem services in New Zealand agro-ecosystems: A literature review. *Ecosyst. Serv.* **2014**, *9*, 115–132. [CrossRef]
60. Dominati, E.J.; Maseyk, F.J.F.; Mackay, A.D.; Rendel, J.M. Farming in a changing environment: Increasing biodiversity on farm for the supply of multiple ecosystem services. *Sci. Total Environ.* **2019**, *662*, 703–713. [CrossRef]
61. Hermann, A.; Schleifer, S.; Wrbka, T. The concept of ecosystem services regarding landscape research: A review. *Living Rev. Landsc. Res.* **2011**, *5*, 1–37. [CrossRef]
62. Kienast, F.; Wildi, O.; Ghosh, S. *A Changing World: Challenges for Landscape Research*; Springer Science & Business Media: Dordrecht, The Netherlands, 2007; Volume 8.
63. Nyerges, T.; Ballal, H.; Steinitz, C.; Canfield, T.; Roderick, M.; Ritzman, J.; Thanatemaneerat, W. Geodesign dynamics for sustainable urban watershed development. *Sustain. Cities Soc.* **2016**, *25*, 13–24. [CrossRef]
64. Newman, G.; Malecha, M.; Yu, S.Y.; Qiao, Z.X.; Horney, J.A.; Lee, J.; Kim, Y.J.; Lee, R.J.; Berke, P. Integrating a resilience scorecard and landscape performance tools into a Geodesign process. *Landsc. Res.* **2019**. [CrossRef]
65. Gu, Y.; Deal, B. Coupling systems thinking and geodesign processes in land-use modelling, design, and planning. *J. Dig. Landsc. Arch.* **2018**, *3*, 51–59.
66. Pettit, C.; Hawken, S.; Ticzon, C.; Nakanishi, H. Geodesign—A Tale of Three Cities. In Proceedings of the International Conference on Computers in Urban Planning and Urban Management, Adelaide, Australia, 11–14 July 2017; pp. 139–161.
67. Zandvoort, M.; Van der Vlist, M.J. The Multi-Layer Safety Approach and Geodesign: Exploring Exposure and Vulnerability to Flooding. In *Geodesign by Integrating Design and Geospatial Sciences;* Springer: Cham, Switzerland, 2014; pp. 133–148.
68. Sophronides, P.; Steenbruggen, J.; Scholten, H.; Giaoutzi, M. Geodesign the multi-layered water safety. *Res. Urban Ser.* **2016**, *4*, 113–138.
69. Xie, Y.Q.; Runck, B.C.; Shekhar, S.; Kne, L.; Mulla, D.; Jordan, N.; Wiringa, P. Collaborative Geodesign and Spatial Optimization for Fragmentation-Free Land Allocation. *ISPRS Int. J. Geo-Inf.* **2017**, *6*, 226. [CrossRef]
70. Raumer, H.-G.S.-v.; Jörg, J.; Alfiky, M. Respecting the Role of Agriculture for an Untegrated Landscape Development at the Urban-rural Fringe Using Geodesign Tools. *J. Digit. Landsc. Archit.* **2016**, 327–336. [CrossRef]

71. Jordan, N.R.; Mulla, D.J.; Slotterback, C.; Runck, B.; Hays, C. Multifunctional agricultural watersheds for climate adaptation in Midwest USA: Commentary. *Renew. Agric. Food Syst.* **2018**, *33*, 292–296. [CrossRef]
72. Shannon McElvaney, D.R. Geodesign and the Future of Planning. Available online: https://www.planning.org/publications/document/9121411/ (accessed on 27 March 2020).
73. Esri GeoDesign Team. GeoDesign Defined: A Framework and Process for a More Sustainable World. Available online: https://storymaps.arcgis.com/stories/ebef9d80309b4812bea11e2f39e3d357 (accessed on 27 March 2020).
74. ESRI. Geodesign Summit 2019. Available online: https://www.esri.com/videos/?event=5cb9e8b83d9d867608dcab88&title=Geodesign%20Summit (accessed on 20 November 2019).
75. Minnesota, U.O. Seven Mile Creek Geodesign. Available online: https://maps.umn.edu/geodesign/ (accessed on 23 January 2020).
76. Iowa State University. Geodesigning the Farm of the Future: Parameterizing for Climate Change. Available online: https://www.youtube.com/watch?time_continue=417&v=lnXomp9Pf-I&feature=emb_title (accessed on 20 April 2020).
77. Claessens, L.; Schoorl, J.; Veldkamp, A. Modelling the location of shallow landslides and their effects on landscape dynamics in large watersheds: An application for Northern New Zealand. *Geomorphology* **2007**, *87*, 16–27. [CrossRef]
78. Borrelle, S.B.; Buxton, R.T.; Jones, H.P.; Towns, D.R. A GIS-based decision-making approach for prioritizing seabird management following predator eradication. *Restor. Ecol.* **2015**, *23*, 580–587. [CrossRef]
79. Longdill, P.C.; Healy, T.R.; Black, K.P. An integrated GIS approach for sustainable aquaculture management area site selection. *Ocean Coast. Manag.* **2008**, *51*, 612–624. [CrossRef]
80. Singh, S.K.; Zeddies, M.; Shankar, U.; Griffiths, G.A. Potential groundwater recharge zones within New Zealand. *Geosci. Front.* **2019**, *10*, 1065–1072. [CrossRef]
81. MacMillan, H.; Moore, A.B.; Augé, A.A.; Chilvers, B.L. GIS-based multi-criteria analysis of breeding habitats for recolonising species: New Zealand sea lions. *Ocean Coast. Manag.* **2016**, *130*, 162–171. [CrossRef]
82. Eagle Technology. Geodesign—Ancient Concept, Universal Applications, Modern Tools. Available online: https://www.eagle.co.nz/geodesign (accessed on 27 August 2019).
83. Moore, A.; Johnson, M.; Gbolagun, J.; Miller, A.; Rombouts, A.; van der Ven, L.; Lord, J.; Coutts, S.; Pagan, M.; Hall, G.B. Integrating agroecology and sustainable tourism: Applying geodesign to farm management in Aotearoa New Zealand. *J. Sustain. Tour* **2018**, *26*, 1543–1561. [CrossRef]
84. Eikelboom, T.; Janssen, R. Collaborative use of geodesign tools to support decision-making on adaptation to climate change. *Mitig. Adapt. Strat. Glob. Chang.* **2017**, *22*, 247–266. [CrossRef] [PubMed]
85. Hansen, R.; Pauleit, S. From multifunctionality to multiple ecosystem services? A conceptual framework for multifunctionality in green infrastructure planning for urban areas. *AMBIO* **2014**, *43*, 516–529. [CrossRef] [PubMed]
86. Von Haaren, C.; Warren-Kretzschmar, B.; Milos, C.; Werthmann, C. Opportunities for design approaches in landscape planning. *Landsc. Urban Plan.* **2014**, *130*, 159–170. [CrossRef]
87. Huang, L.; Xiang, W.N.; Wu, J.G.; Traxler, C.; Huang, J.Z. Integrating GeoDesign with Landscape Sustainability Science. *Sustainability* **2019**, *11*, 833. [CrossRef]
88. Li, W.M.; Milburn, L.A. The evolution of geodesign as a design and planning tool. *Landsc. Urban Plan.* **2016**, *156*, 5–8. [CrossRef]
89. Kastuari, A.; Suwardhi, D.; Hanan, H.; Wikantika, K. State of the art of the landscape architecture spatial data model from a geospatial perspective. In Proceedings of the 11th 3D Geoinfo Conference, Athens, Greece, 20–21 October 2016.
90. ESRI. Geodesign in Practice: Designing a Better World. 2013. Available online: https://www.esri.com/~{}/media/Files/Pdfs/library/ebooks/geodesign-in-practice.pdf (accessed on 10 December 2019).
91. Soini, K. Exploring human dimensions of multifunctional landscapes through mapping and map-making. *Landsc. Urban Plan.* **2001**, *57*, 225–239. [CrossRef]
92. Blaschke, P.M.; Trustrum, N.A.; Derose, R.C. Ecosystem Processes and Sustainable Land-Use in New-Zealand Steeplands. *Agric. Ecosyst. Environ.* **1992**, *41*, 153–178. [CrossRef]
93. Molloy, L. *Soils in the New Zealand Landscape: The Living Mantle*; Kings Time Printing Press Ltd.: Hong Kong, China; New Zealand Society of Soil Science: Lincoln, New Zealand, 1998; p. 253.

94. Lynn, I.; Manderson, A.; Page, M.; Harmsworth, G.; Eyles, G.; Douglas, G.; Mackay, A.; Newsome, P. Land use capability survey handbook: A New Zealand handbook for the classification of land. In *Landcare Research*; AgResearch: Lincoln, New Zealand, 2009.
95. Jones, H.; Clough, P.; Hock, B.; Phillips, C. Economic costs of hill country erosion and benefits of mitigation in New Zealand: Review and recommendation of approach. In *Scion*; Forest Res. Inst.: Rotorua, New Zealand, 2008.
96. Hedley, C.; Manderson, A.; Mudge, P.; Roudier, P.; Fraser, S.; Parfitt, R.; Smaill, S.; Schipper, L.; Kelliher, F. *Improved Measurements of Hill Country Soil Carbon to Assist Carbon Change Studies*; Ministry for Primary Industries: Wellington, New Zealand, 2015.
97. Kerr, G. Why a hill country symposium? In *Proceedings of Hill Country Symposium*; NZ Grassland Association: Rotorua, New Zealand, 2016; pp. 7–9.
98. Beef and Lamb. 2018 Annual Report. Available online: https://beeflambnz.com/sites/default/files/B%2BLNZ_AR_2018_web-compressed.pdf (accessed on 26 August 2019).
99. Norton, D.; Pannell, J. *Desk-Top Assessment of Native Vegetation on New Zealand Sheep and Beef Farms*; School of Forestry, University of Canterbury and Institute for Applied Ecology, Auckland University of Technology: Auckland, New Zealand, 2018; pp. 1–32.
100. Beef and Lamb. Hill Country Sheep and Beef Farms. Available online: https://beeflambnz.com/knowledge-hub/PDF/FS077-hill-country-sheep-beef (accessed on 10 August 2019).
101. Rutledge, D.T.; Cameron, M.P.; Briggs, C.M.; Elliott, S.; Fenton, T.; Hurkens, J.; McDonald, G.; McBride, G.; Phyn, D.; Poot, J.; et al. *WISE: Waikato Integrated Scenario Explorer: Technical Specifications, Version 1.4*; Waikato Regional Council: Hamilton, New Zealand, 2016.
102. Phillips, C.; Marden, M.; Basher, L.R. Geomorphology and forest management in New Zealand's erodible steeplands: An overview. *Geomorphology* **2018**, *307*, 107–121. [CrossRef]
103. Harding, J.S. Historic deforestation and the fate of endemic invertebrate species in streams. *N. Z. J. Mar. Fresh* **2003**, *37*, 333–345. [CrossRef]
104. Basher, L.R. Erosion processes and their control in New Zealand. In *Proceedings of Ecosystem Services in New Zealand–Conditions and Trends*; Lincoln University: Lincoln, New Zealand, 2013; pp. 363–374.
105. Ministry for Primary Industries. Funding Programmes for Tree Planting and Research. Available online: https://www.mpi.govt.nz/funding-and-programmes/forestry/ (accessed on 10 December 2019).
106. Marden, M.; Herzig, A.; Basher, L. Erosion process contribution to sediment yield before and after the establishment of exotic forest: Waipaoa catchment, New Zealand. *Geomorphology* **2014**, *226*, 162–174. [CrossRef]
107. Evison, D. *Climate Change Mitigation in New Zealand-What is the Role of New Planted Forests?* University of Canterbury: Christchurch, New Zealand, 2018.
108. Yao, R.T.; Barry, L.E.; Wakelin, S.J.; Harrison, D.R.; Magnard, L.A.; Payn, T.W. *Planted Forests*; Landcare Research: Lincoln, New Zealand, 2013; pp. 62–78.
109. Lang, S. Future Challenges and Opportunities for Hill Country Farming on the East Coast. Available online: https://ruralleaders.co.nz/files/Lang_S_Future_challenges_and_opportunities_for_hill_country_farming_on_the_East_Coast_Final.pdf (accessed on 15 August 2019).
110. Funk, J.M.; Field, C.B.; Kerr, S.; Daigneault, A. Modeling the impact of carbon farming on land use in a New Zealand landscape. *Environ. Sci. Policy* **2014**, *37*, 1–10. [CrossRef]
111. Funk, J.; Kerr, S. Restoring forests through carbon farming on Māori land in New Zealand/Aotearoa. *Mt. Res. Dev.* **2007**, *27*, 202–206. [CrossRef]
112. Ministry for the Environment and Stats NZ. *New Zealand's Environmental Reporting Series: Our Land 2018*; Ministry for the Environment and Stats NZ: Wellington, New Zealand, 2018.
113. McIvor, I.; Douglas, G.; Dymond, J.; Eyles, G.; Marden, M. Pastoral Hill Slope Erosion in New Zealand and the Role of Poplar and Willow Trees in Its Reduction. Available online: https://books.google.ae/books?hl=zh-CN&lr=&id=SheQDwAAQBAJ&oi=fnd&pg=PA257&dq=Pastoral+Hill+Slope+Erosion+in+New+Zealand+and+the+Role+of+Poplar+and+Willow+Trees+in+Its+Reduction&ots=MBxa6CXS46&sig=Bz-Y4BLXDFe8yE3unK7cJvIJTyM&redir_esc=y#v=onepage&q&f=false (accessed on 20 May 2020).
114. Rodda, H.; Stroud, M.; Shankar, U.; Thorrold, B. A GIS based approach to modelling the effects of land-use change on soil erosion in New Zealand. *Soil Use Manag.* **2001**, *17*, 30–40. [CrossRef]

115. Marden, M. Effectiveness of reforestation in erosion mitigation and implications for future sediment yields, East Coast catchments, New Zealand: A review. *N. Z. Geogr.* **2012**, *68*, 24–35. [CrossRef]
116. Ministry for the Environment. Estimated Long-Term Soil Erosion. Available online: http://archive.stats.govt.nz/browse_for_stats/environment/environmental-reporting-series/environmental-indicators/Home/Land/long-term-soil-erosion.aspx (accessed on 23 August 2019).
117. National Water and Soil Conservation Organisation. *Wise Land Use and Community Development*; National Water and Soil Conservation Organisation: Wellington, New Zealand, 1970.
118. Piegay, H.; Walling, D.E.; Landon, N.; He, Q.P.; Liebault, F.; Petiot, R. Contemporary changes in sediment yield in an alpine mountain basin due to afforestation (the upper Drome in France). *Catena* **2004**, *55*, 183–212. [CrossRef]
119. Nottage, R.; Wilcocks, L.; Lee, J.; Burke, C.; Kalaugher, E.; Roche, J.; Beukes, P.; Lieffering, M.; Newton, P.; Li, F. *Impacts of Climate Change on Land-Based Sectors and Adaptation Options*; Ministry for Primary Industries: Wellington, New Zealand, 2012.
120. Pearce, H.G.; Clifford, V. Fire weather and climate of New Zealand. *N. Z. J. For.* **2008**, *53*, 13–18.
121. Mullan, B.; Porteous, A.; Wratt, D.; Hollis, M. *Changes in Drought Risk with Climate Change*; National Institute of Water & Atmospheric Research Ltd.: Wellington, New Zealand, 2005.
122. Moot, D.J.; Mills, A.; Pollock, K.M. Natural resources for Canterbury agriculture. In *Proceedings of the New Zealand Grassland Association*; NZ Grassland Association: Dunedin, New Zealand, 2006.
123. Hendy, J.; Kerr, S.; Halliday, A.; Owen, S.; Deans, N.; Dickie, B.; Hale, J. *Drought and Climate Change Adaptation: Impacts and Projections*; Motu Economic and Public Policy Research: Wellington, New Zealand, 2018.
124. Grafton, M.; Manning, M. Establishing a Risk Profile for New Zealand Pastoral Farms. *Agriculture* **2017**, *7*, 81. [CrossRef]
125. Climate Change Adaptation Technical Working Group. *Adapting to Climate Change in New Zealand*; Climate Change Adaptation Technical Working Group, Ministry for the Environment: Wellington, New Zealand, 2017.
126. New Zealand Agricultural Greenhouse Gas Research Centre. Impacts of Global Climate Change on New Zealand Agriculture. Available online: https://www.nzagrc.org.nz/fact-sheets,listing,94,impacts-of-global-climate-change-on-new-zealand-agriculture.html (accessed on 6 May 2019).
127. McDonald, H.; Kerr, S. Why do New Zealanders care about agricultural emissions? *Policy Q.* **2012**, *8*. [CrossRef]
128. Ministry for the Environment. *New Zealand Action on Climate Change.*; Ministry for the Environment: Wellington, New Zealand, 2016.
129. Cortés-Acosta, S.; David, A.; Fleming, L.H.; Lou, E.; Owen, S.; Small, B. *Identifying Barriers to Adoption of "no-cost" Greenhouse Gas Mitigation Practices in Pastoral Systems*; Motu Working Paper 19-10; Motu Economic and Public Policy Research: Wellington, New Zealand, 2019.
130. Haggerty, J.; Campbell, H.; Morris, C. Keeping the stress off the sheep? Agricultural intensification, neoliberalism, and 'good'farming in New Zealand. *Geoforum* **2009**, *40*, 767–777. [CrossRef]
131. Dodd, M.; Wedderburn, M.; Parminter, T.; Thorrold, B.; Quinn, J. Transformation toward agricultural sustainability in New Zealand hill country pastoral landscapes. *Agric. Syst.* **2008**, *98*, 95–107. [CrossRef]
132. Peden, R. Farming in the Economy—The Golden Years, 1950s to 1980s'. Te Ara—The Encyclopedia of New Zealand. Available online: https://teara.govt.nz/en/farming-in-the-economy/page-7 (accessed on 5 April 2020).
133. Moller, H.; MacLeod, C.J.; Haggerty, J.; Rosin, C.; Blackwell, G.; Perley, C.; Meadows, S.; Weller, F.; Gradwohl, M. Intensification of New Zealand agriculture: Implications for biodiversity. *N. Z. J. Agric. Res.* **2008**, *51*, 253–263. [CrossRef]
134. Parfitt, R.; Mackay, A.; Ross, D.; Budding, P. Effects of soil fertility on leaching losses of N, P and C in hill country. *N. Z. J. Agric. Res.* **2009**, *52*, 69–80. [CrossRef]
135. Betterige, K.; Kawamura, K.; Costall, D.; Ganesh, S.; Luo, D.; Koolaard, J.; Yoshitoshi, R. Intensive Livestock Farming on New Zealand Hill Country Farms Creates Critical Source Areas of Potential Pollution. *J. Integr. Field Sci.* **2017**, *14*, 103.
136. Parliamentary Commissioner for the Environment. *Climate Change and Agriculture: Understanding the Biological Greenhouse Gases*; Parliamentary Commissioner for the Environment: Wellington, New Zealand, 2016.

137. Swaffield, S. Sustainability practices in New Zealand agricultural landscapes under an open market policy regime. *Landsc. Res.* **2014**, *39*, 190–204. [CrossRef]
138. Tayler, M.; Donnelly, L.; Frater, P.; Stocker, N. Lorne Peak Station–achieving sustainable profitability in challenging Southland hill country. In *Proceedings of Hill Country Symposium—Grassland Research Practice Series*; NZ Grassland Association: Rotorua, New Zealand, 2016; pp. 101–108.
139. Ford, J.G.; Ford, C.M. Highlands Station—A story of economic and environmental sustainability. In *Proceedings of Hill Country Symposium—Grassland Research and Practice Series*; NZ Grassland Association: Rotorua, New Zealand, 2016; pp. 149–150.
140. Hoogendoorn, C.; Lambert, M.; Devantier, B.; Theobald, P.; Park, Z. Nitrogen fertiliser application rates and nitrogen leaching in intensively managed sheep grazed hill country pastures in New Zealand. *N. Z. J. Agric. Res.* **2017**, *60*, 154–172. [CrossRef]
141. Fraser, D.; Vesely, E. Connecting north island hill country farmers nutrient requirements with soil mapping units. In *Proceedings of FLRC Workshop; Massey Univeristy*; Massey Univeristy: Palmerston North, New Zealand, 2011.
142. White, T.; Snow, V.; King, W.M. Intensification of New Zealand beef farming systems. *Agric. Syst.* **2010**, *103*, 21–35. [CrossRef]
143. Todd, J.H.; Malone, L.A.; Benge, J.; Poulton, J.; Barraclough, E.I.; Wohlers, M.W. Relationships between management practices and ground-active invertebrate biodiversity in New Zealand kiwifruit orchards. *Agric. For. Entomol.* **2016**, *18*, 11–21. [CrossRef]
144. Sandhu, H.S.; Wratten, S.D.; Cullen, R. Organic agriculture and ecosystem services. *Environ. Sci. Policy* **2010**, *13*, 1–7. [CrossRef]
145. Sandhu, H.S.; Wratten, S.D.; Cullen, R.; Case, B. The future of farming: The value of ecosystem services in conventional and organic arable land. An experimental approach. *Ecol. Econ.* **2008**, *64*, 835–848. [CrossRef]
146. Hahner, J.L.; Robinson, B.H.; Zhong, H.T.; Dickinson, N.M. The Phytoremediation Potential of Native Plants on New Zealand Dairy Farms. *Int. J. Phytoremediat.* **2014**, *16*, 719–734. [CrossRef] [PubMed]
147. Struik, P.C.; Kuyper, T.W. Sustainable intensification in agriculture: The richer shade of green. A review. *Agron. Sustain. Dev.* **2017**, *37*, 39. [CrossRef]
148. Gillingham, A.; Morton, J.; Gray, M. The role of differential fertiliser application in sustainable management of hill pastures. In *Proceedings of the New Zealand Grassland Association*; NZ Grassland Association: Palmerston North, New Zealand, 2003; pp. 253–257.
149. Yadav, R.; Pathak, G.S. Young consumers' intention towards buying green products in a developing nation: Extending the theory of planned behavior. *J. Clean. Prod.* **2016**, *135*, 732–739. [CrossRef]
150. Ghvanidze, S.; Velikova, N.; Dodd, T.H.; Oldewage-Theron, W. Consumers' environmental and ethical consciousness and the use of the related food products information: The role of perceived consumer effectiveness. *Appetite* **2016**, *107*, 311–322. [CrossRef] [PubMed]
151. McDermott, A.; Scrimgeour, F. Consumers, farmers and the future of New Zealand hill country farming. In *Proceedings of Hill Country Symposium*; Grassland Research and Practice Series; NZ Grassland Association: Rotorua, New Zealand, 2016; pp. 95–100.
152. Fennessy, P.; Glennie, S.; McCorkindale, A. Innovations behind the farm gate that will influence performance of hill farming. In *Proceedings of Hill Country Symposium: Grassland Research Practice Series*; NZ Grassland Association: Rotorua, New Zealand, 2016; pp. 15–20.
153. Samarasinghe, O.; Greenhalgh, S. *Economic and Bio-Physical Models for New Zealand*; Landcare Research: Lincoln, New Zealand, 2011.
154. Beef and Lamb. Land and Environment Planning. Available online: https://beeflambnz.com/knowledge-hub/PDF/land-and-environment-plan-brochure (accessed on 8 August 2019).
155. McNally, S.R.; Beare, M.H.; Curtin, D.; Meenken, E.D.; Kelliher, F.M.; Calvelo Pereira, R.; Shen, Q.; Baldock, J. Soil carbon sequestration potential of permanent pasture and continuous cropping soils in New Zealand. *Glob. Chang. Biol.* **2017**, *23*, 4544–4555. [CrossRef]
156. Dymond, J.R.; Vale, S.S. An event-based model of soil erosion and sediment transport at the catchment scale. *Geomorphology* **2018**, *318*, 240–249. [CrossRef]
157. Burkitt, L.; Bretherton, M.; Singh, R.; Hedley, M. Comparing Nutrient Loss Predictions Using OVERSEER and Stream Water Quality in a Hill Country Sub-Catchment. In *Proceedings of Integrated Nutrient Water Management for Sustainable Farming*; Massey University: Palmerston North, New Zealand, 2016; p. 9.

158. Higham, C.D.; Horne, D.; Singh, R.; Kuhn-Sherlock, B.; Scarsbrook, M.R. Water use on nonirrigated pasture-based dairy farms: Combining detailed monitoring and modeling to set benchmarks. *J. Dairy Sci.* **2017**, *100*, 828–840. [CrossRef]
159. Rendel, J.; Mackay, A.; Smale, P. The value of legumes to a Whanganui hill country farm. *J. N. Z. Grassl.* **2017**, *79*, 27–34.
160. Keller, E.; Baisden, W.; Timar, L. Adapting the Biome-BGC model to New Zealand pastoral agriculture: Climate change and land-use change. In *Proceedings of AGU Fall Meeting Abstracts*; American Geophysical Union: Washington, DC, USA, 2011.
161. Risk, J.; Old, A.; Peyroux, G.; Brown, M.; Yoswara, H.; Wheeler, D.; Lucci, G.; McDowell, R. MITAGATOR™-In action solutions for managing Nitrogen, Phosphorus, sediment and E. coli loss. In *Proceedings of FLRC Workshop*; Massey Univeristy: Palmerston North, New Zealand, 2011.
162. Bryant, J.; Ogle, G.; Marshall, P.; Glassey, C.; Lancaster, J.; Garcia, S.; Holmes, C. Description and evaluation of the Farmax Dairy Pro decision support model. *N. Z. J. Agric. Res.* **2010**, *53*, 13–28. [CrossRef]
163. Morgan, F.J.; Daigneault, A.J. Estimating Impacts of Climate Change Policy on Land Use: An Agent-Based Modelling Approach. *PLoS ONE* **2015**, *10*, e0127317. [CrossRef] [PubMed]
164. Landcare Research. *New Zealand Forest and Agriculture Regional Model (NZ-FARM)*; Landcare Research: Lincoln, New Zealand, 2013.
165. AgResearch. *It's Everybody's Business: Whole Farm Plans—A Vehicle for Implementing Policy*; AgResearch: Hamilton, New Zealand, 2016.
166. Cameron, D. Sustaining the productivity of New Zealand's hill country–A land manager's view. In *Proceedings of Hill Country Symposium: Grassland Research Practice Series*; NZ Grassland Association: Rotorua, New Zealand, 2016; pp. 151–156.
167. Tran, T.V.; Tran, D.X.; Myint, S.W.; Huang, C.-y.; Pham, H.V.; Luu, T.H.; Vo, T.M. Examining spatiotemporal salinity dynamics in the Mekong River Delta using Landsat time series imagery and a spatial regression approach. *Sci. Total Environ.* **2019**, *687*, 1087–1097. [CrossRef] [PubMed]
168. Stevens, D.; Casey, M.; Cousins, K. Farming systems research: Purpose, history and impact in New Zealand hill country. In *Proceedings of Hill Country Symposium: Grassland Research Practice Series*; NZ Grassland Association: Rotorua, New Zealand, 2016; pp. 67–86.
169. Van den Belt, M.; Forgie, V.; Bremer, S.; McDonald, G.; Montes de Oca, O.; Joy, M. *Modelling Tools for Integrated, Adaptive Management: A Case Study of New Zealand Regional Authorities*; Massey University: Palmerston North, New Zealand, 2010.
170. Synge, K.; MacKay, A.; Palmer, A. An evaluation of the Land and Environment Planning Toolkit for advancing soil and nutrient management on sheep and beef farms. In *Proceedings of the New Zealand Grassland Association*; NZ Grassland Association: Tauranga, New Zealand, 2013; pp. 91–96.
171. Sugumaran, R.; Degroote, J. *Spatial Decision Support Systems: Principles and Practices*; Crc Press, Taylor & Francis Group: Boca Raton, FL, USA, 2010.
172. McGlone, M. The Polynesian settlement of New Zealand in relation to environmental and biotic changes. *N. Z. J. Ecol.* **1989**, *12*, 115–129.
173. Ministry for the Environment. *Pressures on the Land*; Ministry for the Environment: Wellington, New Zealand, 1997.
174. Beef and Lamb. Farm Facts 2018. Available online: https://beeflambnz.com/knowledge-hub/PDF/compendium-farm-facts (accessed on 18 August 2019).
175. Rodriguez, J.P.; Beard, T.D.; Bennett, E.M.; Cumming, G.S.; Cork, S.J.; Agard, J.; Dobson, A.P.; Peterson, G.D. Trade-offs across space, time, and ecosystem services. *Ecol. Soc.* **2006**, *11*, 28. [CrossRef]
176. Caruso, B.S.; O'Sullivan, A.D.; Faulkner, S.; Sherratt, M.; Clucas, R. Agricultural Diffuse Nutrient Pollution Transport in a Mountain Wetland Complex. *Water Air Soil Pollut.* **2013**, *224*. [CrossRef]
177. Schon, N.; Mackay, A.; Gray, R.; Minor, M. Influence of phosphorus inputs and sheep treading on soil macrofauna and mesofauna in hill pastures. *N. Z. J. Agric. Res.* **2011**, *54*, 83–96. [CrossRef]
178. Laliberte, E.; Tylianakis, J.M. Cascading effects of long-term land-use changes on plant traits and ecosystem functioning. *Ecology* **2012**, *93*, 145–155. [CrossRef]

179. Wakelin, S.A.; Barratt, B.I.; Gerard, E.; Gregg, A.L.; Brodie, E.L.; Andersen, G.L.; DeSantis, T.Z.; Zhou, J.; He, Z.; Kowalchuk, G.A. Shifts in the phylogenetic structure and functional capacity of soil microbial communities follow alteration of native tussock grassland ecosystems. *Soil Biol. Biochem.* **2013**, *57*, 675–682. [CrossRef]
180. Landis, D.A. Designing agricultural landscapes for biodiversity-based ecosystem services. *Basic Appl. Ecol.* **2017**, *18*, 1–12. [CrossRef]
181. Jorgensen, S.E.; Fath, B.D. *Encyclopedia of Ecology*; Elsevier, B.V: Oxford, UK, 2008.
182. Selman, P. Planning for landscape multifunctionality. *Sustainability* **2009**, *5*, 45–52. [CrossRef]
183. Killion, A.K.; Dixon, A.; Gilbert, J.; Torralba, M.; Greiner, P.T.; Behrer, A.P. Designing spatiotemporal multifunctional landscapes to support dynamic wildlife conservation. *J. Land Use Sci.* **2018**, *13*, 615–630. [CrossRef]
184. Lovell, S.T.; Taylor, J.R. Supplying urban ecosystem services through multifunctional green infrastructure in the United States. *Landsc. Ecol.* **2013**, *28*, 1447–1463. [CrossRef]
185. Biggs, R.; Schlüter, M.; Biggs, D.; Bohensky, E.L.; BurnSilver, S.; Cundill, G.; Dakos, V.; Daw, T.M.; Evans, L.S.; Kotschy, K. Toward principles for enhancing the resilience of ecosystem services. *Annu. Rev. Environ. Resour.* **2012**, *37*, 421–448. [CrossRef]
186. Scherr, S.J.; McNeely, J.A. Biodiversity conservation and agricultural sustainability: Towards a new paradigm of 'ecoagriculture' landscapes. *Philos. Trans. R. Soc. B* **2008**, *363*, 477–494. [CrossRef]
187. Kato, S.; Ahern, J. Multifunctional landscapes as a basis for sustainable landscape development. *J. Jpn. Inst. Landsc. Archit.* **2009**, *72*, 799–804. [CrossRef]
188. Kremen, C.; Iles, A.; Bacon, C. Diversified farming systems: An agroecological, systems-based alternative to modern industrial agriculture. *Ecol. Soc.* **2012**, *17*, 44. [CrossRef]
189. Cumming, G.S.; Buerkert, A.; Hoffmann, E.M.; Schlecht, E.; von Cramon-Taubadel, S.; Tscharntke, T. Implications of agricultural transitions and urbanization for ecosystem services. *Nature* **2014**, *515*, 50–57. [CrossRef]
190. Kirchner, M.; Schmidt, J.; Kindermann, G.; Kulmer, V.; Mitter, H.; Prettenthaler, F.; Rüdisser, J.; Schauppenlehner, T.; Schönhart, M.; Strauss, F. Ecosystem services and economic development in Austrian agricultural landscapes—the impact of policy and climate change scenarios on trade-offs and synergies. *Ecol. Econ.* **2015**, *109*, 161–174. [CrossRef]
191. Ahern, J. From Landscape Research to Landscape Planning. In *Aspects of Integration, Education Application*; Springer: Dordrecht, The Netherlands, 2006; pp. 119–131.
192. Dallimer, M.; Strange, N. Why socio-political borders and boundaries matter in conservation. *Trends Ecol. Evol.* **2015**, *30*, 132–139. [CrossRef]
193. Bergsten, A.; Galafassi, D.; Bodin, Ö. The problem of spatial fit in social-ecological systems: Detecting mismatches between ecological connectivity and land management in an urban region. *Ecol. Soc.* **2014**, *19*, 6. [CrossRef]
194. Schonewald, C.M. Introduction to boundary space. *Complexity* **2000**, *6*, 41–57. [CrossRef]
195. NZ Landcare Trust. Catchment Groups. Available online: https://www.landcare.org.nz/completed-project-item/catchment-groups (accessed on 12 December 2019).
196. Beef and Lamb. Catchment Community Group Programme. Available online: https://beeflambnz.com/your-levies-work/community-catchment-group-programme (accessed on 14 December 2019).
197. Cerretelli, S.; Poggio, L.; Gimona, A.; Yakob, G.; Boke, S.; Habte, M.; Coull, M.; Peressotti, A.; Black, H. Spatial assessment of land degradation through key ecosystem services: The role of globally available data. *Sci. Total Environ.* **2018**, *628*, 539–555. [CrossRef] [PubMed]
198. Arowolo, A.O.; Deng, X.; Olatunji, O.A.; Obayelu, A.E. Assessing changes in the value of ecosystem services in response to land-use/land-cover dynamics in Nigeria. *Sci. Total Environ.* **2018**, *636*, 597–609. [CrossRef]
199. Yi, H.; Güneralp, B.; Filippi, A.M.; Kreuter, U.P.; Güneralp, İ. Impacts of land change on ecosystem services in the San Antonio River Basin, Texas, from 1984 to 2010. *Ecol. Econ.* **2017**, *135*, 125–135. [CrossRef]
200. Crespin, S.J.; Simonetti, J.A. Loss of ecosystem services and the decapitalization of nature in El Salvador. *Ecosyst. Serv.* **2016**, *17*, 5–13. [CrossRef]
201. Mendoza-González, G.; Martínez, M.L.; Lithgow, D.; Pérez-Maqueo, O.; Simonin, P. Land use change and its effects on the value of ecosystem services along the coast of the Gulf of Mexico. *Ecol. Econ.* **2012**, *82*, 23–32. [CrossRef]

202. Kreuter, U.P.; Harris, H.G.; Matlock, M.D.; Lacey, R.E. Change in ecosystem service values in the San Antonio area, Texas. *Ecol. Econ.* **2001**, *39*, 333–346. [CrossRef]
203. Polasky, S.; Nelson, E.; Pennington, D.; Johnson, K.A. The impact of land-use change on ecosystem services, biodiversity and returns to landowners: A case study in the state of Minnesota. *Environ. Resour. Econ.* **2011**, *48*, 219–242. [CrossRef]
204. Cabral, P.; Feger, C.; Levrel, H.; Chambolle, M.; Basque, D. Assessing the impact of land-cover changes on ecosystem services: A first step toward integrative planning in Bordeaux, France. *Ecosyst. Serv.* **2016**, *22*, 318–327. [CrossRef]
205. Maestre, F.T.; Quero, J.L.; Gotelli, N.J.; Escudero, A.; Ochoa, V.; Delgado-Baquerizo, M.; García-Gómez, M.; Bowker, M.A.; Soliveres, S.; Escolar, C. Plant species richness and ecosystem multifunctionality in global drylands. *Science* **2012**, *335*, 214–218. [CrossRef]
206. Maseyk, F.J.; Dominati, E.J.; Mackay, A.D. Change in ecosystem service provision within a lowland dairy landscape under different riparian margin scenarios. *Int. J. Biodivers. Sci. Ecosyst. Serv. Manag.* **2018**, *14*, 17–31. [CrossRef]
207. Ausseil, A.G.; Herzig, A.; Dymond, J. Optimising Land Use for Multiple Ecosystem Services Objectives: A Case Study in the WaitakiCatchment, New Zealand. In *Proceedings of 6th International Congress on Environmental Modelling and Software*; Helmholtz Centre for Environmental Research: Leipzig, Germany, 2012.
208. Willemen, L. *Mapping and Modelling Multifunctional Landscapes*; Wageningen University: Wageningen, The Netherlands, 2010.
209. Muir, R. *The Evaluation of Landscape. In Approaches to Landscape*; Palgrave: London, UK, 1999; pp. 182–211.
210. Wu, J. Landscape sustainability science: Ecosystem services and human well-being in changing landscapes. *Landsc. Ecol.* **2013**, *28*, 999–1023. [CrossRef]
211. Baños-González, I.; Martínez-Fernández, J.; Esteve-Selma, M. Using dynamic sustainability indicators to assess environmental policy measures in Biosphere Reserves. *Ecol. Indic.* **2016**, *67*, 565–576. [CrossRef]
212. Albert, C.; Galler, C.; Hermes, J.; Neuendorf, F.; Von Haaren, C.; Lovett, A. Applying ecosystem services indicators in landscape planning and management: The ES-in-Planning framework. *Ecol. Indic.* **2016**, *61*, 100–113. [CrossRef]
213. Niemeijer, D.; De Groot, R.S. A conceptual framework for selecting environmental indicator sets. *Ecol. Indic.* **2008**, *8*, 14–25. [CrossRef]
214. Heink, U.; Hauck, J.; Jax, K.; Sukopp, U. Requirements for the selection of ecosystem service indicators–The case of MAES indicators. *Ecol. Indic.* **2016**, *61*, 18–26. [CrossRef]
215. Almenar, J.B.; Rugani, B.; Geneletti, D.; Brewer, T. Integration of ecosystem services into a conceptual spatial planning framework based on a landscape ecology perspective. *Landsc. Ecol.* **2018**, *33*, 2047–2059. [CrossRef]
216. Ekroos, J.; Olsson, O.; Rundlöf, M.; Wätzold, F.; Smith, H.G. Optimizing agri-environment schemes for biodiversity, ecosystem services or both? *Biol. Conserv.* **2014**, *172*, 65–71. [CrossRef]
217. Pohekar, S.; Ramachandran, M. Application of multi-criteria decision-making to sustainable energy planning—A review. *Renew. Sustain. Energy Rev.* **2004**, *8*, 365–381. [CrossRef]
218. Bretagnolle, V.; Berthet, E.; Gross, N.; Gauffre, B.; Plumejeaud, C.; Houte, S.; Badenhausser, I.; Monceau, K.; Allier, F.; Monestiez, P.; et al. Towards sustainable and multifunctional agriculture in farmland landscapes: Lessons from the integrative approach of a French LTSER platform. *Sci. Total Environ.* **2018**, *627*, 822–834. [CrossRef]
219. Kimberly, A. *Essential of Landscape Ecology*; Oxford University Press: Oxford, UK, 2019.
220. Milestad, R.; Björklund, J. Strengthening the adaptive capacity of rural communities: Multifunctional farms and village action groups. In Proceedings of the 8th European IFSA Symposium, Clermont-Ferrand, France, 6–10 July 2008; pp. 6–10.
221. Hassink, J.; Agricola, H.; Thissen, J. Participation rate of farmers in different multifunctional activities in the Netherlands. *Outlook Agric.* **2016**, *45*, 192–198. [CrossRef]
222. Wu, J.G. Linking landscape, land system and design approaches to achieve sustainability. *J. Land Use Sci.* **2019**. [CrossRef]
223. McWilliam, W.; Fukuda, Y.; Moller, H.; Smith, D. Evaluation of a dairy agri-environmental programme for restoring woody green infrastructure. *Int. J. Agric. Sustain.* **2017**, *15*, 350–364. [CrossRef]
224. Bradley, B. Integrated Crop. In *Management Systems in the EU: Amended Final Report for European Commission, DG Environment*; Agra CEAS Consulting Ltd.: Ashford, UK, 2002.

225. Mouchet, M.A.; Rega, C.; Lasseur, R.; Georges, D.; Paracchini, M.-L.; Renaud, J.; Stürck, J.; Schulp, C.J.; Verburg, P.H.; Verkerk, P.J. Ecosystem service supply by European landscapes under alternative land-use and environmental policies. *Int. J. Biodivers. Sci. Ecosyst. Serv. Manag.* **2017**, *13*, 342–354. [CrossRef]
226. Asquith, N.M.; Vargas, M.T.; Wunder, S. Selling two environmental services: In-kind payments for bird habitat and watershed protection in Los Negros, Bolivia. *Ecol. Econ.* **2008**, *65*, 675–684. [CrossRef]

land MDPI

Article

Using the Ecosystem Services Concept to Assess Transformation of Agricultural Landscapes in the European Alps

Uta Schirpke [1,2], Erich Tasser [2], Georg Leitinger [1] and Ulrike Tappeiner [1,2,*]

[1] Department of Ecology, University of Innsbruck, 6020 Innsbruck, Austria; Uta.Schirpke@uibk.ac.at (U.S.); Georg.Leitinger@uibk.ac.at (G.L.)
[2] Institute for Alpine Environment, Eurac Research, 39100 Bolzano, Italy; erich.tasser@eurac.edu
* Correspondence: ulrike.tappeiner@uibk.ac.at

Abstract: Mountain farming sustains human well-being by providing various ecosystem services (ES). In the last decades, socio-economic developments have led to worldwide changes in land-use/cover (LULC), but the related effects on ES have not been fully explored. This study aimed at assessing the impacts of the transformation of agricultural land on ES in the European Alps. We mapped 19 ES within the agriculturally used areas in the year 2000 and analyzed LULC changes by 2018. We compared eight regions with a similar development, regarding social–ecological characteristics, to outline contrasting trends. Our results indicate that the ES decreased most strongly in regions with a massive abandonment of mountain grassland, while ES in the 'traditional agricultural region' remained the most stable. In regions with an intensification of agriculture, together with urban sprawl, ES had the lowest values. Across all regions, a shift from ES that are typically associated with mountain farming towards forest-related ES occurred, due to forest regrowth. By relating differing trends in ES to social–ecological developments, we can discuss our findings regarding new landscapes and farming systems across the European Alps. Our quantitative and spatially explicit findings provide a valuable basis for policy development, from the regional to the international/EU level, and for adopting sustainable management strategies.

Keywords: social–ecological system; mountain region; spatial analysis; land-use change; farming

Citation: Schirpke, U.; Tasser, E.; Leitinger, G.; Tappeiner, U. Using the Ecosystem Services Concept to Assess Transformation of Agricultural Landscapes in the European Alps. *Land* **2022**, *11*, 49. https://doi.org/10.3390/land11010049

Academic Editor: Monika Kopecká

Received: 17 November 2021
Accepted: 28 December 2021
Published: 30 December 2021

1. Introduction

The IPBES (Intergovernmental Platform on Biodiversity and Ecosystem Services) conceptual framework names three interactions between human societies and the non-human world: nature, nature's benefits to humans, and a good quality of life. To value NCPs (nature's contributions to people; defined here as any positive contribution or benefit, and occasionally negative contributions, losses, or degradations, that humans receive from nature), the concept of ecosystem services (ES) is often used. Since, NCPs are consistent with the original use of the term ES in the Millennium Ecosystem Assessment [1], we define ES as the contributions ecosystems make to human well-being, including the goods and benefits that people subsequently derive from them. The IPBES further reinforces the need for initiatives at the science–society interface, aiming at sustainable futures in the light of global change [2]. Our study applies the ES concept to value the transformation of landscapes in this context, contributing to a possible sustainable adaptation of land-use/cover changes (LULC). Here, we focus on agricultural landscapes, as they are particularly affected by global change, with wide-ranging consequences for society [3–5]. Agricultural ecosystems contribute to a variety of ES, such as food and fodder provision, soil conservation, erosion protection, climate regulation, habitat provision, aesthetics, and recreation [6–8]. In particular, organic or traditional farming systems provide high levels of multiple ES, while conventional farming systems are focused on food production [9–13]. In mountain regions, small-scale farming systems and sustainable management practices have been developed over centuries to cope with the challenging topographic and climatic conditions [14]. This has shaped appealing mountain landscapes, which are rich in biodiversity and provide many ES to local people, tourists, and adjacent lowland populations [15–24]. However, institutional and political drivers, socio-economic shifts, urbanization, and technical developments have reduced the competitiveness of these marginal areas and induced a massive abandonment of alpine pastures and meadows in European mountain regions during the last century [25–32]. At the same time, less steep areas in the valley bottoms with a favorable climate and easy access have been intensified, often managed by larger and more specialized farms [33,34]. Such changes have led to still ongoing transformations in agricultural landscapes, with implications for biodiversity and manifold ES [22,23,35–37]. For example, the intensification of agricultural land causes a decline in water quality, because of higher nutrient input, and a reduction in pollination, due to the use of pesticides and a habitat loss [6,38,39]. In addition to a decline in forage provision, the abandonment of alpine pastures and meadows leads to a loss of many cultural ES. In contrast, the provision of timber and non-wood products, and the regulation of the climate and protection from hazards increase due to forest regrowth [33,36]. Hence, previous conditions and past processes not only show an impact on current landscape patterns and functions, but can also determine, to a great extent, future pathways of landscape change [40].

Such developments require the attention of decision-makers and land managers, to foster a sustainable development of mountain regions and maintain high levels of multiple ES provision [41–44]. There is growing evidence that the concept of ES, acknowledging the human benefits obtained from the interaction with ecosystems, provides a valuable basis to support landscape planning and management, in various ways [41,45]. This may include raising the awareness of stakeholders, developing management strategies, and taking decisions [46]. In particular, ES maps can be supportive for identifying developments of ES over space and time [36,38,47]. They can be used for revealing synergies and trade-offs among multiple ES [24,48,49], and consequently, for setting priorities in land-use decisions; for example, how intensive agricultural use maximizes provisioning ES, while reducing other ES [50,51]. Maps can also help to identify the spatial separation between farming activities and consumers, which is responsible for trading agricultural products globally [19,52,53]. On this basis, decision-makers can develop nature-based solutions, such as promoting dietary shifts, to strengthen the consumption of local products.

However, quantitative and spatially explicit information on the impacts from the transformation of agricultural land is often not available [54,55]. One reason is that studies on LULC in agricultural landscapes are often not sufficiently linked to the concept of ES [51]. Although there is an increasing number of studies dealing with ES in mountain regions in general, many studies have not considered changes in ES over time [56] or did not specifically analyze agricultural landscapes [51]. Regarding spatial coverage, most studies concentrated on the local level, e.g., [28,57–59], or, if carried out at regional or national level, largely neglected social–ecological differences within and across regions [21]. Consequently, national or even regional policies fail to consider diverging local developments, which occur due to the high complexity of large mountain ranges, such as the European Alps, that include a high variety of climatic, topographic, socio-cultural, and political conditions [14]. Furthermore, many studies focused on a limited number of ES [60]. Data on ES that are not directly linked to land-use/land-cover (LULC) or that are more difficult to assess (e.g., many cultural ES) are largely lacking. Therefore, ES are still rarely integrated into policies and decision-making [60].

To contribute to a more comprehensive understanding of recent developments in the European Alps, this study was aimed at assessing the impacts of the transformation of agricultural landscapes on 19 ES. By differentiating eight regions with distinct social–ecological characteristics, our findings illustrate contrasting developments in ES and highlight diverse pathways for agricultural landscapes in mountain regions.

2. Materials and Methods

2.1. Study Area

The European Alps are the highest mountain chain in Europe and contain a large variety of landscapes, species, and cultures. They count about 14 million inhabitants and stretch over eight different countries, including parts of France, Switzerland, Italy, Austria, Germany, and Slovenia, as well as the countries Liechtenstein and Monaco. The Alps are a hot spot of biodiversity, and the complex topography influences the natural distribution of soil, the typology of land, and habitat variety. About 49% of the area is covered by forest, followed by agricultural land (27%), high mountain landscapes with shrubs, natural grasslands and rocks (19%), artificial surfaces (3.7%), and water (1.1%).

Due to the high variety of social–ecological conditions across the Alps, the analyses of this study are based on eight regions with different economic and social structures or environmental situations [61]. These regions were identified by Tappeiner et al. [61] through cluster analysis (Ward method, squared Euclidian distance), based on 21 indicators that reflect the three pillars of sustainability in equal measure (Table S1). The classification refers to data between 2000 and 2008, as well as between 1990 and 2002 for change indicators. An updated classification is currently not available. The eight regions (Figure 1) can be summarized as follows:

- 'Employment hubs' are municipalities to which many employed persons commute daily. They have a good transport infrastructure and offer a good range of jobs in the secondary and tertiary sectors.
- 'Residential municipalities' are typical residential and dormitory municipalities located around major employment hubs. Daily commuting is possible without great loss of time, due to the above-average traffic infrastructure. The residential environment in these municipalities is attractive, and land prices are affordable, which leads to increased urban sprawl.
- 'Important tourist centers' have very well-developed accommodation facilities; the employment situation is better than average in the Alps. Most of them are rural municipalities with largely intact agriculture and an attractive landscape.
- 'Dynamic rural areas' are characterized by a rural location and a dynamic labor market. The employment of women and older persons in particular has improved significantly here, not least due to the positive development of tourism. Agriculture in these areas is largely intact. Of concern, however, is the above-average emigration of employed persons.
- 'Standard Alpine regions' reach average values for the Alps in all aspects. Typical of these are low tourism intensity, a negative commuter balance, and a decline in agriculture. Balanced migration and birth rates, however, prevent excessive over-aging in these municipalities.
- 'Traditional agricultural regions' are characterized by a severe over-aging of society, poor traffic infrastructure, and a moderate retreat of, mostly extensive, agriculture from the area. The poor employment situation in these regions is likely to contribute to the fact that the number of abandoned farms is limited. Overall, this results in a rich, traditional landscape.
- 'Rural retreats' are characterized by good traffic infrastructure, which residents use to commute to work while keeping their center of life in the rural hinterland. Agriculture has largely retreated from the area, creating a slightly fragmented and highly diverse landscape.
- 'Forgotten rural areas' are characterized by significant over-aging and a particularly strong abandonment of agriculture. A major reason for this is remoteness and poor traffic infrastructure. The areas show great economic weakness and are threatened by depopulation.

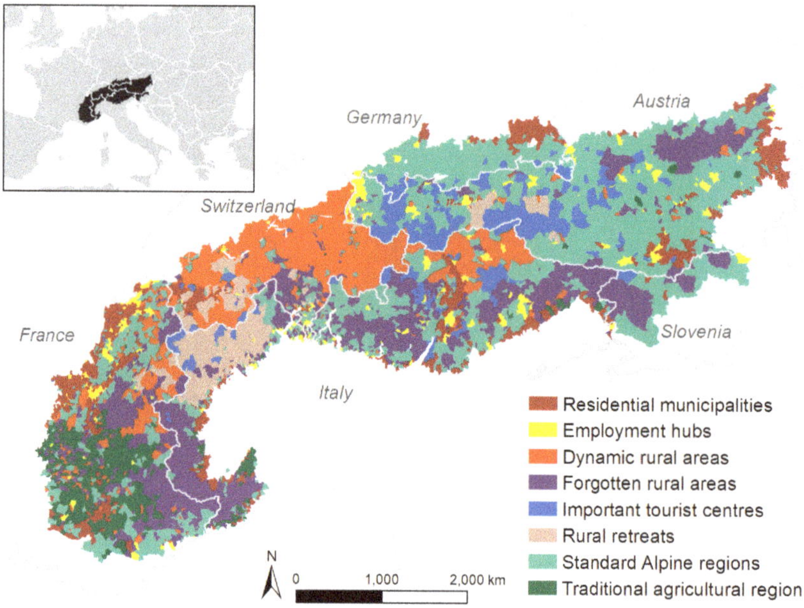

Figure 1. Location of the European Alps in Europe (small map), and the eight social–ecological regions in the study area, as identified by Tappeiner et al. [61] (large map). Authors own elaboration.

2.2. Analysis Steps

We analyzed changes in agricultural land between 2000 and 2018 in the European Alps for the eight above described social–ecological regions, based on LULC distribution, and related these developments to 19 ES. The ES mapping and impact analysis in this study comprised the following three steps (Figure 2):

1. Aggregation of LULC types: We used CORINE Land Cover data (CLC) in raster format with a spatial resolution of 100 × 100 m for the years 2000 [62] and 2018 [63]. We aggregated the 44 CLC classes to 11 LULC types (Table S2), mainly representing the first and second level of thematic detail, according to the hierarchical nomenclature of CLC [62,63]. Based on the LULC distribution in 2000, we selected four agricultural LULC types (crop cultivation, permanent culture, fertilized grassland, unfertilized grassland), which we used to extract the aggregated LULC maps in 2000 and 2018 to the same spatial extent, focusing on agricultural areas.
2. Calculation of ES values: We created ES raster maps by relating the LULC types in 2000 and 2018 to ES values (Table S3). Moreover, we distinguished raster cells with slope $<$ and $\geq 30°$ to distinguish flat areas that do not need 'protection from hazards (R1)' due to the presence of steep areas. ES values represent the ES supply, which was weighted by socio-cultural preferences [50]. Tasser et al. [50] and Schirpke et al. [25] derived the ES supply from an extensive literature review on ES-relevant ecosystem processes and functions related to water, soil, plants, animals, microorganisms, agricultural production, and landscape structure. Socio-cultural preferences (from 1 = low to 5 = high) were obtained from surveys [18,64]. Hence, ES that were more preferred obtained higher final ES values than those ES of lower importance (for details, see Tasser et al. [50]). Final ES values are expressed as a dimensionless index, ranging from 0 to 5, and were used to map ES based on the aggregated LULC types, i.e., each raster cell of a specific LULC type was associated with the respective ES value of Table S3.

3. Impact analysis: To identify differing trends in LULC and ES across the eight regions with differing social–ecological characteristics, we spatially overlaid the raster maps (aggregated LULC, ES values) with the eight regions (Figure 1). We calculated area-weighted mean values for each region in 2000 and 2018, which were used to map and evaluate changes in LULC and ES values.

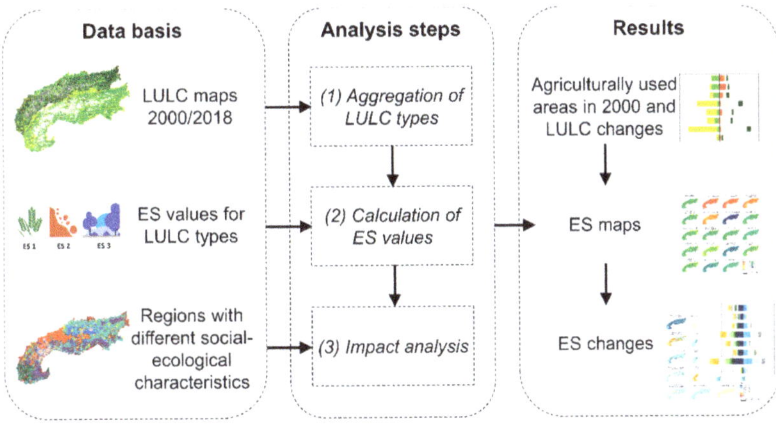

Figure 2. Conceptual steps for examining trends in ES values across different regions with distinct social–ecological characteristics in the European Alps, only considering areas that were agriculturally used in 2000. Authors own elaboration.

3. Results

3.1. LULC Changes

The composition of agricultural land in 2000 varied across the eight regions (Figure 3). The residential municipalities, employment hubs, the standard Alpine region, and the traditional agricultural region comprised mostly intensively used LULC types, such as crop cultivation, permanent cultures, and fertilized grassland, whereas unfertilized grassland prevailed in the other four regions. LULC changes between 2000 and 2018 generally consisted in the abandonment of fertilized and unfertilized grassland, and in an increase in settlement area, forest, and abandoned grassland. Change rates, however, greatly differed across the eight regions (Figure 3). The smallest changes occurred in the traditional agricultural region, while the largest changes happened in rural retreats and forgotten rural areas. Residential municipalities and employment hubs had the largest increases in settlement areas, while forest increased above average in forgotten rural areas and rural retreats. Agriculturally used grasslands were frequently abandoned, especially in the latter two regions, but also in the tourist centers and in the dynamic rural areas, resulting in forested areas or succession stages towards forest, such as dwarf-shrub habitats and bushland. In addition, crop cultivation and permanent cultures slightly increased around the main settlement regions, mainly in residential municipalities and standard Alpine regions.

3.2. Changes in ES Values

Considering only agricultural LULC types (i.e., crop cultivation, permanent culture, fertilized grassland, unfertilized grassland) that were present in 2000, ES values varied across the eight regions (Figures 4 and A1). The lowest ES values occurred in the economically prosperous employment hubs, including the suburbanization region (residential municipalities), mainly due to less ecosystems with high ES supply and below-average values for cultural and regulating ES. In contrast, regions with a high increase in forest or a high share of unfertilized grassland, including alpine pastures and traditional agro-forestry systems, had the highest ES values; in particular, wood production (P5), occurrence of mushrooms and wild berries (P4), protection against natural hazards (R1), availability of usable water (R2), preservation of valuable habitats and species (R3, R4, and R5), positive impact on climate (R9), opportunities for leisure activities (C1), aesthetic experiences (C4), and cultural heritage (C5). Regions with a high proportion of intensive agricultural land (i.e., crop cultivation, permanent crops, and fertilized grassland) had lower ES values for regulating and cultural ES.

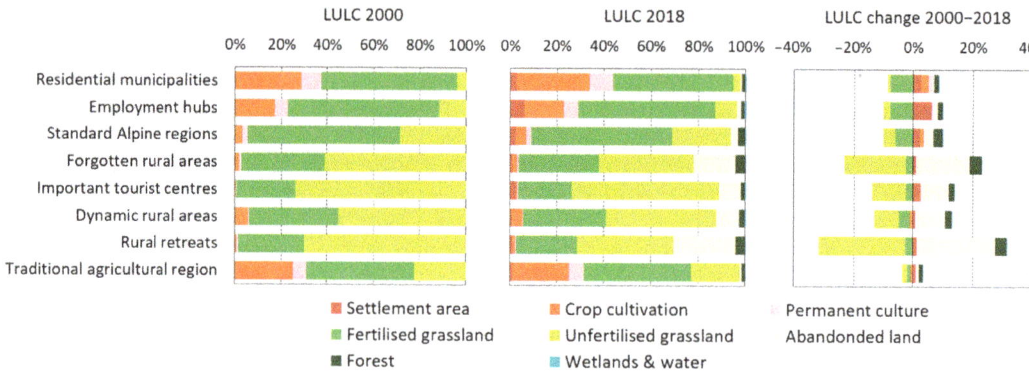

Figure 3. LULC distribution across eight social–ecological regions in the European Alps in 2000 (**left**) and 2018 (**center**), as well as LULC changes between 2000 and 2018 (**right**), only referring to the area that was covered in 2000 by agricultural LULC types (i.e., crop cultivation, permanent culture, fertilized grassland, and unfertilized grassland). Wetlands, rivers, lakes are summarized as 'wetlands & water'. Authors own elaboration.

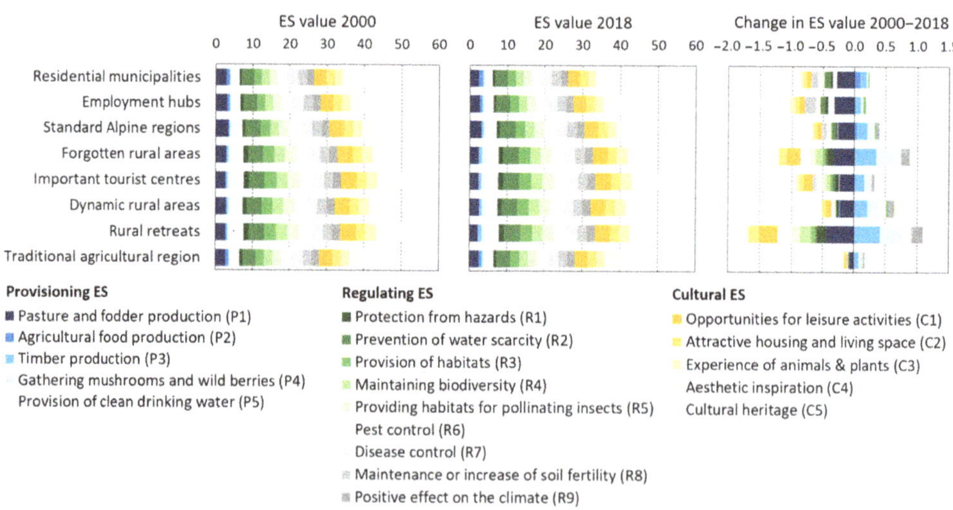

Figure 4. ES values in 2000 (**left**), 2018 (**center**), and change in ES value between 2000 and 2018 (**right**). Authors own elaboration.

ES values between 2000 and 2018 mostly declined, especially, regulating and cultural ES, mainly due to LULC changes of agricultural land towards other LULC types, including abandoned land, forest, and settlement areas. On the other hand, provisioning ES increased except for fodder production (P1), but the changes in ES values varied across the eight regions (Figures 4 and 5). Corresponding to the small LULC changes, the smallest changes in ES values occurred in the traditional agricultural region. Changes in employment hubs and residential municipalities were also below average, but there was a further decline in provisioning ES, due to the increasing urban sprawl. Rural retreats had a particularly strong decrease in many cultural and regulating ES values, with the exception of the positive effect on the climate (R9), due to an increase of forests and abandoned land (including heathlands, transitional woodlands, and shrub) on former agricultural land; however, provisioning ES also increased above average, apart from fodder production (P1) and agricultural food production (P2). Across all ES, positive trends only prevailed over negative ones in the dynamic rural areas and the traditional agricultural regions. In spatial terms, the greatest changes occurred in the Southern Alps in Italy and Slovenia, and the Western Alps were more affected by changes than the Eastern Alps.

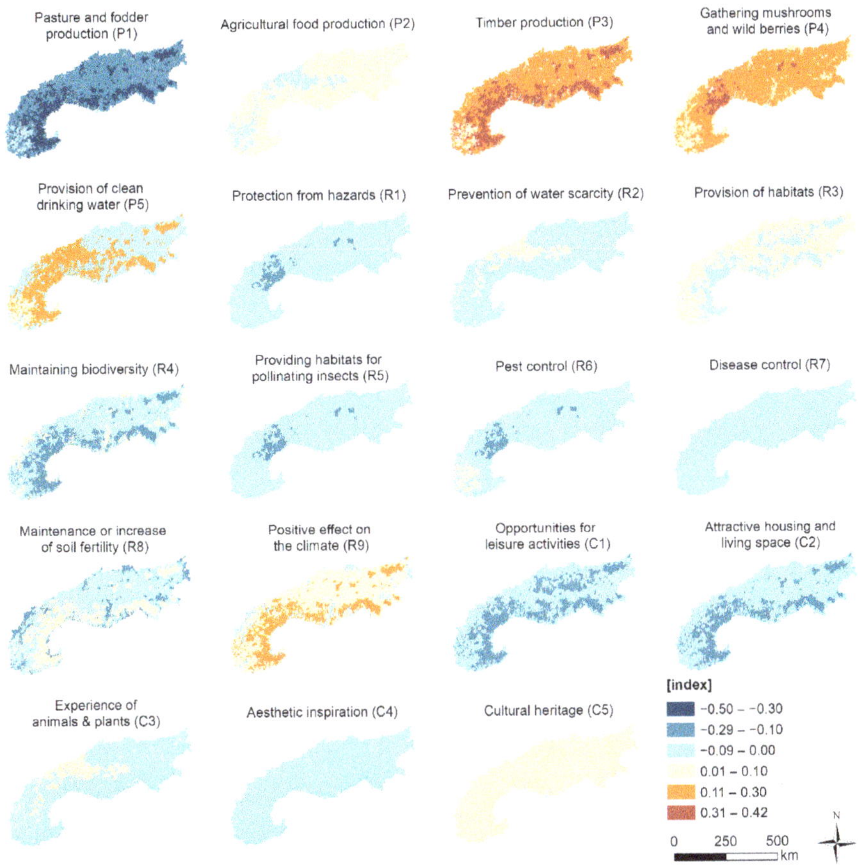

Figure 5. Changes in ES values between 2000 and 2018 across the European Alps. Only agricultural LULC types (crop cultivation, permanent culture, fertilized grassland, unfertilized grassland) that were present in 2000 were considered. Thus, these maps illustrate the changes in agricultural land to other LULC types (including abandoned land, forest, and settlement areas). Authors own elaboration.

4. Discussion

4.1. Current Trends in Alpine Agricultural Landscapes

European land management has not been evolving unidirectionally, following predefined trajectories, but rather as path-dependent processes affected by technological, institutional, economic, and social drivers, including sudden events [31]. This is also true for the European Alps [30,31]. Since the beginning of the 20th century, the number of farms in most Alpine regions has decreased by more than 50%, and the share of the population employed in agriculture has decreased from about 70% to less than 5% [34], while employment has strongly increased in the secondary, and later in the tertiary, sectors [34,65]. Due to unfavorable growing conditions, such as short growing seasons, steep slopes, and small property parcels, which necessitate expensive management practices, while having low productivity, mountain farming cannot compete in national and international markets [31,32,59]. Today, many farmers generate their main income outside their own farm, e.g., in business parks, industrial facilities, shopping centers, and tourism, and the share of part-time farmers is about 70% in the Alps [61,66]. Therefore, the agricultural area decreased on average by about 20%, and in some areas up to even 70% [23]. Our results indicate that this trend is still ongoing. The abandoned grassland areas are currently subject to a natural succession process towards site-typical climax vegetation (forest up to the natural timberline, with dwarf shrubs and alpine grassland above) with impacts on ecosystem structure and processes [67]. Our results also indicate that land use has been intensified in favorable locations, mainly through conversion to permanent crops or transformation of agroecosystems into urban or suburban areas. The extent to which these trends will continue depends not only on socio-economic drivers, but increasingly also on climate change. At high altitudes, climate warming will lead to a rise in the timberline, from 300 m (at +2°K) to 800 m (at +5°K), resulting in a decrease in alpine grassland [67]. However, the temperature increase will cause only a marginal expansion of forested area in 84.3% of Alpine municipalities, because they do not have areas in the alpine and nival belts [68]. Although climate change does have impacts on land use below 2000 m a.s.l., economic impacts override climate effects [59,69]. As a result of temperature increase and regionally lower precipitation, agricultural use will shift from grassland toward arable farming and permanent crops at lower elevations, whereas grassland farming will intensify at higher elevations [12,59].

Our results indicate that these transformations of agricultural use can jeopardize ES provision and may simultaneously aggravate associated disservices, such as increased leaching of soil nutrients or pests [6,70]. Many ES have declined in recent decades on land formerly used for agriculture, due to the intensification of use or urban sprawl, resulting in LULC types that produce fewer regulating and cultural ES, in particular (see also [25,50]). In addition, some provisioning ES, such as food and fodder production, decreased. In contrast, if forest growth occurs on formerly used grassland (above all in the Italian and Slovenian Alpine regions), timber production will increase, but provision of drinking water (i.e., streamflow) could decrease [71].

4.2. Implications for Management and Decision-Making

Our results show that ES values are reduced in most of the selected regions, but with different expressions when divided into provisioning, regulating, and cultural ES. This suggests that increases in ES value can be achieved through targeted regional planning, which also conserves landscape and species diversity, as well as powerful bundles of ES [72]. Moreover, abandoned land can contribute to sustainable land use transitions, providing opportunities to nature-based solutions based on biodiversity, cultural, and regulating services [73]. For management and decision-making, a respective framing must be set to comprehensively evaluate the impacts of agricultural strategies (i.e., on environment and economy). Here, the ES concept should be better integrated into existing frameworks such as the sustainable rural livelihood framework [65]. Moreover, a stronger focus on transdisciplinary research, including the development of adaptive pathways would enable stakeholders to translate ES changes into a tangible local or regional agricultural strategy [74]. To highlight the interdependence of different economic sectors and the need for collective action at the local/regional level, to successfully tackle future challenges, the resilience of ES needs to be addressed, in an ecosystem-based approach, in order to duly incorporate the steadily increasing knowledge of changing ecosystem functions and ecosystem processes due to climate change [57,75]. This requires a clear commitment to basic research in the field of global change and the use of promising scientific approaches, especially in topographically complex areas, in order to have results available quickly at the landscape level [76,77].

To complete the picture, an appropriate framework must consider the historical development of agricultural strategies, and socio-economic and landscape developments, which means that 'history' must be part of future strategies. Results such as those shown in our study can form the basis and at the same time the starting point for future development paths, which are also increasingly taken up scientifically in landscape ecology, e.g., [40]. However, shifting to more resilient pathways, i.e., developing innovative and adaptive pathways that can mitigate the negative effects of global change on ES [78], can pose significant challenges, especially if land use decisions are predominantly based on agricultural market values. Farmers and decision makers seem to be 'locked-in' to their production-oriented view [40], disregarding the importance of land-use change in promoting other values such as greenhouse gas emissions and sequestration or recreational use and biodiversity [72].

4.3. Methodological Considerations

In this study, we applied a simple approach for mapping and quantifying ES values based on LULC maps, which is often applied to generate comprehensive information suitable for decision-making, because it sufficiently accounts for underlying mechanisms and directly illustrates possible impacts from LULC changes [79,80]. While this approach is easily replicable, the results contain some uncertainties that need to be considered. One issue regards the LULC types that were used to map ES values, as we differentiated only four types of agricultural use. These are linked to different levels of fertilizer use and have distinct ecological functions and differing species composition [50]. Further levels of fertilization of grassland or differently stocked pastures [12,13], as well as specific types of annual and permanent crops [55,81] could not be distinguished, due to lacking spatial information at a cross-national level. A further refinement could also include a distinction between conventional and organic farming systems for annual crops or permanent cultures [9–11].

There are also limitations with regard to the underlying databases. An updated version of the classification of the eight Alpine regions was not yet available and the classification of some municipalities may have changed, as socio-economic indicators, especially, are less stable than environmental conditions. This may have greater effects on municipalities at low to medium elevations compared to municipalities located higher [82]. Nevertheless, future studies should reclassify the Alpine regions using recent data, which would be particularly important when predicting future agricultural landscapes. Another uncertainty is related to the LULC maps, which originates from methodological issues during the interpretation of different remote sensing data over time, for generating the CLC [83]. To reduce mapping uncertainties, we used the newer versions of CLC. However, in this relatively short time period, only the immediately visible changes from an intensification of use are reflected, while long-term effects such as forest regrowth on abandoned grassland can only be captured over longer monitoring periods [36,67]. Over such short periods as in our study, only transitional stages to forest (e.g., heathlands, transitional woodland, shrub) could be considered. In future studies, the results may be improved by differentiating ES values between young and mature forests. Basing our analysis on earlier time steps with a greater extent of agriculturally used land would have revealed greater transformations of agricultural landscapes and related impacts on ES [36,77,84].

Furthermore, it has to be noted that our results represent the potential ES supply weighted by socio-cultural preferences, that is, the capacity of ecosystems to provide ES independently of their actual use [80]. However, many studies indicated spatial mismatches between ES supply and ES demand, i.e., the demand exceeds the supply at the local or regional level, requiring the transfer of goods or the movement of people [19,52,85,86]. Such dynamics need to be taken into account in the development of sustainable management strategies, and our results should, therefore, be complemented with spatial information on ES demand [20,24].

5. Conclusions

By applying the concept of ES, the consequences for society can be assessed in a comprehensive way, highlighting both the direct impacts on agricultural production and the associated effects on regulating and cultural ES. Our results reveal that the agricultural area in the Alpine region is under massive pressure, as up to 30% of agricultural land in some regions has been abandoned or converted to other uses within the last two decades, despite the efforts made within the framework of the Common Agriculture Policy (CAP) of the European Union (EU). Consequently, ES values mostly declined between 2000 and 2018, especially, regulating and cultural ES, while some forest-related provisioning ES have increased. Our results also indicate that LULC change rates and, hence, changes in ES greatly differed across regions with different social-ecological characteristics. The smallest changes occurred in the traditional agricultural region, while rural retreats and forgotten rural areas were affected by the largest changes.

Such quantitative and spatially explicit information on impacts from the transformation of agricultural land can be used as an information basis for developing sustainable management strategies and for evaluating underlying policies such the CAP. The frequent abandonment of mountain grassland, providing an above-average number of ES, also emphasizes the importance of the Green Deal in the EU, which should be an impulse for an agricultural and food transition. The Green Deal's target of 25% ecologically valuable farmland in agriculture is one of the central and most important targets, and, therefore, particular attention should be paid to the maintenance of mountain grassland in the European Alps. Finally, to support decision-making in adopting tangible local or regional strategies that can maintain cultural landscapes and multiple ES, greater efforts should be put into transdisciplinary research, allowing for the development of adaptive pathways, depending on the historical development of agricultural use, and socio-economic and landscape developments.

Supplementary Materials: The following are available online at https://www.mdpi.com/article/10.3390/land11010049/s1, Table S1: Indicators selected for cluster analysis; Table S2: LULC types aggregated from CLC classes; Table S3: ES values (ES supply weighted with socio-cultural preferences) for different LULC types.

Author Contributions: Conceptualization, U.S. and E.T.; methodology, U.S. and E.T.; formal analysis, U.S. and E.T.; writing—original draft preparation, U.S., E.T., G.L. and U.T. All authors have read and agreed to the published version of the manuscript.

Funding: This research received no external funding.

Institutional Review Board Statement: Not applicable.

Informed Consent Statement: Not applicable.

Data Availability Statement: All data used for analysis are publicly available and data sources are indicated. Further information is included in the Supplementary Material.

Conflicts of Interest: The authors declare no conflict of interest.

Appendix A

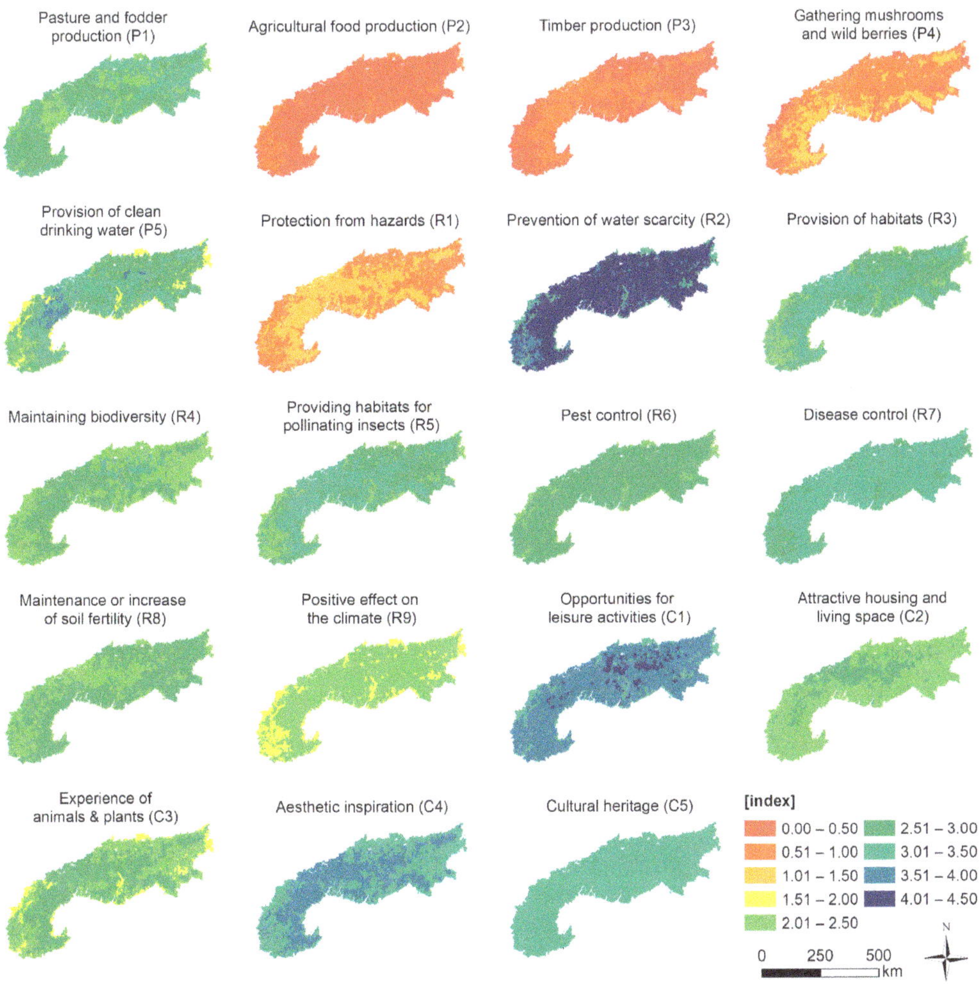

Figure A1. Distribution of ES values in 2018 across the European Alps. Only agricultural LULC types (crop cultivation, permanent culture, fertilized grassland, unfertilized grassland) that were present in 2000 were considered. Thus, these maps illustrate the changes in agricultural land to other LULC types (other agricultural types, abandoned land, forest, settlement areas). Authors own elaboration.

References

1. Assessment, M.E. *Ecosystems and Human Well-Being*; Island Press: Washington, DC, USA, 2005; Volume 5.
2. Pascual, U.; Balvanera, P.; Díaz, S.; Pataki, G.; Roth, E.; Stenseke, M.; Watson, R.T.; Başak Dessane, E.; Islar, M.; Kelemen, E.; et al. Valuing nature's contributions to people: The IPBES approach. *Curr. Opin. Environ. Sustain.* **2017**, *26–27*, 7–16. [CrossRef]
3. Fuhrer, J.; Smith, P.; Gobiet, A. Implications of climate change scenarios for agriculture in alpine regions—A case study in the Swiss Rhone catchment. *Sci. Total Environ.* **2014**, *493*, 1232–1241. [CrossRef] [PubMed]
4. Baude, M.; Meyer, B.C.; Schindewolf, M. Land use change in an agricultural landscape causing degradation of soil based ecosystem services. *Sci. Total Environ.* **2019**, *659*, 1526–1536. [CrossRef] [PubMed]
5. García-Llamas, P.; Geijzendorffer, I.R.; García-Nieto, A.P.; Calvo, L.; Suárez-Seoane, S.; Cramer, W. Impact of land cover change on ecosystem service supply in mountain systems: A case study in the Cantabrian Mountains (NW of Spain). *Reg. Environ. Chang.* **2019**, *19*, 529–542. [CrossRef]

6. Zhang, W.; Ricketts, T.H.; Kremen, C.; Carney, K.; Swinton, S.M. Ecosystem services and dis-services to agriculture. *Ecol. Econ.* **2007**, *64*, 253–260. [CrossRef]
7. Swinton, S.M.; Lupi, F.; Robertson, G.P.; Hamilton, S.K. Ecosystem services and agriculture: Cultivating agricultural ecosystems for diverse benefits. *Ecol. Econ.* **2007**, *64*, 245–252. [CrossRef]
8. Palm, C.; Blanco-Canqui, H.; DeClerck, F.; Gatere, L.; Grace, P. Conservation agriculture and ecosystem services: An overview. *Agric. Ecosyst. Environ.* **2014**, *187*, 87–105. [CrossRef]
9. Kremen, C.; Miles, A. Ecosystem Services in Biologically Diversified versus Conventional Farming Systems: Benefits, Externalities, and Trade-Offs. *Ecol. Soc.* **2012**, *17*, 40. [CrossRef]
10. Schrama, M.; de Haan, J.J.; Kroonen, M.; Verstegen, H.; Van der Putten, W.H. Crop yield gap and stability in organic and conventional farming systems. *Agric. Ecosyst. Environ.* **2018**, *256*, 123–130. [CrossRef]
11. Boone, L.; Roldán-Ruiz, I.; Van Linden, V.; Muylle, H.; Dewulf, J. Environmental sustainability of conventional and organic farming: Accounting for ecosystem services in life cycle assessment. *Sci. Total Environ.* **2019**, *695*, 133841. [CrossRef]
12. Jäger, H.; Peratoner, G.; Tappeiner, U.; Tasser, E. Grassland biomass balance in the European Alps: Current and future ecosystem service perspectives. *Ecosyst. Serv.* **2020**, *45*, 101163. [CrossRef]
13. Wezel, A.; Stöckli, S.; Tasser, E.; Nitsch, H.; Vincent, A. Good Pastures, Good Meadows: Mountain Farmers' Assessment, Perceptions on Ecosystem Services, and Proposals for Biodiversity Management. *Sustainability* **2021**, *13*, 5609. [CrossRef]
14. Troeger-Weiß, G. *Die Alpen—Geschichte und Zukunft Einer Europäischen Kulturlandschaft*; C.H.Beck: Munich, Germany, 2015; Volume 51, ISBN 9783406681837.
15. Scolozzi, R.; Schirpke, U.; Detassis, C.; Abdullah, S.; Gretter, A. Mapping Alpine Landscape Values and Related Threats as Perceived by Tourists. *Landsc. Res.* **2015**, *40*, 451–465. [CrossRef]
16. Schirpke, U.; Meisch, C.; Marsoner, T.; Tappeiner, U. Revealing spatial and temporal patterns of outdoor recreation in the European Alps and their surroundings. *Ecosyst. Serv.* **2018**, *31*, 336–350. [CrossRef]
17. Tenerelli, P.; Demšar, U.; Luque, S. Crowdsourcing indicators for cultural ecosystem services: A geographically weighted approach for mountain landscapes. *Ecol. Indic.* **2016**, *64*, 237–248. [CrossRef]
18. Zoderer, B.M.; Tasser, E.; Carver, S.; Tappeiner, U. An integrated method for the mapping of landscape preferences at the regional scale. *Ecol. Indic.* **2019**, *106*, 105430. [CrossRef]
19. Schirpke, U.; Tappeiner, U.; Tasser, E. A transnational perspective of global and regional ecosystem service flows from and to mountain regions. *Sci. Rep.* **2019**, *9*, 6678. [CrossRef]
20. Grêt-Regamey, A.; Weibel, B. Global assessment of mountain ecosystem services using earth observation data. *Ecosyst. Serv.* **2020**, *46*, 101213. [CrossRef]
21. Schirpke, U.; Wang, G.; Padoa-Schioppa, E. Editorial: Mountain landscapes: Protected areas, ecosystem services, and future challenges. *Ecosyst. Serv.* **2021**, *49*, 101302. [CrossRef]
22. Marini, L.; Klimek, S.; Battisti, A. Mitigating the impacts of the decline of traditional farming on mountain landscapes and biodiversity: A case study in the European Alps. *Environ. Sci. Policy* **2011**, *14*, 258–267. [CrossRef]
23. Zimmermann, P.; Tasser, E.; Leitinger, G.; Tappeiner, U. Effects of land-use and land-cover pattern on landscape-scale biodiversity in the European Alps. *Agric. Ecosyst. Environ.* **2010**, *139*, 13–22. [CrossRef]
24. Schirpke, U.; Candiago, S.; Egarter Vigl, L.; Jäger, H.; Labadini, A.; Marsoner, T.; Meisch, C.; Tasser, E.; Tappeiner, U. Integrating supply, flow and demand to enhance the understanding of interactions among multiple ecosystem services. *Sci. Total Environ.* **2019**, *651*, 928–941. [CrossRef] [PubMed]
25. Schirpke, U.; Tasser, E. Trends in Ecosystem Services across Europe Due to Land-Use/Cover Changes. *Sustainability* **2021**, *13*, 7095. [CrossRef]
26. Nadal-Romero, E.; Rubio, P.; Kremyda, V.; Absalah, S.; Cammeraat, E.; Jansen, B.; Lasanta, T. Effects of agricultural land abandonment on soil organic carbon stocks and composition of soil organic matter in the Central Spanish Pyrenees. *CATENA* **2021**, *205*, 105441. [CrossRef]
27. Munteanu, C.; Kuemmerle, T.; Boltiziar, M.; Butsic, V.; Gimmi, U.; Halada, L.; Kaim, D.; Király, G.; Konkoly-Gyuró, É.; Kozak, J.; et al. Forest and agricultural land change in the Carpathian region-A meta-analysis of long-term patterns and drivers of change. *Land Use Policy* **2014**, *38*, 685–697. [CrossRef]
28. Huber, L.; Schirpke, U.; Marsoner, T.; Tasser, E.; Leitinger, G. Does socioeconomic diversification enhance multifunctionality of mountain landscapes? *Ecosyst. Serv.* **2020**, *44*, 101122. [CrossRef]
29. Flury, C.; Huber, R.; Tasser, E. Future of mountain agriculture in the Alps. In *The Future of Mountain Agriculture*; Springer: Berlin/Heidelberg, Germany, 2013; pp. 105–126. ISBN 9783642335846.
30. Hinojosa, L.; Tasser, E.; Rüdisser, J.; Leitinger, G.; Schermer, M.; Lambin, E.F.; Tappeiner, U. Geographical heterogeneity in mountain grasslands dynamics in the Austrian-Italian Tyrol region. *Appl. Geogr.* **2019**, *106*, 50–59. [CrossRef]
31. Jepsen, M.R.; Kuemmerle, T.; Müller, D.; Erb, K.; Verburg, P.H.; Haberl, H.; Vesterager, J.P.; Andrič, M.; Antrop, M.; Austrheim, G.; et al. Transitions in European land-management regimes between 1800 and 2010. *Land Use Policy* **2015**, *49*, 53–64. [CrossRef]
32. Dax, T.; Schroll, K.; Machold, I.; Derszniak-Noirjean, M.; Schuh, B.; Gaupp-Berghausen, M. Land Abandonment in Mountain Areas of the EU: An Inevitable Side Effect of Farming Modernization and Neglected Threat to Sustainable Land Use. *Land* **2021**, *10*, 591. [CrossRef]

33. Egarter Vigl, L.; Schirpke, U.; Tasser, E.; Tappeiner, U. Linking long-term landscape dynamics to the multiple interactions among ecosystem services in the European Alps. *Landsc. Ecol.* **2016**, *31*, 1903–1918. [CrossRef]
34. Tasser, E.; Schermer, M.; Siegl, G.; Tappeiner, U. *Wir Landschaftmacher. Vom Sein und Werden der Kulturlandschaft in Nord-, Ost- und Südtirol*; Athesia: Bozen/Bolzano, Italy, 2012; ISBN 8882669068.
35. Schirpke, U.; Zoderer, B.M.; Tappeiner, U.; Tasser, E. Effects of past landscape changes on aesthetic landscape values in the European Alps. *Landsc. Urban Plan.* **2021**, *212*, 104109. [CrossRef]
36. Schirpke, U.; Tscholl, S.; Tasser, E. Spatio-temporal changes in ecosystem service values: Effects of land-use changes from past to future (1860–2100). *J. Environ. Manag.* **2020**, *272*, 111068. [CrossRef]
37. Egarter Vigl, L.; Marsoner, T.; Schirpke, U.; Tscholl, S.; Candiago, S.; Depellegrin, D. A multi-pressure analysis of ecosystem services for conservation planning in the Alps. *Ecosyst. Serv.* **2021**, *47*, 101230. [CrossRef]
38. Stürck, J.; Schulp, C.J.E.; Verburg, P.H. Spatio-temporal dynamics of regulating ecosystem services in Europe- The role of past and future land use change. *Appl. Geogr.* **2015**, *63*, 121–135. [CrossRef]
39. Marando, F.; Salvatori, E.; Sebastiani, A.; Fusaro, L.; Manes, F. Regulating Ecosystem Services and Green Infrastructure: Assessment of Urban Heat Island effect mitigation in the municipality of Rome, Italy. *Ecol. Modell.* **2019**, *392*, 92–102. [CrossRef]
40. Tappeiner, U.; Leitinger, G.; Zariņa, A.; Bürgi, M. How to consider history in landscape ecology: Patterns, processes, and pathways. *Landsc. Ecol.* **2020**, *36*, 2317–2328. [CrossRef]
41. Guerry, A.D.; Polasky, S.; Lubchenco, J.; Chaplin-Kramer, R.; Daily, G.C.; Griffin, R.; Ruckelshaus, M.; Bateman, I.J.; Duraiappah, A.; Elmqvist, T.; et al. Natural capital and ecosystem services informing decisions: From promise to practice. *Proc. Natl. Acad. Sci. USA* **2015**, *112*, 7348–7355. [CrossRef]
42. Huber, R.; Rigling, A.; Bebi, P.; Brand, F.S.; Briner, S.; Buttler, A.; Elkin, C.; Gillet, F.; Grêt-Regamey, A.; Hirschi, C.; et al. Sustainable land use in mountain regions under global change: Synthesis across scales and disciplines. *Ecol. Soc.* **2013**, *18*, 36. [CrossRef]
43. Brunner, S.H.; Grêt-Regamey, A. Policy strategies to foster the resilience of mountain social-ecological systems under uncertain global change. *Environ. Sci. Policy* **2016**, *66*, 129–139. [CrossRef]
44. Martín-López, B.; Leister, I.; Lorenzo Cruz, P.; Palomo, I.; Grêt-Regamey, A.; Harrison, P.A.; Lavorel, S.; Locatelli, B.; Luque, S.; Walz, A. Nature's contributions to people in mountains: A review. *PLoS ONE* **2019**, *14*, e0217847. [CrossRef] [PubMed]
45. McKenzie, E.; Posner, S.; Tillmann, P.; Bernhardt, J.R.; Howard, K.; Rosenthal, A. Understanding the Use of Ecosystem Service Knowledge in Decision Making: Lessons from International Experiences of Spatial Planning. *Environ. Plan. C Gov. Policy* **2014**, *32*, 320–340. [CrossRef]
46. Hauck, J.; Görg, C.; Varjopuro, R.; Ratamäki, O.; Maes, J.; Wittmer, H.; Jax, K. "Maps have an air of authority": Potential benefits and challenges of ecosystem service maps at different levels of decision making. *Ecosyst. Serv.* **2013**, *4*, 25–32. [CrossRef]
47. Paprotny, D.; Terefenko, P.; Giza, A.; Czapliński, P.; Vousdoukas, M.I. Future losses of ecosystem services due to coastal erosion in Europe. *Sci. Total Environ.* **2021**, *760*, 144310. [CrossRef]
48. Spake, R.; Lasseur, R.; Crouzat, E.; Bullock, J.M.; Lavorel, S.; Parks, K.E.; Schaafsma, M.; Bennett, E.M.; Maes, J.; Mulligan, M.; et al. Unpacking ecosystem service bundles: Towards predictive mapping of synergies and trade-offs between ecosystem services. *Glob. Environ. Chang.* **2017**, *47*, 37–50. [CrossRef]
49. Turkelboom, F.; Leone, M.; Jacobs, S.; Kelemen, E.; García-Llorente, M.; Baró, F.; Termansen, M.; Barton, D.N.; Berry, P.; Stange, E.; et al. When we cannot have it all: Ecosystem services trade-offs in the context of spatial planning. *Ecosyst. Serv.* **2018**, *29*, 566–578. [CrossRef]
50. Tasser, E.; Schirpke, U.; Zoderer, B.M.; Tappeiner, U. Towards an integrative assessment of land-use type values from the perspective of ecosystem services. *Ecosyst. Serv.* **2020**, *42*, 101082. [CrossRef]
51. Geneletti, D.; Scolozzi, R.; Adem Esmail, B. Assessing ecosystem services and biodiversity tradeoffs across agricultural landscapes in a mountain region. *Int. J. Biodivers. Sci. Ecosyst. Serv. Manag.* **2018**, *14*, 188–208. [CrossRef]
52. Ala-Hulkko, T.; Kotavaara, O.; Alahuhta, J.; Hjort, J. Mapping supply and demand of a provisioning ecosystem service across Europe. *Ecol. Indic.* **2019**, *103*, 520–529. [CrossRef]
53. González-García, A.; Palomo, I.; González, J.A.; López, C.A.; Montes, C. Quantifying spatial supply-demand mismatches in ecosystem services provides insights for land-use planning. *Land Use Policy* **2020**, *94*, 104493. [CrossRef]
54. Maes, J.; Egoh, B.; Willemen, L.; Liquete, C.; Vihervaara, P.; Schägner, J.P.; Grizzetti, B.; Drakou, E.G.; La Notte, A.; Zulian, G.; et al. Mapping ecosystem services for policy support and decision making in the European Union. *Ecosyst. Serv.* **2012**, *1*, 31–39. [CrossRef]
55. Lasseur, R.; Vannier, C.; Lefebvre, J.; Longaretti, P.-Y.; Lavorel, S. Landscape-scale modeling of agricultural land use for the quantification of ecosystem services. *J. Appl. Remote Sens.* **2018**, *12*, 046024. [CrossRef]
56. Mengist, W.; Soromessa, T.; Legese, G. Ecosystem services research in mountainous regions: A systematic literature review on current knowledge and research gaps. *Sci. Total Environ.* **2020**, *702*, 134581. [CrossRef] [PubMed]
57. Lavorel, S.; Grigulis, K.; Leitinger, G.; Kohler, M.; Schirpke, U.; Tappeiner, U. Historical trajectories in land use pattern and grassland ecosystem services in two European alpine landscapes. *Reg. Environ. Chang.* **2017**, *17*, 2251–2264. [CrossRef] [PubMed]
58. Sauter, I.; Kienast, F.; Bolliger, J.; Winter, B.; Pazúr, R. Changes in demand and supply of ecosystem services under scenarios of future land use in Vorarlberg, Austria. *J. Mt. Sci.* **2019**, *16*, 2793–2809. [CrossRef]

59. Briner, S.; Elkin, C.; Huber, R. Evaluating the relative impact of climate and economic changes on forest and agricultural ecosystem services in mountain regions. *J. Environ. Manag.* **2013**, *129*, 414–422. [CrossRef]
60. Grêt-Regamey, A.; Sirén, E.; Brunner, S.H.; Weibel, B. Review of decision support tools to operationalize the ecosystem services concept. *Ecosyst. Serv.* **2017**, *26*, 306–315. [CrossRef]
61. Tappeiner, U.; Borsdorf, A.; Tasser, E.I. *Alpenatlas: Society, Economy, Environment = Atlas des Alpes = Atlante delle Alpi = Atlas Alp = Mapping the Alps*; Spektrum Akademischer Verlag: Heidelberg, Germany, 2008; ISBN 978-3-8274-2004-6.
62. European Environment Agency Corine Land Cover (CLC) 2000, Version 2020_20u1. Available online: https://land.copernicus.eu/pan-european/corine-land-cover/clc-2000 (accessed on 26 April 2021).
63. European Environment Agency Corine Land Cover (CLC) 2018, Version 2020_20u1. Available online: https://land.copernicus.eu/pan-european/corine-land-cover/clc2018 (accessed on 26 April 2021).
64. Pecher, C.; Bacher, M.; Tasser, E.; Tappeiner, U. Agricultural landscapes between intensification and abandonment: The expectations of the public in a Central-Alpine cross-border region. *Landsc. Res.* **2018**, *43*, 428–442. [CrossRef]
65. Stotten, R.; Ambrosi, L.; Tasser, E.; Leitinger, G. Social-ecological resilience in remote mountain communities: Toward a novel framework for an interdisciplinary investigation. *Ecol. Soc.* **2021**, *26*, 29. [CrossRef]
66. Niedermayr, J.; Hoffmann, C.; Stawinoga, A.; Streifeneder, T. Agro-structural patterns in the Alps 2000–2010. *Jahrb. Osterr. Ges. Agrar.* **2015**, *24*, 275–284.
67. Tasser, E.; Leitinger, G.; Tappeiner, U. Climate change versus land-use change—What affects the mountain landscapes more? *Land Use Policy* **2017**, *60*, 60–72. [CrossRef]
68. Pecher, C.; Tasser, E.; Tappeiner, U. Definition of the potential treeline in the European Alps and its benefit for sustainability monitoring. *Ecol. Indic.* **2011**, *11*, 438–447. [CrossRef]
69. Huber, R.; Briner, S.; Bugmann, H.; Elkin, C.; Hirschi, C.; Seidl, R.; Snell, R.; Rigling, A. Inter- and transdisciplinary perspective on the integration of ecological processes into ecosystem services analysis in a mountain region. *Ecol. Process.* **2014**, *3*, 9. [CrossRef]
70. Shackleton, C.M.; Ruwanza, S.; Sinasson Sanni, G.K.; Bennett, S.; De Lacy, P.; Modipa, R.; Mtati, N.; Sachikonye, M.; Thondhlana, G. Unpacking Pandora's Box: Understanding and Categorising Ecosystem Disservices for Environmental Management and Human Wellbeing. *Ecosystems* **2016**, *19*, 587–600. [CrossRef]
71. Strasser, U.; Förster, K.; Formayer, H.; Hofmeister, F.; Marke, T.; Meißl, G.; Nadeem, I.; Stotten, R.; Schermer, M. Storylines of combined future land use and climate scenarios and their hydrological impacts in an Alpine catchment (Brixental/Austria). *Sci. Total Environ.* **2019**, *657*, 746–763. [CrossRef]
72. Bateman, I.J.; Harwood, A.R.; Mace, G.M.; Watson, R.T.; Abson, D.J.; Andrews, B.; Binner, A.; Crowe, A.; Day, B.H.; Dugdale, S.; et al. Bringing ecosystem services into economic decision-making: Land use in the United Kingdom. *Science* **2013**, *341*, 45–50. [CrossRef] [PubMed]
73. Fayet, C.M.J.; Reilly, K.H.; Van Ham, C.; Verburg, P.H. What is the future of abandoned agricultural lands? A systematic review of alternative trajectories in Europe. *Land Use Policy* **2022**, *112*, 105833. [CrossRef]
74. Schermer, M.; Stotten, R.; Strasser, U.; Meißl, G.; Marke, T.; Förster, K.; Formayer, H. The Role of Transdisciplinary Research for Agricultural Climate Change Adaptation Strategies. *Agronomy* **2018**, *8*, 237. [CrossRef]
75. Kohler, M.; Devaux, C.; Grigulis, K.; Leitinger, G.; Lavorel, S.; Tappeiner, U. Plant functional assemblages as indicators of the resilience of grassland ecosystem service provision. *Ecol. Indic.* **2017**, *73*, 118–127. [CrossRef]
76. Schirpke, U.; Kohler, M.; Leitinger, G.; Fontana, V.; Tasser, E.; Tappeiner, U. Future impacts of changing land-use and climate on ecosystem services of mountain grassland and their resilience. *Ecosyst. Serv.* **2017**, *26*, 79–94. [CrossRef]
77. Schirpke, U.; Leitinger, G.; Tasser, E.; Rüdisser, J.; Fontana, V.; Tappeiner, U. Functional spatial units are fundamental for modelling ecosystem services in mountain regions. *Appl. Geogr.* **2020**, *118*, 102200. [CrossRef]
78. Lavorel, S.; Colloff, M.J.; Locatelli, B.; Gorddard, R.; Prober, S.M.; Gabillet, M.; Devaux, C.; Laforgue, D.; Peyrache-Gadeau, V. Mustering the power of ecosystems for adaptation to climate change. *Environ. Sci. Policy* **2019**, *92*, 87–97. [CrossRef]
79. Jacobs, S.; Burkhard, B.; Van Daele, T.; Staes, J.; Schneiders, A. 'The Matrix Reloaded': A review of expert knowledge use for mapping ecosystem services. *Ecol. Modell.* **2015**, *295*, 21–30. [CrossRef]
80. Burkhard, B.; Kroll, F.; Nedkov, S.; Müller, F. Mapping ecosystem service supply, demand and budgets. *Ecol. Indic.* **2012**, *21*, 17–29. [CrossRef]
81. Rega, C.; Short, C.; Pérez-Soba, M.; Luisa Paracchini, M. A classification of European agricultural land using an energy-based intensity indicator and detailed crop description. *Landsc. Urban Plan.* **2020**, *198*, 103793. [CrossRef]
82. Hinojosa, L.; Napoléone, C.; Moulery, M.; Lambin, E.F. The "mountain effect" in the abandonment of grasslands: Insights from the French Southern Alps. *Agric. Ecosyst. Environ.* **2016**, *221*, 115–124. [CrossRef]
83. García-Álvarez, D.; Camacho Olmedo, M.T. Changes in the methodology used in the production of the Spanish CORINE: Uncertainty analysis of the new maps. *Int. J. Appl. Earth Obs. Geoinf.* **2017**, *63*, 55–67. [CrossRef]
84. Locatelli, B.; Lavorel, S.; Sloan, S.; Tappeiner, U.; Geneletti, D. Characteristic trajectories of ecosystem services in mountains. *Front. Ecol. Environ.* **2017**, *15*, 150–159. [CrossRef]
85. Serna-Chavez, H.M.; Schulp, C.J.E.; Van Bodegom, P.M.; Bouten, W.; Verburg, P.H.; Davidson, M.D. A quantitative framework for assessing spatial flows of ecosystem services. *Ecol. Indic.* **2014**, *39*, 24–33. [CrossRef]
86. Syrbe, R.U.; Grunewald, K. Ecosystem service supply and demand–the challenge to balance spatial mismatches. *Int. J. Biodivers. Sci. Ecosyst. Serv. Manag.* **2017**, *13*, 148–161. [CrossRef]

land

MDPI

Article

Emergent Properties of Land Systems: Nonlinear Dynamics of Scottish Farming Systems from 1867 to 2020

Richard Aspinall [1,*], Michele Staiano [2] and Diane Pearson [3]

1 James Hutton Institute, Craigiebuckler, Aberdeen AB15 8QH, UK
2 Statistics Technology and Analysis of Data, Department of Industrial Engineering, University of Naples Federico II, 80125 Napoli, Italy; mstaiano@unina.it
3 Farmed Landscape Research Centre, School of Agriculture and Environment, College of Sciences, Massey University, Palmerston North 4442, New Zealand; d.pearson@massey.ac.nz
* Correspondence: rjaspinall10@gmail.com

Abstract: Dynamics of arable and pastoral farming systems in Scotland over the period 1867–2020 are documented using time series analysis methods, including for nonlinear dynamical systems. Results show arable and pastoral farming, at a national scale, are dynamic over a range of timescales, with medium- and short-term dynamics associated with endogenous system forces and exogenous factors, respectively. Medium-term dynamics provide evidence of endogenous systems-level feedbacks between farming sectors responding to change in world and national cereal prices as an economic driver, and act to dampen impacts of exogenous shocks and events (weather, disease). Regime shifts are identified in national cereal prices. Results show change and dynamics as emergent properties of system interactions. Changes in dynamics and strength of endogenous dampening over the duration of the study are associated with dynamical changes from major governmental policy decisions that altered the boundary conditions for interdependencies of arable and pastoral farming.

Keywords: land system dynamics; emergent properties; time series analysis; nonlinear dynamics; Recurrence Plots; Scotland

Citation: Aspinall, R.; Staiano, M.; Pearson, D. Emergent Properties of Land Systems: Nonlinear Dynamics of Scottish Farming Systems from 1867 to 2020. *Land* **2021**, *10*, 1172. https://doi.org/10.3390/land10111172

Academic Editor: Marta Debolini

Received: 20 September 2021
Accepted: 28 October 2021
Published: 1 November 2021

Publisher's Note: MDPI stays neutral with regard to jurisdictional claims in published maps and institutional affiliations.

1. Introduction

Much research in land systems science has focused on process–response (cause and effect) relationships of changes in land use and land cover with a variety of drivers of change as causal factors [1–4]. Many of these studies have focused on dynamics defined by change resulting from either land conversion (changes in type of cover and/or use) or land modification (land use intensification, land degradation, land abandonment) using snapshots in time and simple differencing between dates to elucidate patterns in observed changes. Observed changes are hypothesized to be caused by a variety of drivers and processes. In a whole systems context, however, dynamics are the set of behaviours exhibited as a result of the interactions of the elements that define land as a system [5]. The nature of land systems as complex systems with dynamic and emergent behaviours is recognized in, for example, land use transitions and the causal roles of endogenous forces and exogenous factors [6] and in calls for identification of regime shifts [7,8]. To date, however, despite recognizing land systems as inherently coupled systems, relatively few studies of land systems beyond agent-based models [9] have attempted to interpret dynamics as a function of the structures, interactions, and feedback mechanisms that define land as a coupled system. Additionally, as Turner and colleagues note, despite wide recognition of land as an exemplar of coupled human-environment systems [10], these explanations typically invoke one of the human or environment subsystem explanations in more detail, and few are rooted in the interactions of human and environment systems [1].

There are three main limitations on the current description and understanding of dynamics in land systems. First, the short time spans of studies provide a limited set of system states recording change, which can allow changes to be mis-characterized [8]. Second, drivers are treated as constants over the short time span of interest, with little attempt to describe change in drivers over time or feedbacks from the system to the drivers; this can preclude identification of regime shifts and other dynamic system responses, particularly if change occurs as a punctuated equilibrium process [7,8] and may lead to misidentification of mechanisms generating change [11]. Third, in the context of whole systems approaches to land, land systems are not only dynamic, but also dynamical, in that their state can change over time even in the absence of changes in use or cover. This has implications for studies of process–response relationships in land change, since change may be a result of dynamical responses that are endogenous to the land system, in addition to being caused by drivers and other exogenous forcing variables.

The dynamic and dynamical nature of land systems are central to understanding land as a complex, coupled human-environment system. Land conversion and modification are undoubtedly important, and central to the scientific and applied needs for understanding land system change and its impacts [12–14]; understanding dynamics and dynamical changes in land systems is essential for describing functional behaviours of the whole land system. In dynamic and dynamical systems with many feedbacks and interactions the linear and sequential distinction between cause and effect becomes weak since each variable is both a cause and an effect. In this context, explanation of dynamics is based on system functioning via interaction of structures through uncovering endogenous forces and measurement of system-level responses to exogenous factors. This is related to the problem of equifinality, in which there are multiple plausible explanations for an outcome, a phenomenon well known in environmental science [15], and with an importance recognized for policy advice and developing models for socio-ecological systems [9].

Description of the dynamics of a system requires data that describe the structures, funds, flows (inputs, outputs, changes in funds) over time, as well as frameworks that organize the funds, flows and feedbacks into a system, and mathematical and other kinds of models that encapsulate functional dynamics of the system. Erb et al., and Kuemmerle et al., describe potential inputs, outputs, and structures for use to describe land intensity [16,17], responding to knowledge gaps that limit understanding and characterization of dynamics and patterns of land use intensity, but their conceptual framing is limited in description of the structure of the land system itself, and no attempt is made to incorporate feedbacks or land system funds beyond the biophysical structures they identify. Rahim et al., (2017) describe a causal loop model for analysis of supply and demand in Malaysian rice production as a complex system but have yet to quantify the model [18]. Elements defining land systems include state descriptors and system drivers, such as the structures and funds that comprise the system, but also the interactions of these elements, associated with connections between structures and funds through flows and feedbacks.

The aim of this study is to analyse long-, medium- and short-term patterns in land system changes, to understand the dynamics associated with interaction of systems at these different scales. Few studies have attempted formally to characterize multi-scale dynamics or analyse long time-series of land system data. The study also demonstrates some of the techniques available for this type of analysis, using a case study for Scotland.

This paper analyses dynamics of farming systems in Scotland, using data describing farming at a national (Scotland-wide) scale from the last third of the 19th century until 2020. The record of farming in Scotland over this period is well known from studies contemporary with the changes observed (see, for example, annual publications of the Transactions of the Royal Highland and Agricultural Society (1790–1969) and Scottish Agricultural Economics (1950–1960), and reviews and audits of the history [19–25]. The difference here is that the analysis develops from the perspective of farming as a land system and spans the full period from 1867 to 2020 within a single quantitative analysis. Time series analysis, including methods for nonlinear dynamical systems, is used (i) to examine dynamics of the system over the full timespan and (ii) to characterize internal feedbacks and coupling of farming as a system at the national scale. These analyses reveal some system characteristics and behaviours associated with the evolution of the system itself. The results also identify regime shifts using analytical and graphical tools for study of nonlinear dynamical systems.

Analysis is based on the contention that the time series of data recording the history of land use for farming in Scotland from 1867 to 2020 contain a record of the impacts of all long-, medium-, and short-term dynamics associated with both endogenous system forces and exogenous factors that have influenced the land system. Just as spatial patterns embed all the processes, from many spatial and temporal scales, that are involved in the production of landscape patterns (Dobson, 1990, 1992), so the temporal record of land systems contains an embedded record of both the effects of processes acting over long-, medium- and short- timescales and system responses. Because of this, the temporal scale at which a land system is studied should be made explicit, as the factors needed for explanation of changes and dynamics will vary with the time scale of interest. The case study shows that arable and pastoral farming, at a national scale, are dynamic over a range of timescales, but that throughout much of the timespan of the study the system has maintained a pattern of changes consistent with endogenous systems-level feedbacks between sectors that act to dampen the impacts of exogenous drivers. Changes in these system dynamics over the timespan are associated with policy changes that altered the interaction of arable and pastoral farming.

The rest of this paper is structured as follows:

1. Terminology, assumptions, and organizing principles for the data and analyses presented are defined. These represent pre-analytical decisions and definitions that prescribe the purpose of analyses and the specific hypotheses investigated.
2. Methods for addressing dynamics in time series data, including for nonlinear dynamics are outlined.
3. The Data section describes the detailed time series of data used; these define dimensions of farming land systems for Scotland as the basis for describing dynamics. The data represent funds, flows, and drivers of the land system, and data summarize aspects of the history of farming. The data are used for decomposition of hierarchically structured, time-related changes in both (a) the selected drivers of system dynamics and (b) system funds, structures, and feedbacks.
4. The Results section
    (a) reports the results of analysis of time series data to identify trends, cycles, and random elements at long-, medium-, and short timescales. The short-term component in the data is treated mathematically and operationally as statistical noise, but in practice reflects the impacts of real exogenous shocks, and other perturbations at specific times during the period of interest. This analysis shows the capacity of time series analyses to reveal the variety of long-, medium-, and short-term patterns recorded within the data, and ways in which system variables interact when viewed over various time spans.
    (b) analyses time series data for cereal prices, area planted with cereals, and number of sheep using methods from nonlinear dynamics. This analysis is based on understanding the interactions of arable and pastoral farming at a national scale in Scotland over the 19th and 20th Century. Arable and pastoral farming typically are treated as separate land uses and receive separate economic treatment as relatively distinct farming sectors in contemporary studies; this reflects increasing specialization in farming associated with intensification and modernization [26], and land cover and land management differences. However, this separation has not always been the case, and the interaction of pastoral and arable farming has long been widely recognized [21,27–29].
5. Discussion about the results and implications and opportunities of the approaches used for study of dynamics in land systems.

2. Terminology, Assumptions, and Organization: Pre-Analytical Definitions

In the discussion that follows, land system implicitly refers to a coupled human-environment system. The description of the system and definition of system elements are central to analysis; this forms a necessary and fundamental pre-analytical stage for subsequent data collection and analysis. An underpinning conceptual model for land systems has been described elsewhere [5], using funds and flows to define the system structures and their interactions. This conceptual representation of the land system describes a series of sub-systems that are associated with both driving factors, that operate as system processes, and different types of capital (human, social, financial, physical/manufactured, and natural). A suite of human and social factors that influence individual and group decisions and decision-making is also included within a decision-making subsystem. Funds are linked by flows, as fluxes and changes in funds. The conceptual model in Aspinall and Staiano (2017) does not quantify the funds and flows, although it does indicate some of the time scales over which the fluxes and changes in the system elements operate, from days, months, and seasons to years and decades, and longer. The model has been used to underpin an accounting approach for analysis of supply of provisioning services and the dynamics of agricultural land use in Scotland between 1940 and 2016 [26].

The timespan considered is 154 years (1867 to 2020, inclusive). The time step is 1 year. The geographic unit used is the aggregate national land use in arable (cereals) and pastoral (sheep) farming in Scotland. In the example here we are primarily interested in exploring the nature of change in the land system and in the ways in which structural measures of the system as well as drivers have changed over the long-, medium-, and short-term, using data for the period from the last third of the 19th Century to 2020. The long run and annual time step allow us to measure long-, medium-, and short-term patterns within these data.

Hierarchy theory is helpful in conceptual organization of hypotheses about scale (Wu, 2013), including the interaction of different time scales, and the use of time series analyses. Allen and Starr (1982) define hierarchies as a process-oriented framework, and Allen (2009) lists some general principles for ordering levels in ecological hierarchies. These include:

1. higher levels operate more slowly and at a lower frequency than lower levels;
2. higher levels exert constraints on lower levels;
3. higher levels function as a context to lower levels.

This hierarchical organization of process embeds and defines relationships between processes at different levels, based on timescale of processes from fast to slow (short to long), with dynamics of slower processes at higher levels appearing as a constant at lower levels, and dynamics of faster processes at lower levels appearing as noise at higher levels. Hierarchy theory provides, therefore, a coherent conceptual architecture for addressing complexity, ordering levels by rate of processes, and defining coupling of system components across and within timescales such that they can be decomposed for description, analysis, and understanding (Wu, 2013). Nonlinear dynamics also offers potential, particularly in the interaction of fast and slow processes [30] and understanding the consequences of managing resources based on one over the other [31], leading to fragility in system resilience [32]. A hierarchical structure of process dynamics from fast to slow is embedded in the conceptualization and analysis of the system dynamics for farming used to inform interpretation.

3. Methods for Addressing Dynamics

Dynamics are changes or motion in systems that reflect the nature and interactions of system elements, including system states, feedbacks, and evolution. As such, dynamics are characterized in temporal changes in system drivers and states and in the operation of interactions between system elements. Techniques from time series analysis and for analysis of nonlinear dynamical systems are used to describe, extract, and understand observed dynamical behaviours and patterns from noisy time series. This section outlines the methods used in the context of some of the time-varying characteristics of system funds and drivers; the methods are then used to describe and understand the dynamics of arable and pastoral system behaviours.

Dynamics are examined using two sets of methods:

(i) Time series analysis, including lag plots and decomposition of time series into long-, medium-, and short-term components

(ii) Recurrence plots and Recurrence Quantification Analysis.

3.1. Time Series Analysis

Time series analysis is described in a number of standard texts [33,34] and implemented in numerous statistical and mathematical packages. Representation of change in the time domain is straightforward, using plots of data against time. Lag plots and decomposition of time series are used to identify long-, medium-, and short-term patterns in the time series, specifically identifying long-, medium-, and short-term patterns using detrending, smoothing, and calculation of residuals respectively. Least-squares regression is used to model exponential growth and other long-term changes in data over the complete timespan of the data. Deviations from these trends are modelled with smoothing splines to describe cycles over medium-term time scales. The residuals from the trends and cycles describe short-term variation. Results are reported both as absolute values and, for comparison between variables, as normalized values using z scores. The separate patterns and values of long-, medium-, and short-term components (trends, cycles, and residuals) for multiple variables can be analysed further to assess possible influences and feedbacks between variables at different timescales.

To identify and describe trends and cycles in time series of prices, and to link prices in Scotland to global prices, we use the Christiano-Fitzgerald Filter, a bandpass filter designed to identify patterns in data that lie within a specific range of frequencies [35,36]. The Christiano-Fitzgerald Filter has been used to identify long-term trends, cycles, and boom/bust episodes for world price data for commodities [37]. The filter characterizes time series as the sum of periodic functions, using a bandpass to accommodate trends without restrictions on the distribution of the underlying data. The method allows filtered series to be extracted for the duration of the full span of time in the data, without discarding data from the beginning or end of the series, as observations from the beginning of the period can be filtered with future values and from the end with past values. In the use here, the filter identifies cycles in prices data, allowing comparison of data for Scotland with world prices for commodities.

3.2. Recurrence Plots and Recurrence Quantification Analysis

Further analysis of the time series data is carried out using Recurrence Plots (RP) and Recurrence Quantification Analysis (RQA). RP and RQA are nonlinear dynamics methods for analysis and visualization of time series data [38,39]. Analysis is based on phase space reconstruction [40,41], a method for discovery of deterministic structure present in real-world dynamical systems using time series data of a single variable [42–44]. The approaches can be applied to coupled variables through cross and joint recurrence plots [45,46]. Detailed descriptions of the approach can be found in literature from physics and mathematics [38,45,47] and complexity science [39]; and examples of applications found in a variety of disciplines including economics [48,49], ecology [50–52], psychology [53,54], epidemiology [55], atmospheric science [56–58], and geosciences [59–62].

Using time series of values for structural or state variables as basic building blocks, we define X, a vector set of elements describing system structure and fund at a series of discrete time intervals, to represent the land system. Change in elements of this set over time (t) formally can be represented with a standard equation describing change over time:

$$\frac{dX}{dt} = f(X) \tag{1}$$

Continuous and discrete representations of change are unified by time-scale formulation [63].

Data are analysed by plotting values of the time series for X against lagged values of X, a standard procedure for time series that reveals changes from time $(t-k)$ to time (t) over the period of the lag (k), and that graphically identifies the patterns of changes. We use this method to identify annual changes of magnitude greater or less than 2.0 SDs from the probability density function of all observed differences, using a lag of 1 year to represent annual decisions in the farming calendar over time.

Specializing Equation (1) to derive the matrix of first order time derivatives of X for all lags across the discrete time series defines a new space and a Jacobian matrix:

$$\mathbf{J} = \{x(t),\ x(t-\Delta t),\ x(t-2\Delta t), \ldots x(t-n\Delta t)\} \tag{2}$$

This matrix describes vectors of delay space coordinates that estimate the original phase space generating the dynamics of X [40,41]. The eigenvalues that can be calculated from the Jacobian matrix are local Lyapunov exponents, used in diagnostic analysis of chaotic systems [64], including in geomorphology [65] and ecology [51].

In practice, the matrix of delay space coordinates is calculated for time lags up to the embedding dimension and each column is a vector of coordinates $\vec{x_t}$

$$\vec{x_t} = \{x(t),\ x(t-\Delta t),\ x(t-2\Delta t), \ldots x(t-(d_\epsilon - 1)\Delta t)\} \tag{3}$$

where:

Δ_t is the time delay or lag between data

d_ϵ is the embedding dimension or dimension of the space required to recover the dynamics.

The embedding dimension is estimated for a time series using the method proposed by Cao [66].

The delay space matrix is a representation of the phase space and used for phase space reconstruction. Recurrence plots [38] and Recurrence Quantification Analysis [47] are used to evaluate the dynamics of the time series from the delay space coordinates. These methods are robust, RP and RQA being independent of limiting constraints, such as data set size, non-stationarity, and assumptions about underlying statistical distributions of data. RP and RQA can also identify thresholds in datasets, and have been proposed as a nonlinear time series analysis method for detection of environmental thresholds [50]. In the context of land systems, RP and RQA offer potential for both characterizing and identifying complex dynamics and identification of thresholds and regime shifts.

A Recurrence Plot is a graphical tool for interpretation of delay space. The plot is based on computation of a matrix of distances $\mathbb{R}$ between the vectors of reconstructed points in the delay space, identifying when transitions in the delay space revisit the same value:

$$\mathbb{R}_{ij} = \begin{cases} 1 : \vec{x_i} \approx \vec{x_j} \\ 0 : \vec{x_i} \not\approx \vec{x_j} \end{cases} \quad i,j = 1,\ldots,N \tag{4}$$

where N is the number of considered states and $\vec{x_i} \approx \vec{x_j}$ means equivalence up to a distance r, a radius threshold identifying proximity in the delay space. The RP is hence sensitive to the value of r. A value that is too small will result in a sparse RP with little to no information. Similarly, a value that is too large will fill the RP, again providing little to no information. A number of criteria guide selection of r [45]:

(a) it should not exceed 10% of the mean or the maximum phase space diameter [47,67,68]
(b) it should be defined so that the recurrence point density in the RP is about 1% [69]
(c) to avoid problems related to noise, r has to be chosen such that it is five time larger than the standard deviation of the observational noise, i.e., $r > 5\sigma$ [70].

To generate RPs for each of the variables in this study we set the initial value of r to the minimum of (a) 10% of the maximum value of the phase space diameter and (b) 5σ where σ is initially estimated from the standard deviation of the annual changes at lag $t = 1$ over the duration of the time series. We then iterate from this value of r to reduce the point density in the RP, examining changes in the RP as r changes. The value of r used here for each variable is the smallest that retains pattern in the RP.

The RP is a square, symmetrical plot of the $N \times N$ matrix $\mathbb{R}$. In the analysis, N is the number of time points under study. Values from Equation (4) are plotted, 1 being coloured black, 0 being white. Black points highlight the recurrences of the dynamical system (defined by the radius r), the patterns in the RP giving insight into periodic structures and clustering properties within the data that do not show up in the original time series. The main diagonal is the identity line. RPs reveal structures in the data which can be single dots, diagonal, horizontal, and vertical lines, and blocks. Infrequent states are represented as isolated dots. Diagonal lines are the result of the system visiting the same region of state space several times. Horizontal and vertical lines represent periods when the system remains in the same state for a while; the lengths of lines represent the time the system is in the state. A threshold, or regime shift, will appear as a two (or more) separate square areas along the diagonal. White noise produces a uniformly distributed structure, and periodic oscillations produce a regular pattern within the RP. White bands are caused by abrupt changes in the dynamics and by extreme events, facilitating identification of extreme and rare events [45].

Analysis of the structures in the RP use methods from Recurrence Quantification Analysis [39,47] providing several measures indicative of the dynamics [45]. We use (i) recurrence rate, which is the percentage of points in the RP and indicates the probability that a specific state recurs; (ii) laminarity, which is the percentage of recurrence points that form vertical lines and is a measure of the presence of laminar phases in the system, and (iii) entropy, the Shannon entropy of the probability distribution of diagonal line lengths, indicative of the deterministic coupling of the system. RQA can be applied to a single RP for the full time series and to sliding windows traversing the time series. The RQA values for sliding windows of different sizes are computed to build up a picture of the dynamical properties over time between 1867 and 2020 across different time periods within the timespan of the data. A RP for the full time series is constructed to identify any regime shifts.

4. Data

The data used are time series of annual records describing the farming system in Scotland from 1867 to 2020. Specifically, data are

(i) finance and economics (national and global prices of cereals and other commodities),
(ii) descriptors of funds in arable and pastoral farming land use, namely area of cereal crops and numbers of sheep, and their change over time.

Additional data include time series for environment (weather data describing rainfall and temperature), and events (disease outbreaks, wars, introduction of legislation, trade agreements). These data are distilled to identify key events relevant to explanation of patterns and residuals in the analysis of the time series (see Figure 1).

A condition for use for all data is that each time series is required to be complete, with no missing data values within the timespan covered. Data describing structural elements of Scottish agriculture are collated as time series compiled from the Annual Agricultural (June) Censuses of Scotland which have been published over the last 154 years (Transactions of the Royal Highland and Agricultural Society of Scotland (1867–1910), Board of Agriculture for Scotland (1911–1927), Department of Agriculture for Scotland (1928–1958), Department of Agriculture and Fisheries for Scotland (1959–89), Scottish Office Agriculture and Fisheries Department (1990–1995), The Scottish Office Agriculture, Environment and Fisheries Department (1996–1999), Scottish Executive Rural Affairs Department (2000), Scottish Executive Environment and Rural Affairs Department (2001–2007), Scottish Government Rural and Environment Research and Analysis Directorate (2008–2010), Scottish Government Rural Payments and Inspections Directorate (2011), Scottish Government Environment and Forestry Directorate (2012–2013), Scottish Government Directorate for Environment and Forestry (2014–2017), Scottish Government Rural and Environmental Science & Analytical Services Division (2018–2020)—see, for example, [71–77].

The Agricultural (June) Census started in 1866, but many of the entries for 1866 are officially considered to be unrepresentative because of incomplete returns [23], and 1867 is used as the start for the time series. The total planted area in cereals is used to represent the arable sector; the number of sheep are used to represent the pastoral sector [22]. Together, wheat, barley, and oats are over 99% of the area of cereals grown in Scotland and have not been less than 97% in the period since 1867. In 2018, cereals contributed about 14% of the annual value of Scottish agriculture (about £430 million); sheep were a further 7% of the total value of agriculture (£236 m) [78].

Financial data on prices of cereals are from multiple sources. Wheat, barley, and oats prices for Scotland are from the annual reports on Agricultural Statistics (1912–1978), Economic Reports on Scottish Agriculture (1980–2020), Scottish Agricultural Economics (1950–1959), and records of prices from Fiars Markets around Scotland from the Transactions of the Royal Highland and Agricultural Society (1790–1969), themselves a continuation of a longer series of records from Fiars Markets from 1550–1780 [79]. All prices data are converted to pounds sterling per tonne (£/tonne), from a variety of source price and weight units (viz. Scots and English pounds, shillings, and pence (£/s/d), GB Pounds after decimalization in 1971, and weight (e.g., boll, bushel, cwt, ton, tonne) in use at the time of original data collection. The prices and their trends for the three cereals over time are similar. Barley price is used for price of cereals since barley is the cereal grown over the largest area since the mid-1960s; oats were the major cereal crop by area until the mid-1960s [22].

Potential links between these data as descriptors of structure and drivers of Scottish farming are derived from the literature describing farming in Scotland [21,22], the UK [29], and from the conceptual model of land systems described by Aspinall and Staiano (2017). We focus on cereal area (km^2), numbers of sheep (millions), and prices of cereals (barley £/tonne). There are no single or sets of equations that relate these variables, since relationships between them would not only have to deal with their different metrics and scaling, but are also linked within a social and environmental system with limited potential for description with invariant or deterministic mathematical functions. Although marginal cost and other models for agriculture attempt to inform decisions e.g., [80,81], these are developed for specific times and circumstances, and there are no existing universal models describing the relationships between prices, crop areas, and livestock numbers that could be considered to apply over the timespan of our study. Even with this, however, some hypotheses about the general nature and direction of relationships can be formulated to describe the general process we examine, based on principles and market signals resulting from interactions of supply, demand, and price.

Traditionally, farming systems across the UK and its component countries have combined arable and pastoral activities within many of the different environments farmed [27]. Historically, grass was grown as part of crop and land use rotations in Scotland, helping rest and fertilize land prior to the next cycle of crop production [21]. Rotation grass grazing also serves as a low maintenance land use during periods when pressures reduce the capacity or potential for cereal farming [21,22]. Bowers and Cheshire compare sheep numbers with wheat price for England and Wales from 1893 to 1940, using five year means for the variables and a lag of three years for sheep numbers, revealing a negative linear relationship between wheat price and sheep numbers [29]. Our hypotheses about the cereal, grassland, and sheep farming system in Scotland is similar to Bowers and Cheshire's argument. An increase in price of a commodity produced by farming (barley price), itself linked to increased demand or insufficient supply (or both) for the commodity, might be expected to result in an increase in effort to produce that commodity (increased area for cereals). In turn, interdependence between parts of the system means that this increase in production will be associated with a reduction in land for other uses and a decline in livestock (sheep). Conversely, as prices or demand fall, and supply increases, the area of cereals will decrease, and other land uses (sheep farming) will increase. Both cereals and sheep are, therefore, expected to show a correlation with barley price, and cereal area with sheep numbers. Additionally, trends in barley price in Scotland are expected to be aligned with global trends in prices.

In land system terms, this set of relationships and feedbacks links economic/financial, arable (cereal) and pastoral land uses, and livestock (sheep) components of the land system. Fortunately, even in the absence of mathematical models for the relationships and feedbacks, it is possible to interpret the patterns in time-series data against the expectations of these general hypotheses about the directions of relationships and feedbacks. In addition to the hypothesized relationships between prices, cereal area, and sheep numbers, which reflect endogenous relationships within the farming system, there are a variety of exogenous variables that can also be described with time series data. For some, such as plant and animal diseases, war, and legislation and regulation the variable is binary. For others, such as weather variables, the data provide a continuous measure that can be interpreted using time series analysis to separate trend, cycles, and extremes (as statistical 'noise'). These exogenous events present shocks to the system that must be accommodated, depending on the specific nature of the event, over time scales from short- to longer terms.

5. Results

Figure 1a shows plots of the price of barley (£/t), number of sheep (millions), and area of cereals (km^2) for Scotland from 1867–2020. The plot is scaled with natural logarithms. The plots show that area of cereals and number of sheep, as funds, and barley price, as a driver, vary with superficially different patterns. As noted above, these data contain a record of changes at many temporal scales over the last century and a half.

Figure 1b shows some major legislative, trade, commodity price, weather, war, and disease events from 1867–2020. These provide not only boundary conditions for farming but also events that coincide with many of the more extreme values identified in the time series data. For example, government policy during and after the two world wars, and membership of the EEC/EU Common Agricultural Policy (CAP) provide system boundary conditions that shape agriculture in the short and medium terms. The 1947 Agriculture Act and associated support mechanisms (price guarantees, deficiency payments, marketing boards, investment in R&D etc.) set the context for farming from 1947 until the UK formerly joined the EEC CAP in 1973. Similarly, the influence of events such as wars, disease outbreaks, extreme weather, and financial crashes were short-term shocks that have clear signals in the short-term results ('noise') in the time series.

5.1. National and World Crop Prices

Long-, medium-, and short-term trends, cycles and patterns are identified in each of the cereal price datasets for Scotland, and in global prices data for cereals [37]. Prices for wheat, barley, and oats are shown in Figure 2a. Analysis of trend, cycles, and noise in data for prices of wheat, barley, and oats in Scotland shows that all have similar long-, medium-, and short-term trends. The long-term trend is exponential growth for each of the prices of wheat, barley, and oats (Figure 2a), the exponents in the equations being quite similar for the three (Table 1). Deviations from the long-term trend are modelled with a smoothing spline, revealing four main cycles over the 154-year period (Figure 2b). The residuals after removal of the long-term trend show high prices in 1918, 1942, the early 1980s, mid 1990s, 2007/8, 2012, and 2018, and low prices in 1972, 2005, 2009, and 2014–16 (Figure 2c). The concentration of high and low residuals in the post-1972 period is unsurprising given the increase in absolute values of prices due to inflation, but fluctuations are also latterly related to payments from CAP being in Euros, prices thus being subject to exchange rate variations in addition to inflation [82].

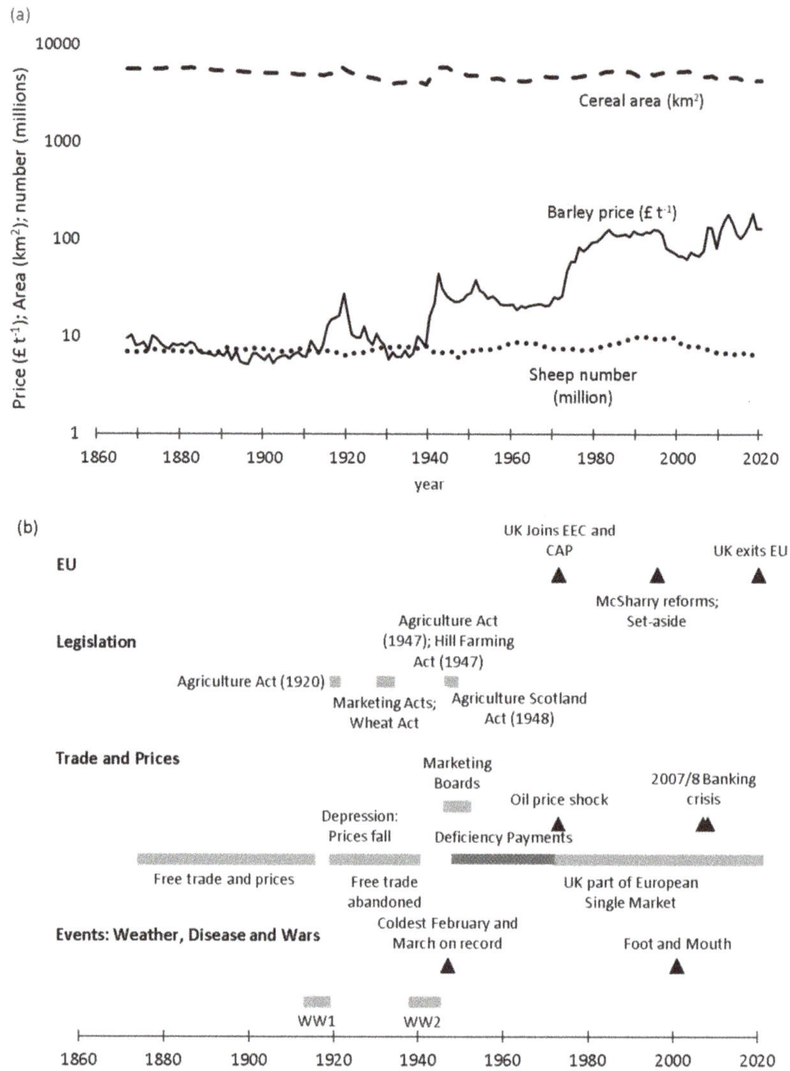

Figure 1. (**a**) Price of barley (£/t), number of sheep (millions), and area of cereals (km^2) for Scotland from 1867–2020. The plot is scaled with natural logarithms. (**b**) selected major legislative, trade, commodity price, weather, war, and disease events from 1867–2020.

Applying the Christiano-Fitzgerald filter to cereal prices in Scotland and world cereal prices from datasets used by Jacks [37,83] shows that the medium-term cycles revealed in cereal prices in Scotland are synchronized with cycles in prices for grain crops (wheat, barley, corn, and rye) for world data (Figure 2d).

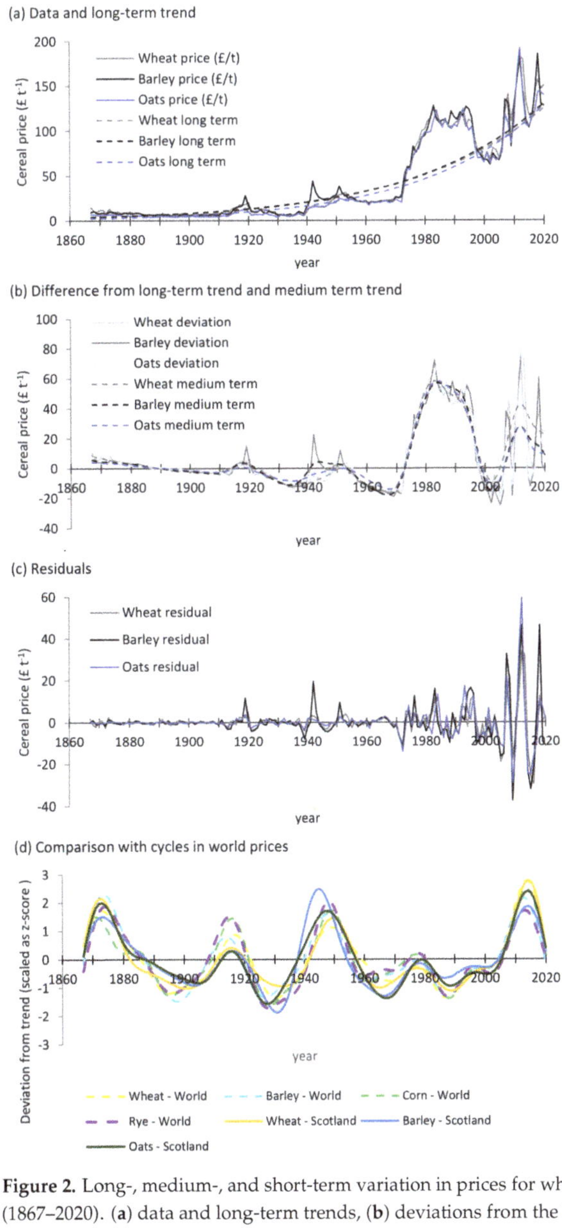

Figure 2. Long-, medium-, and short-term variation in prices for wheat, barley, and oats in Scotland (1867–2020). (**a**) data and long-term trends, (**b**) deviations from the long-term trend and smoothing spline model, (**c**) residuals after removal of the long- and medium-term trends. (**d**) comparison of cycles for prices of wheat, barley, and oats in Scotland with grain crops (wheat, barley, corn, and rye) in world data from Jacks [37,83]. Note: the prices shown in Figure 2a,b, and c are prices for the year. In Figure 2d the cycles comparing prices in Scotland with world prices are based on these prices adjusted for inflation and indexed to a specific year so that they match the indexing for prices for the world dataset used (see Jacks [83]). Price of barley in the year in question, not adjusted for inflation, is used in the analysis of price with other data describing the farming system, since this is the price data available for each individual year when decisions are being made within the farming system.

Table 1. Regression models for prices of cereals (1867–2020).

Cereal Price (£/Tonne)	Exponential Growth Equation	*t*-Test for Exponent	F-Test	Variance Explained (R2)
Wheat price	$y = a_w e^{1.0227x}$	t = 23.90 $p < 0.0001$	F = 571.1 df: 1 and 152 $p < 0.0001$	78.9%
Barley price	$y = a_b e^{1.0235x}$	t = 26.41 $p < 0.0001$	F = 687.2 df: 1 and 152 $p < 0.0001$	82.1%
Oats price	$y = a_o e^{1.0228x}$	t = 29.28 $p < 0.0001$	F = 687.2 df: 1 and 152 $p < 0.0001$	84.9%
$a_w = 3.91;\ a_b = 3.66;\ a_o = 2.51$ for x 1..154 (1867..2020)				

5.2. Lag Plots and Recurrence Plots

Lag plots and RPs for the three variables are shown in Figure 3. Lag plots of barley price (Figure 3a) show a decrease in price of more than 2 SDs in the year-on-year difference in 1920 and increases in 1919, 1940, 1942, 1973, and 2007. The 1920 decrease reflects adjustment after increased prices during the first world war [21]; the increases in 1940 and 1942 reflect UK government decisions to guarantee prices for farmers during the second world war. The 1973 increase is associated with both UK accession to CAP and the world oil price crisis [84], and the 2007 increase with the world food and financial crises [85]. The RP for barley price (Figure 3b) shows clear evidence of regime shifts, with areas of black along the diagonal of the plot and virtually no recurrence points outside those boxes. The regime shift in the 1970s is clear in the time series plot (Figure 1), but the RP shows there was a further shift starting in the 1950s during post war recovery and lasting into the 1970s. The white bars coinciding with the world wars indicate extreme variability in barley price; high variability since 1973 is also revealed in the absence of recurrence points.

The lag plot for area of cereals (Figure 3c) shows that 1919 was a decrease in area planted of greater than 2SDs of the long-term annual changes, while 1918, 1940, 1941, 1942, and 1993 were increases in area of more than 2SDs. The decline in 1919 represents a return to pre-war farming practices after the focus on cereals during the First War [19,21], offsetting the 1918 increase. The 1940–1942 increases represent the intense and sustained efforts to increase crop production during the Second World War [20]. The increase in area in 1993 coincided with the introduction of set-aside [86]. Set-aside was a policy to reduce the area of cereals, but payments under the scheme were based on the registered cultivated area. The RP for cereals shows long term cycles in the area, with variability during the wars. The RP also shows increased cycles in the period since 1973.

The lag plot for sheep (Figure 3e) shows that most year-on-year changes in the numbers of sheep are relatively small. Three years, 1941, 1947, and 2001, were decreases of greater than 2 SDs, while 1948 is an increase of more than 2 SDs from the previous year. Of these, 1941 represents government policy to reduce the sheep flock to allow crop production to be increased during the Second World War [21]; the number of sheep was reduced by over 1 million in 1941, and total sheep numbers were reduced by about 25% from pre-war levels during the years of the war [20]. February and March 1947 were extremely cold and snowy (see below), the timing additionally coinciding with lambing that led to high sheep mortality (almost 1 million sheep fewer in 1947 than in 1946). In 2001 an outbreak of foot-and-mouth led to 962,000 sheep being culled in Scotland to control the disease [87]. In 1948 over 700,000 sheep were added to the Scottish total through efforts to recover from the 1947 winter. The RP for sheep shows very clear evidence of cycles in the number of sheep, with regular pattern of recurrences spaces about 30 years apart, three cycles being evident since 1950. Numbers were more stable in the latter part of the 19th century.

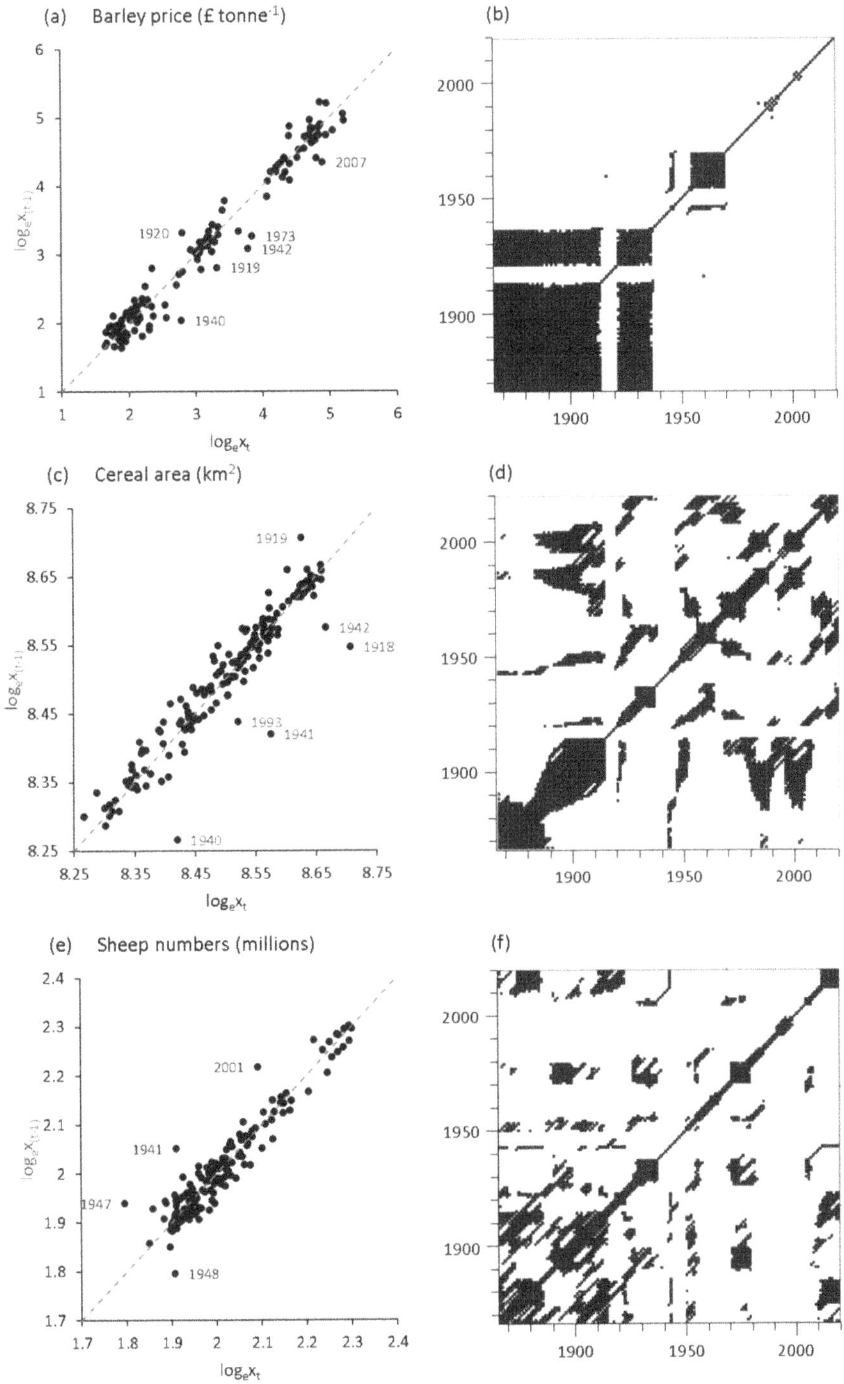

Figure 3. Lag plots and Recurrence Plots for barely price, cereal area, and sheep numbers in Scotland (1867–2020). (**a**) lag plot and (**b**) RP for barley price, (**c**) lag plot, and (**d**) RP for cereal area, and (**e**) lag plot, and (**f**) RP for sheep numbers.

5.3. Time Series Analysis and Recurrence Quantification Analysis

Figures 4–6 show the time series analysis results and RQA results for barley price, area of cereals, and number of sheep respectively.

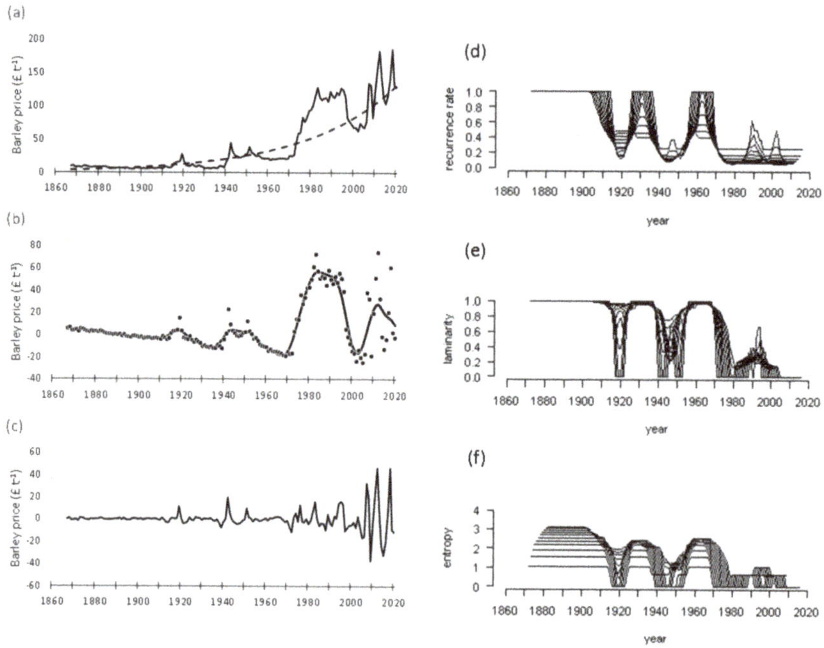

Figure 4. Time series analysis results and RQA results for barley price in Scotland (1867–2020): (**a**) data and long-term trend, (**b**) deviation from long-term trend and smoothing spline, (**c**) residuals from long- and medium-term trends, (**d**) recurrence rate, (**e**) laminarity, and (**f**) entropy from RQA using sliding windows of 10–30 years duration in 2-year increments.

The time series analysis for barley price has been described above but is shown in Figure 4a–c for comparison with the results of the RQA for barley price. In the RQA for barley price (Figure 4d–f) the recurrence rate (Figure 4d), laminarity (Figure 4e), and entropy (Figure 4f) all show that barley price was relatively stable until the first world war, from the mid-1920s and through the 1930s, and from the mid-1950s to about 1970. The low values of recurrence rate, laminarity, and entropy since 1973 reflect increasing variability and volatility in price.

The total area planted with cereals in Scotland has varied between 3900 km^2 and 6000 km^2 over the period from 1867 to 2020, with major changes in both the cereals planted and yields. The long-term trend is an annual decline in area planted of 0.14%, accumulating to a total of about 19% over the 154-year period (Figure 5a). The yearly difference between annual data and the long-term trend ranges from -1000 to $+1000$ km^2, and shows four cycles superimposed on the long-term trend, with greater areas planted in the 1870s and 1880s, during the two world wars, and again in the 1980s (Figure 5b). Negative deviations from the long-term trend are in the 1920s and 1930s, mid-1950s to mid-1960s, and mid-1990s and late 2000s. The residuals, after removing the long-term trend and medium-term cycles, are high in 1918 and 1942, and low in 1939, 1993 and 1994, and 2006 and 2007 (Figure 5c), similar to the results of the lag plot (Figure 3c). The recurrence rate (Figure 5d), laminarity (Figure 5e), and entropy (Figure 5f) show that cereal area changes gradually for most of the 154 years, although was more dynamic during the two wars, and also in the early-mid 1990s, coinciding with the onset of the policy and practice of set-aside.

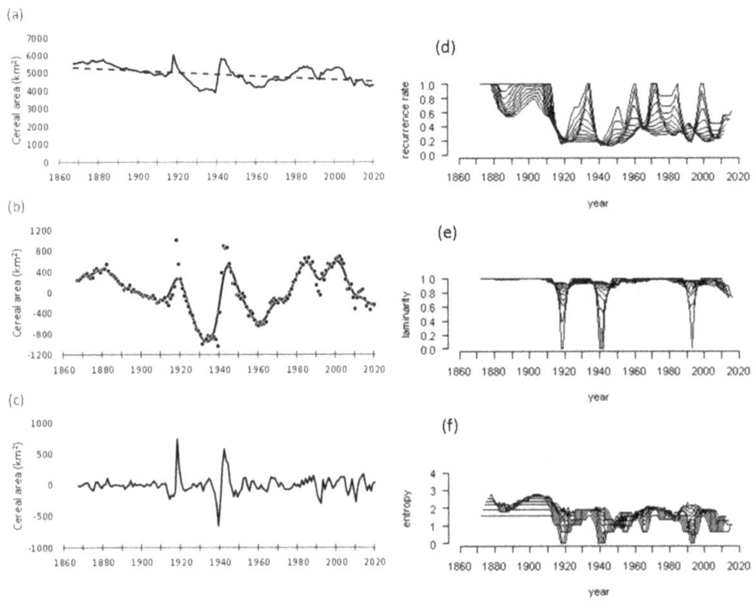

Figure 5. Time series analysis results and RQA results for cereal area in Scotland (1867–2020): (**a**) data and long-term trend, (**b**) deviation from long-term trend and smoothing spline, (**c**) residuals from long- and medium-term trends, (**d**) recurrence rate, (**e**) laminarity, and (**f**) entropy from RQA using sliding windows of 10–30 years duration in 2-year increments.

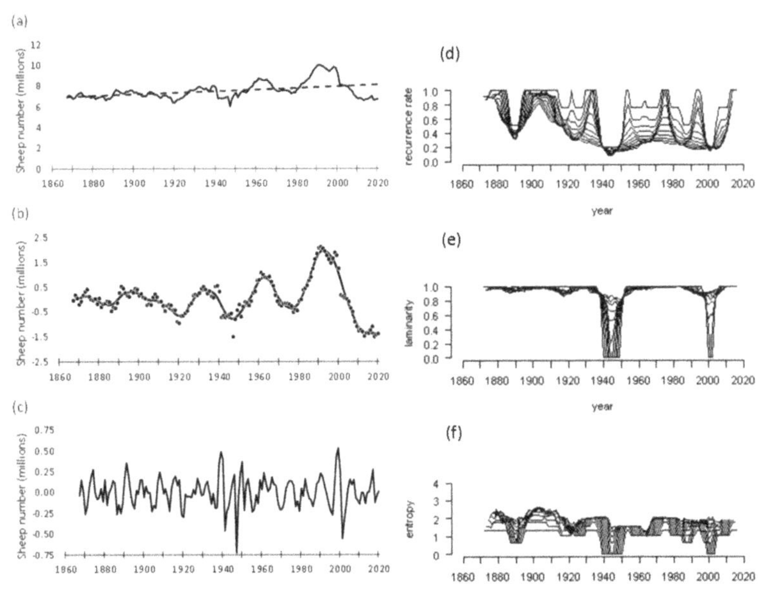

Figure 6. Time series analysis results and RQA results for sheep numbers in Scotland (1867–2020): (**a**) data and long-term trend, (**b**) deviation from long-term trend and smoothing spline, (**c**) residuals from long- and medium-term trends, (**d**) recurrence rate, (**e**) laminarity, and (**f**) entropy from RQA using sliding windows of 10–30 years duration in 2-year increments.

Sheep numbers in Scotland have varied between 6 and 10 million over the period from 1867 to 2020. The long-term trend is of increasing numbers at a rate of 0.1% per year, accumulating to a total of about 17% over the 154-year period (Figure 6a). The yearly difference between annual data and the long-term trend ranges from about −1.5 million to +1.5 million and shows five cycles (Figure 6b) superimposed on the long-term trend, with maxima in the late 1890s, 1930s, 1960s, and late 1980s and early 1990s, and minima in the 1880s, 1919/20, late 1940s, 1970s, and 2010s. The residuals, after removing the long-term trend and the medium-term cycles show peaks in 1937–1939, 1950, 1998, and 1999, and lows in 1947 and 2001 (Figure 6c). RQA results for sheep numbers show that although the recurrence rate is low for much of the 154 years (Figure 6d), the laminarity shows only two periods of extreme change (Figure 6f), these being during the 1940s, corresponding to the second world war and subsequent high mortality of sheep in the cold spring of 1947 [21], and in the early 2000s, coinciding with the outbreak of foot and mouth disease that reduced sheep flocks [87].

5.4. Dynamics from Interdependencies in the System

Interdependencies among the three data series are assessed from correspondence between the time series of medium-term changes with long-term trends and short-term noise removed. The medium-term patterns of variation for barley price, cereal area, and sheep numbers, expressed as time series and as x-y plots in Figures 7–9; all variables are normalized with their mean and standard deviation to account for differences in scaling between the variables. The correlation coefficients r and percent r2 for the pairs of variables for 1867–1947, 1947–1972, and 1973 to 2020 are shown in Table 2. All coefficients except two (marked by n.s. in the table) are significant at $p \leq 0.001$. The signs of the correlations correspond to expectations for associations between the variables.

The sequencing of cycles is of interest since this indicates their timings relative to one another and is indicative of the influences we posit in our general model (see above). Cycles for barley price and cereal area are synchronized and in phase until the late 1940s after which they become less synchronized (Figure 7a). This can also be seen in the x-y plot (Figure 7b), where data for 1867–1947 are tightly clustered along a line with positive slope, and data from 1960 onwards, and particularly from the early 1970s have a different trajectory in the x-y space. The change in trajectory in 1992, coinciding with the introduction of set-aside and lasting until 2012, is particularly evident.

The medium-term trends for cereal area and sheep number are also synchronized, but with a lag that places the peaks for sheep at the minima for cereal area, and vice versa for the period from 1867 to the early 1970s (Figure 8a). After the early 1970s this synchronization weakens (Figure 8a). The x-y plot of the medium-term trends (Figure 8b) shows the switch in emphasis to sheep from cereals during the 1920s and the agricultural depression of that period, the increase in area of cereals and decline in sheep numbers between the mid-1930s and particularly from 1939–1944. The general negative association between cereal area and sheep numbers from 1867–1972 contrasts with the positive association in the trajectory after 1973 when sheep and cereals became decoupled under CAP.

Medium-term trends and cycles in barley price and sheep numbers are also synchronized, although there is a lag between the two cycles (Figure 9a), as is expected from the associations already described (Figures 7 and 8). The x-y plot of these trends shows close association between 1867 and the 1950s, before the trajectory of the data changes to a peak for price and sheep numbers between 1985 and 1992 (Figure 9b). This reflects the decoupling of sheep numbers and cereal prices.

These associations are also apparent in the correlation coefficients (Table 2). Between 1867–1946 and 1947–1972, barley price has a positive correlation with cereal area and negative with sheep numbers (Table 2). After 1973, the correlation coefficient between cycles for cereal area and sheep number changes to +0.764, from negative correlations prior to 1973. The correlations between barley price and both cereal area and sheep numbers are not significant for the period 1973–2020. In summary, the different trajectories of each of the plots in Figures 7–9 for the 1973–2020 period compared with 1867–1946 and 1947–1972 are marked, the close associations between the variables in 1867–1946 and 1947–1972 being shown by a trace over time that clusters along a positive or negative line through the x-y space (Figures 7–9) and the 1973–2020 trace departing from these patterns.

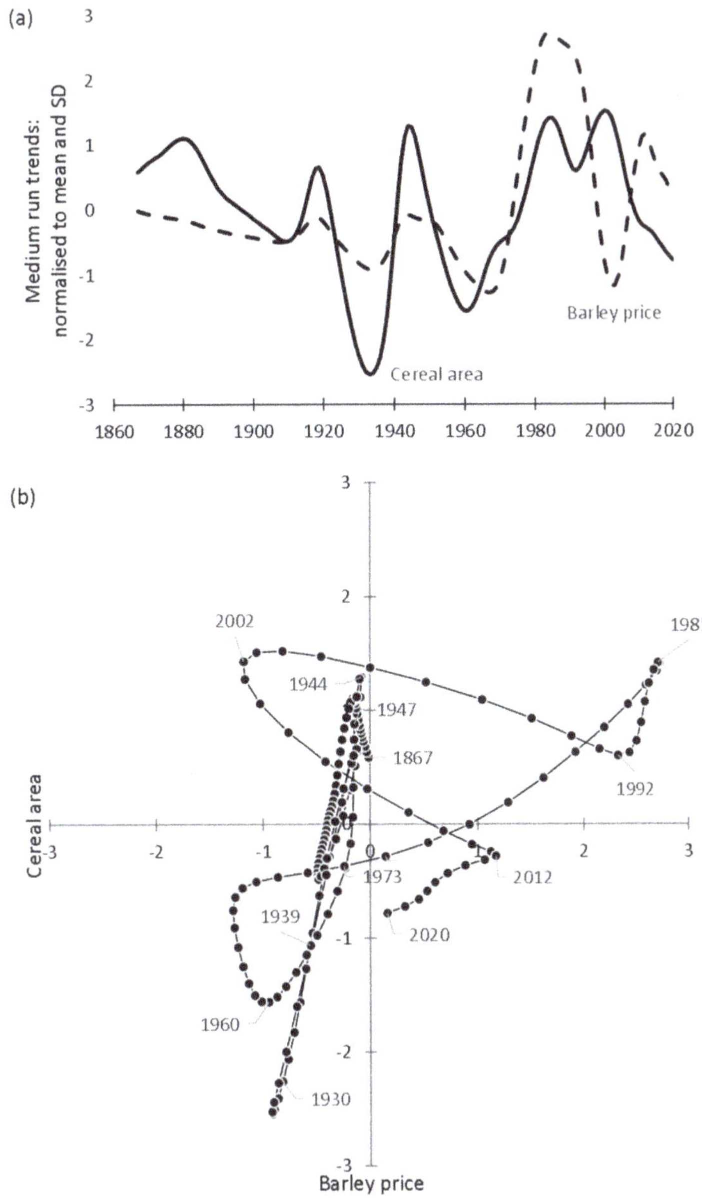

Figure 7. Medium-term patterns of variation for barley price and cereal area in Scotland (1867–2020) as (**a**) time series and (**b**) phase plots. Note: the variables are normalized to account for differences in magnitude.

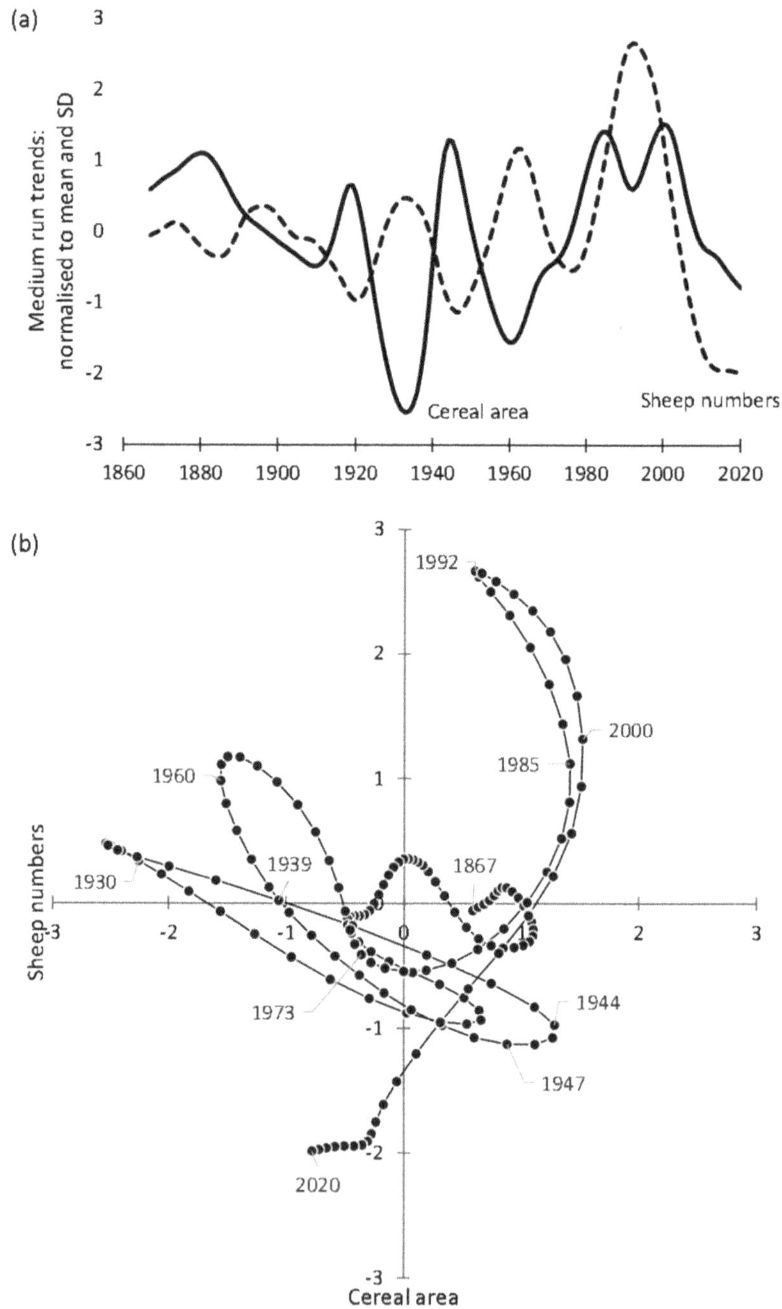

Figure 8. Medium-term patterns of variation for cereal area and sheep numbers in Scotland (1867–2020) as (**a**) time series and (**b**) phase plots. Note: the variables are normalized to account for differences in magnitude.

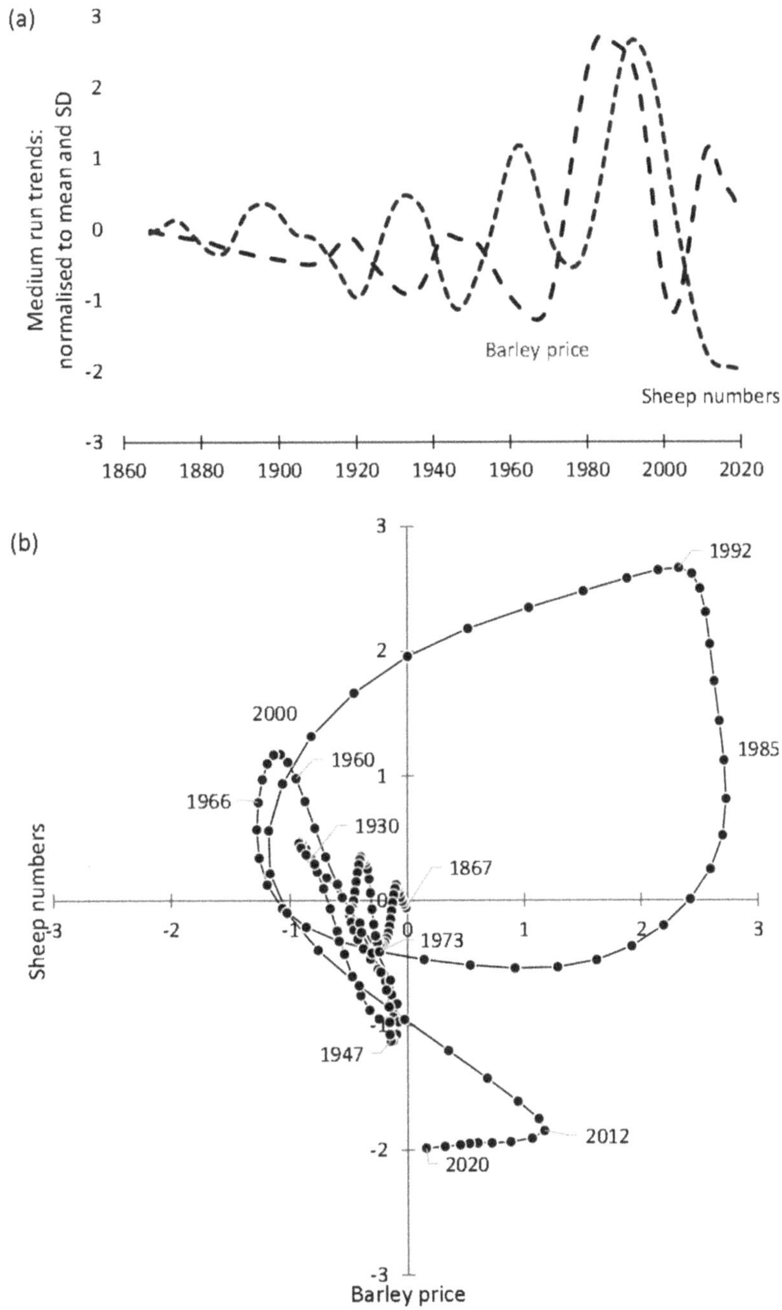

Figure 9. Medium-term patterns of variation for barley price and sheep numbers in Scotland (1867–2020) as (**a**) time series and (**b**) phase plots. Note: the variables are normalized to account for differences in magnitude.

Table 2. Correlation coefficients (upper right quadrant) and r^2 (lower left quadrant) between medium-term cycles for 1867–1946, 1947–1972, and 1973–2020.

	Barley Price	Cereal Area	Sheep Numbers
	1867–1946		
Barley price		0.967	−0.528
Cereal area	93.6%		−0.503
Sheep numbers	27.9%	25.3%	
	1947–1972		
Barley price		0.653	−0.857
Cereal area	42.6%		−0.905
Sheep numbers	73.4%	82.0%	
	1973–2020		
Barley price		0.157 n.s.	0.358 n.s.
Cereal area	2.5%		0.764
Sheep numbers	12.8%	58.4%	

6. Discussion

6.1. Dynamics of Scottish Farming Systems

Trends and cycles over different timespans and timescales identified within the data using time series analysis, as well as RP and RQA, characterize long-, medium-, and short-term dynamics of cereal and sheep farming and cereal prices. Irregular cycles are evident in each of barley price, cereal area, and sheep numbers, the cycles being synchronized with each other but with phase shifts. The period of these cycles is between 15 and 40 years. Cycles for cereal area and barley price are synchronized and in phase up to the early 1970s (Table 2); both barley price and cereal area are negatively correlated with sheep numbers until 1972 (Table 2), particularly under the policies that operated from 1947–1972. The changes in synchronization and correlations following 1973 reflect decoupling of arable and sheep farming sectors under the provisions of the EU CAP. The long-term trends and patterns of cycles, as well as the year-to-year variability superimposed on the long- and medium-term trends, for farming, reveal the multiscale nature of temporal variation in changes to farming systems. The RP and RQA also help to identify regime shifts. The RP (Figure 3b) and RQA (Figure 4) for Barley price shows clear evidence of regime shifts, with one regime over the period from 1867 to the late-1930s (interrupted by World War One), and two further shifts in about 1950 and 1970; since 1970 the price has been highly volatile. Regime shifts are not apparent for cereal area and sheep numbers (Figures 3, 5 and 6).

Results of analysis of changes in Scottish farming over a century and a half show the signal of endogenous system dynamics. Domestic cereal prices are linked to changes in world prices (Figure 2d), and to national and international policies and events (Figure 1b), but behind the influences of these exogenous factors, there is evidence from the period from 1867 to 1972 for the dampening influence of endogenous dynamics associated with the (loose) coupling of components of Scottish farming systems. From 1973–2020 system feedback and interaction at a national aggregate scale has been weakened as sheep and cereal farming have been decoupled. The dampening feedback provided resilience to Scottish farming as long-term trends and medium-term changes in the world and domestic economies, and short-term events influenced farming. The importance of system dynamics for description and explanation of changes in system funds, and the presence of long-term trends and medium-term cycles also challenges analysis of changes based on data that cover only a restricted timespan. The results show higher level interdependencies between arable and pastoral sectors, dependencies that have themselves changed during the course of the twentieth century as boundary conditions are changed by events, policies (e.g., the Agriculture Act of 1947, the UK's accession to the EEC/EU CAP in 1973), that are important for understanding both arable and pastoral farming, development of policy, and land management. The lessons from the period studied, despite much of it being historic remain important for strategic decisions about policy regarding farming, land management, and farming livelihoods in Scotland. The results show ways in which dynamic behaviours of farming systems have evolved as policy context has changed. The extent to which the dynamics have become decoupled and less resilient with modernization of farming raises concerns for land use in future especially as new policy is developed following the UK's departure from the EU and CAP in 2020.

6.2. Land System Dynamics and Time Series Analyses

The results and analyses also highlight the variety of ways in which exogenous drivers and endogenous interactions of state variables within the system can influence land system change and dynamics across these time scales. Medium-term trends, revealed as cycles in the data here, are particularly important in this analysis. It is important to note that cycles (in both drivers and funds of the system) are not cycles in a strict mathematical sense, and they are not required to be regular, to have fixed periods or magnitude of oscillation, or to be predictable [83]. Rather, they are medium-term patterns of deviations from long-term trends, with short-term noise filtered out. The focus on cycles is based on the expectation that they reflect behaviours that result from interaction of system factors over the medium-term, and, as such, cycles are of particular interest in characterizing and understanding system dynamics. The interactions and feedbacks of system components over time result in a statistical tendency for cycles, as (irregular) waves, to be found in the data, with values increasing and decreasing as feedbacks propagate through the system (with characteristic, but variable, time scales). Regime shifts are apparent in the RP for barley price, with cycles in the RPs for cereal area and sheep number (Figure 3d,f) and by recurrence rate, laminarity and entropy in the RQA (Figures 4d–f, 5d–f and 6d–f), as well as in the medium-term patterns of the time series (Figures 4b, 5b and 6b). Interpreting these cycles provides insights into system changes (Figures 7–9). Long-term trends represent slow dynamics and secular changes. Short-term changes represent impacts of events, and year-to-year stochastic variability, as well as a range of uncertainties, including inherent uncertainty of environmental and social systems, measurement errors (statistical uncertainty), short-term decision-making (partial controllability of complex systems), and structural uncertainty (the inability to describe the system fully) [88].

The use of time series and nonlinear dynamical systems methods is guided by hypotheses about the nature of farming as a coupled land system integrating human- and environment- drivers through farmer choices and decisions, manifest at the aggregate national and regional scales. Results are informative on the nature of dynamics of farming systems, the relation between dynamics and both endogenous feedbacks and exogenous noise, the influence of different timescales in establishing explanations based on potential processes and drivers, and on the impacts of drivers at multiple scales from farm to international trade, finance, and legislation. Together, the methods reveal aspects of the dynamic nature of drivers that underpin land system change, evolution, and dynamics, as well as the specific nature of dynamics in land systems themselves. The examples also elucidate some fundamental principles and mechanisms for studying land systems as complex coupled human-environment systems; the approaches have application to study and explanation of both dynamics and change.

The short-, medium-, and long-term trends and process relationships embedded within time series' data offer potential for study of not only change in land systems, but also temporal and cross-scale dynamics in system function, leading to improved understanding of coupling between human and environment systems, evolution in land systems over time, and influence and response to changes in land system drivers. The analysis uses a long data series, necessary for identifying long- and medium-term patterns. A snapshot in time cannot reveal these dynamics, and consideration of too short a time span can lead to misinterpretation of change and dynamics, for example by focusing only on increase or decrease [25].

If a central tenet for study of land systems, as exemplars of coupled systems, is that they are dynamic systems because of the functional interactions between the human and environment subsystems, then the dynamics of both system drivers and dynamical behaviours of land systems themselves (based on the interactions and coupling of human and environment) is as much a part of land dynamics as changes in the structures of land systems. Dynamics of both land system structures and drivers are also necessarily embedded in the pervasive impacts of spatial, temporal, and organizational scales, and in both the hierarchical complexity and contingent history of land systems within societal and environmental change more generally. Even in the absence of major categorical conversion in type or intensity of change, land systems operate with complex dynamics, and they require to be understood as dynamical complex systems.

The variety of dynamics represented by the patterns in the data used in this case study emphasizes the need for explicit pre-analytical hypotheses to be constructed about relationships between land system dynamics and changes with potential trends and changes in system drivers. The results also emphasize why hypotheses need to be explicit about the time scale, or scales, of interest, since long-, medium-, and short- time scale patterns are contained within the observed data, and all may be relevant to understanding the variety of land system dynamics. If we accept that time series data represent and reflect all the processes from all scales involved in their formation, then these data potentially provide a source of insight into multi-scale consequences of the actions of drivers. In this context, time series analysis provides a set of mechanisms for distinguishing these temporal patterns at various time scales.

In summary, in systems terms the analysis of the historical record of changes in cereal area and price, and sheep numbers in Scotland reveal a complex pattern of interdependencies and coupling over time and at different scales, combining endogenous system dynamics with short-term variability associated with stochastic events, within a broader set of higher-level interdependencies and boundary conditions for the system. The long time-period of the study also shows that the embedded system dynamics can make farming relatively resilient to changes in policy, exogenous shocks (such as weather events or disease outbreaks), or regime changes and thresholds (as seen here in prices). The whole systems perspective is one that is seldom considered by short-term or sectoral approaches to farming. Although many of the results are not new, the long-term, whole systems perspective shows the evolution of land use in Scottish farming as a dynamic and dynamical system, hence demonstrating that this kind of approach is suitable for study and interpretation using a single analysis. The contribution of time series analysis and the tools of NLDS (RP, RQA) in land systems science is also evident. The long time series of data, and the impacts of historical contingency over the timespan of the study, combined with the coupling and complexity of system-level relationships, weakens the chances that steady-state latent structures would emerge by means of classical modelling. Instead, analysis with time series analysis and methods from NLDS allows exploration of discontinuities (if any) in the system dynamics, allowing abrupt changes and extreme values to be identified, that would be difficult or impossible to capture in steady-state global models. Time series analysis and NLDS also enable exploration of system dynamics at hierarchically nested time scales, moving beyond use of classical models in describing phenomena over short periods that are of little relevance over the longer duration of land use history, as captured in the data used in the case study. As such, the results offer a challenge to the land systems community to address timescales and dynamics explicitly, while demonstrating some approaches and methodologies for achieving this. Authors should discuss the results and how they can be interpreted in perspective of previous studies and of the working hypotheses. The findings and their implications should be discussed in the broadest context possible.

6.3. Directions for Future Research

Further research is needed into dynamics of land systems based on system interdependencies, interactions, and feedbacks. As noted in the Introduction, studies of system dynamics based in system structures and coupling are rare within land system science.

Methods such as time series analysis, for analysis of non-linear dynamic systems, and models for exploring input-output within non-linear dynamic systems such as NARMAX [89], offer tools with potential for use by the land systems community. This study shows the importance of long time-series of data for capturing dynamics, the dynamical behaviours found for Scotland being evident over medium-term cycles of about 30 years duration. The consequences of policy changes for dynamic behaviours are also apparent from this case study, producing dynamical shifts in system coupling, and showing the importance of comparative studies across different socio-political, economic, and other contexts that provide the boundary conditions for land use decisions. Finally, this case study uses aggregate national data. The extent to which this is representative of the behaviours and experience of individual farm or other land use units requires further research.

7. Conclusions

Time series analysis, including methods from analysis of non-linear dynamical systems, are used to separate long-, medium- and short-term dynamics encapsulated within a historical record of land system states for farming in Scotland over the period from 1867 to 2020. The results show that cereal prices in Scotland follow similar trends and cycles to those shown in global prices, and that the dynamics of both the area under cereal cultivation and sheep numbers are linked to the dynamics of barley prices, as well as to each other, particularly for the period from 1867–1972. The relationships revealed in the medium-term trends are weaker since 1973 as prices, and cereal and sheep farming have become decoupled under modernization associated with policies in the EEC/EU CAP and as prices have become more volatile. These medium-term cycles in the data represent the endogenous dynamics of the farming system itself, operating within boundary conditions set by the policy environment. Short-term variability in the data reflect year-to-year variability associated with weather, disease, and other events.

Our results characterize dynamics from internal feedbacks and coupling of farming as a system at the national scale, reveal some system characteristics and behaviours associated with the dynamical evolution of farming as a system, and identify some regime shifts over the full 154-year timespan of the census. Specifically, the results reveal (i) consequences of several exogenous factors as events that had an impact on system states, (ii) show that arable and pastoral farming, at a national scale, are dynamically related over a range of timescales and coupled to global trends, and (iii) that throughout much of the timespan of the study the system has maintained a pattern of changes consistent with endogenous systems-level feedbacks between sectors that act to dampen the impacts of exogenous factors. Changes in system dynamics over the timespan are also associated with policy changes that altered the interaction of arable and pastoral farming.

The analysis is based on the contention that the time series of system states recording the history of land use contain an embedded record of the impacts of long-, medium-, and short-term dynamics associated with both endogenous system forces and exogenous factors that have influenced the land system. Because of this, both the underlying systems framework structuring the land system and the temporal scales at which a land system is studied should be made explicit, as the information needed for explanation of changes and dynamics will vary with the system structure and the time scales of interest. The use of time series analysis and methods from non-linear dynamics forces explicit attention to system structure, time scales, and the multi-scale behaviours of land systems. This demonstration of interdependencies between the prices, and arable and pastoral systems in Scotland shows that farming land use in Scotland has functioned as a complex system and was particularly resilient as a coupled arable-pastoral system prior to 1973, displaying characteristic behaviours of endogenous variables within a nonlinear dynamical system with noise-dampening feedbacks. The cases study illustrates a more general problem. Because of the dominance of studies of land conversion and modification, the prevalence of studies of short timespan [90], and the requirement for long time series of data to support time series and NLDS analyses, there are, correspondingly, still few exemplars or results of analytical approaches applied to land system dynamics found in the land systems literature, beyond those based on change detection. More are needed. Further studies of land systems could usefully attempt to identify emergent properties and behaviours of land systems, developing analyses focusing on dynamics in long-term time-series data, complementing analyses based on spatial snapshots over short time spans.

Author Contributions: Conceptualization, R.A., M.S. and D.P.; Data curation, R.A.; Formal analysis, R.A. and M.S.; Investigation, R.A., M.S. and D.P.; Methodology, R.A., M.S. and D.P.; Software, R.A. and M.S.; Validation, R.A., M.S. and D.P.; Writing—original draft, R.A., M.S. and D.P.; Writing—review & editing, R.A., M.S. and D.P. All authors have read and agreed to the published version of the manuscript.

Funding: R.A. acknowledges financial support from University of Naples for collaboration with the STAD group based at University of Naples Federico II. M.S. acknowledges financial support from the European Union's Horizon 2020 research and innovation programme under grant agreement No 689669. D.P. and R.A. acknowledge financial support from Massey University International Visitor Research Fund. This research received no specific grant from any funding agency in the public, commercial, or not-for-profit sectors.

Acknowledgments: This work reflects the authors' views only.

Conflicts of Interest: The authors declare no conflict of interest.

References

1. Turner, B.; Meyfroidt, P.; Kuemmerle, T.; Müller, D.; Chowdhury, R.R. Framing the search for a theory of land use. *J. Land Use Sci.* **2020**, *15*, 489–508. [CrossRef]
2. Vadjunec, J.M.; Frazier, A.E.; Kedron, P.; Fagin, T.; Zhao, Y. A Land Systems Science Framework for Bridging Land System Architecture and Landscape Ecology: A Case Study from the Southern High Plains. *Land* **2018**, *7*, 27. [CrossRef]
3. Verburg, P.H.; Alexander, P.; Evans, T.; Magliocca, N.R.; Malek, Z.; Rounsevell, M.; van Vliet, J. Beyond land cover change: Towards a new generation of land use models. *Curr. Opin. Environ. Sustain.* **2019**, *38*, 77–85. [CrossRef]
4. Güneralp, B.; Seto, K.C.; Ramachandran, M. Evidence of urban land teleconnections and impacts on hinterlands. *Curr. Opin. Environ. Sustain.* **2013**, *5*, 445–451. [CrossRef]
5. Aspinall, R.; Staiano, M. A Conceptual Model for Land System Dynamics as a Coupled Human–Environment System. *Land* **2017**, *6*, 81. [CrossRef]
6. Lambin, E.F.; Meyfroidt, P. Land use transitions: Socio-ecological feedback versus socio-economic change. *Land Use Policy* **2010**, *27*, 108–118. [CrossRef]
7. Müller, D.; Sun, Z.; Vongvisouk, T.; Pflugmacher, D.; Xu, J.; Mertz, O. Regime shifts limit the predictability of land-system change. *Glob. Environ. Chang.* **2014**, *28*, 75–83. [CrossRef]
8. Ramankutty, N.; Coomes, O. Land-use regime shifts: An analytical framework and agenda for future land-use research. *Ecol. Soc.* **2016**, *21*, 1. [CrossRef]
9. Williams, T.; Guikema, S.; Brown, D.; Agrawal, A. Assessing model equifinality for robust policy analysis in complex socio-environmental systems. *Environ. Model. Softw.* **2020**, *134*, 104831. [CrossRef]
10. Turner, B.; Geoghegan, J.; Lawrence, D.; Radel, C.; Schmook, B.; Vance, C.; Manson, S.; Keys, E.; Foster, D.; Klepeis, P.; et al. Land system science and the social-environmental system: The case of Southern Yucatán Peninsular Region (SYPR) project. *Curr. Opin. Environ. Sustain.* **2016**, *19*, 18–29. [CrossRef]
11. Turchin, P. Dynamical Feedbacks between Population Growth and Sociopolitical Instability in Agrarian States. *Struct. Dyn.* **2005**, *1*, 1. [CrossRef]
12. Crossman, N.D.; Bryan, B.A.; de Groot, R.S.; Lin, Y.-P.; Minang, P.A. Land Science Contributions to Ecosystem Services. *Curr. Opin. Environ. Sustain.* **2013**, *5*, 509–514. [CrossRef]
13. Nielsen, J.Ø.; de Bremond, A.; Chowdhury, R.R.; Friis, C.; Metternicht, G.; Meyfroidt, P.; Munroe, D.; Pascual, U.; Thomson, A. Toward a Normative Land Systems Science. *Curr. Opin. Environ. Sustain.* **2019**, *38*, 1–6. [CrossRef]
14. Schößer, B.; Helming, K.; Wiggering, H.; Schosser, B. Assessing land use change impacts—A comparison of the SENSOR land use function approach with other frameworks. *J. Land Use Sci.* **2010**, *5*, 159–178. [CrossRef]
15. Schumm, S.A. *To Interpret the Earth: 10 Ways to Be Wrong*; Cambridge University Press: Cambridge, UK, 1998.
16. Erb, K.-H.; Haberl, H.; Jepsen, M.R.; Kuemmerle, T.; Lindner, M.; Müller, D.; Verburg, P.H.; Reenberg, A. A conceptual framework for analysing and measuring land-use intensity. *Curr. Opin. Environ. Sustain.* **2013**, *5*, 464–470. [CrossRef]
17. Kuemmerle, T.; Erb, K.; Meyfroidt, P.; Müller, D.; Verburg, P.H.; Estel, S.; Haberl, H.; Hostert, P.; Jepsen, M.R.; Kastner, T.; et al. Challenges and opportunities in mapping land use intensity globally. *Curr. Opin. Environ. Sustain.* **2013**, *5*, 484–493. [CrossRef]
18. Rahim, F.H.A.; Hawari, N.N.; Abidin, N.Z. Supply & Demand of Rice in Malaysia: A System Dynamics Approach. *Int. J. Supply Chain Manag.* **2017**, *6*, 234–240.

19. Jones, D.T.; Duncan, J.F.; Conacher, H.M.; Scott, W.R. *Rural Scotland during the War*; Oxford University Press: London, UK, 1926.
20. Marshall, D. Scottish Agriculture During the War. In *Transactions of the Highland and Agricultural Society of Scotland*; 5th Series; RHASS: Glasgow, UK, 1946; Volume LVIII, pp. 1–77.
21. Symon, J.A. *Scottish Farming Past and Present*; Oliver & Boyd Ltd.: Edinburgh, UK, 1959.
22. Coppock, J.T. *An Agricultural Atlas of Scotland*; John Donald Publishers Ltd.: Edinburgh, UK, 1976.
23. Wood, H.J. *An Agricultural Atlas of Scotland*; George Gill and Sons Ltd: London, UK, 1931.
24. Highland and Agricultural Society. *Report on the Present State of the Agriculture of Scotland 1878. Presented at the International Agricultural Congress at Paris in June 1878*; Highland and Agricultural Society: Edinburgh, UK, 1878.
25. SAC Rural Policy Centre. *Farming's Retreat from the Hills*; SAC: Edinburgh, UK, 2008.
26. Aspinall, R.; Staiano, M. Ecosystem services as the products of land system dynamics: Lessons from a longitudinal study of coupled human-environment systems. *Landsc. Ecol.* **2018**, *34*, 1503–1524. [CrossRef]
27. Coppock, J.T. An Agricultural Geography of Great Britain. *Geogr. J.* **1971**, *137*, 400. [CrossRef]
28. Garrett, R.; Ryschawy, J.; Bell, L.W.; Cortner, O.; Ferreira, J.; Garik, A.V.N.; Gil, J.D.B.; Klerkx, L.; Moraine, M.; Peterson, C.A.; et al. Drivers of decoupling and recoupling of crop and livestock systems at farm and territorial scales. *Ecol. Soc.* **2020**, *25*, 24. [CrossRef]
29. Bowers, J.K.; Cheshire, P. *Agriculture, the Countryside and Land Use: An Economic Critique*; Methuen: London, UK, 1983.
30. Boffetta, G.; Crisanti, A.; Paparella, F.; Provenzale, A.; Vulpiani, A. Slow and fast dynamics in coupled systems: A time series analysis view. *Phys. D Nonlinear Phenom.* **1998**, *116*, 301–312. [CrossRef]
31. Ford, A. Simulating systems with fast and slow dynamics: Lessons from the electric power industry. *Syst. Dyn. Rev.* **2018**, *34*, 222–254. [CrossRef]
32. Briske, D.D.; Illius, A.W.; Anderies, J.M. *Nonequilibrium Ecology and Resilience Theory*; Springer: Berlin/Heidelberg, Germany, 2017; pp. 197–227. [CrossRef]
33. Chatfield, C.; Xing, H. *The Analysis of Time Series: An Introduction with R*, 7th ed.; Chapman and Hall/Crc Texts in Statistical Science; CRC Press: Boca Raton, FL, USA, 2019.
34. Hamilton, J.D. *Time Series Analysis*; Princeton University Press: Princeton, NJ, USA, 1994.
35. Christiano, L.J.; Fitzgerald, T.J. *The Band Pass Filter*; National Bureau of Economic Research: Cambridge, MA, USA, 1999; Volume 73.
36. The Band Pass Filter. *International Economic Review 44*; No. 2; John Wiley & Sons: Hoboken, NJ, USA, 2003; pp. 435–465.
37. Jacks, D.S. From boom to bust: A typology of real commodity prices in the long run. *Cliometrica* **2019**, *13*, 201–220. [CrossRef]
38. Eckmann, J.P.; Oliffson Kamphorst, S.; Ruelle, D. Recurrence Plots of Dynamical Systems. *Europhys. Lett.* **1987**, *4*, 973–977. [CrossRef]
39. Webber, C.L.; Marwan, N. (Eds.) *Recurrence Quantification Analysis: Theory and Best Practices, Understanding Complex Systems*; Springer: London, UK, 2015.
40. Packard, N.H.; Crutchfield, J.P.; Farmer, J.D.; Shaw, R.S. Geometry from a Time Series. *Phys. Rev. Lett.* **1980**, *45*, 712–716. [CrossRef]
41. Takens, F. *Detecting Strange Attractors in Turbulence*; Paper Presented at the Dynamical Systems and Turbulence, Warwick 1980; Springer: Berlin/Heidelberg, Germany, 1981.
42. Broomhead, D.; King, G. Extracting qualitative dynamics from experimental data. *Phys. D Nonlinear Phenom.* **1986**, *20*, 217–236. [CrossRef]
43. Williams, G. *Chaos Theory Tamed*, 1st ed.; CRC Press: London, UK, 1997.
44. King, G.; Stewart, I. Phase space reconstruction for symmetric dynamical systems. *Phys. D Nonlinear Phenom.* **1992**, *58*, 216–228. [CrossRef]
45. Marwan, N.M.; Romano, C.; Thiel, M.; Kurths, J. Recurrence Plots for the Analysis of Complex Systems. *Phys. Rep.* **2007**, *438*, 237–329. [CrossRef]
46. Zhou, W.-X.; Sornette, D. Lead-lag cross-sectional structure and detection of correlated–anticorrelated regime shifts: Application to the volatilities of inflation and economic growth rates. *Phys. A Stat. Mech. Appl.* **2007**, *380*, 287–296. [CrossRef]
47. Zbilut, J.P.; Webber, C.L. Embeddings and delays as derived from quantification of recurrence plots. *Phys. Lett. A* **1992**, *171*, 199–203. [CrossRef]
48. Yao, C.-Z.; Lin, Q.-W. Recurrence plots analysis of the CNY exchange markets based on phase space reconstruction. *N. Am. J. Econ. Financ.* **2017**, *42*, 584–596. [CrossRef]
49. Huffaker, R.G. Phase Space Reconstruction from Economic Time Series Data: Improving Models of Complex Real-World Dynamic Systems. *Int. J. Food Syst. Dyn.* **2010**, *3*, 184–193.
50. Zaldívar, J.-M.; Strozzi, F.; Dueri, S.; Marinov, D.; Zbilut, J.P. Characterization of regime shifts in environmental time series with recurrence quantification analysis. *Ecol. Model.* **2008**, *210*, 58–70. [CrossRef]
51. Ellner, S.; Turchin, P. Chaos in a Noisy World: New Methods and Evidence from Time-Series Analysis. *Am. Nat.* **1995**, *145*, 343–375. [CrossRef]
52. Cooke, B.J.; Roland, J. Early 20th Century Climate-Driven Shift in the Dynamics of Forest Tent Caterpillar Outbreaks. *Am. J. Clim. Chang.* **2018**, *7*, 253–270. [CrossRef]
53. Coco, M.I.; Dale, R. Cross-recurrence quantification analysis of categorical and continuous time series: An R package. *Front. Psychol.* **2014**, *5*, 510. [CrossRef]

54. Wallot, S.; Leonardi, G. Analyzing Multivariate Dynamics Using Cross-Recurrence Quantification Analysis (Crqa), Diagonal-Cross-Recurrence Profiles (Dcrp), and Multidimensional Recurrence Quantification Analysis (Mdrqa)—A Tutorial in R. *Front. Psychol.* **2018**, *9*, 2232. [CrossRef]
55. Kot, M.; Schaffer, W.; Truty, G.; Graser, D.; Olsen, L. Changing criteria for imposing order. *Ecol. Model.* **1988**, *43*, 75–110. [CrossRef]
56. Miranda, F.; Ramos, F.; Von Randow, C.; Dias-Júnior, C.; Chamecki, M.; Fuentes, J.; Manzi, A.; De Oliveira, M.; De Souza, C. Detection of Extreme Phenomena in the Stable Boundary Layer over the Amazonian Forest. *Atmosphere* **2020**, *11*, 952. [CrossRef]
57. Mukherjee, S.; Zawar-Reza, P.; Sturman, A.; Mittal, A.K. Characterizing atmospheric surface layer turbulence using chaotic return map analysis. *Theor. Appl. Clim.* **2013**, *122*, 185–197. [CrossRef]
58. Trauth, M.H.; Asrat, A.; Duesing, W.; Foerster, V.; Kraemer, K.H.; Marwan, N.; Maslin, M.A.; Schaebitz, F. Classifying Past Climate Change in the Chew Bahir Basin, Southern Ethiopia, Using Recurrence Quantification Analysis. *Clim. Dyn.* **2019**, *53*, 2557–2572. [CrossRef]
59. Medvinsky, A.B.; Adamovich, B.V.; Chakraborty, A.; Lukyanova, E.V.; Mikheyeva, T.M.; Nurieva, N.I.; Radchikova, N.P.; Rusakov, A.V.; Zhukova, T.V. Chaos far away from the edge of chaos: A recurrence quantification analysis of plankton time series. *Ecol. Complex.* **2015**, *23*, 61–67. [CrossRef]
60. Spiridonov, A. Recurrence and Cross Recurrence Plots Reveal the Onset of the Mulde Event (Silurian) in the Abundance Data for Baltic Conodonts. *J. Geol.* **2017**, *125*, 381–398. [CrossRef]
61. Spiridonov, A.; Balakauskas, L.; Stankevič, R.; Kluczynska, G.; Gedminienė, L.; Stančikaitė, M. Holocene vegetation patterns in southern Lithuania indicate astronomical forcing on the millennial and centennial time scales. *Sci. Rep.* **2019**, *9*, 14711. [CrossRef]
62. Spiridonov, A.; Stankevič, R.; Gečas, T.; Šilinskas, T.; Brazauskas, A.; Meidla, T.; Ainsaar, L.; Musteikis, P.; Radzevičius, S. Integrated Record of Ludlow (Upper Silurian) Oceanic Geobioevents—Coordination of Changes in Conodont, and Brachiopod Faunas, and Stable Isotopes. *Gondwana Res.* **2017**, *51*, 272–288. [CrossRef]
63. Ahlbrandt, C.D.; Morian, C. Partial Differential Equations on Time Scales. *J. Comput. Appl. Math.* **2002**, *141*, 35–55. [CrossRef]
64. Wolf, A.; Swift, J.B.; Swinney, H.L.; Vastano, J.A. Determining Lyapunov exponents from a time series. *Phys. D Nonlinear Phenom.* **1985**, *16*, 285–317. [CrossRef]
65. Phillips, J.D. *Earth Surface Systems: Complexity, Order and Scale*; Blackwell Publishers: Oxford, UK, 1999.
66. Cao, L. Practical method for determining the minimum embedding dimension of a scalar time series. *Phys. D Nonlinear Phenom.* **1997**, *110*, 43–50. [CrossRef]
67. Mindlin, G.M.; Gilmore, R. Topological Analysis and Synthesis of Chaotic Time Series. *Phys. D Nonlinear Phenom.* **1992**, *58*, 229–242. [CrossRef]
68. Koebbe, M.; Mayer-Kress, G. *Use of Recurrence Plots in the Analysis of Time-Series Data*; Paper Presented at the Nonlinear Modeling and Forecasting, Proceedings of SFI Studies in the Science of Complexity; Santa Fe Institute: Santa Fe, NM, USA, 1992; Volume XXI.
69. Zbilut, J.P.; Zaldivar-Comenges, J.-M.; Strozzi, F. Recurrence quantification based Liapunov exponents for monitoring divergence in experimental data. *Phys. Lett. A* **2002**, *297*, 173–181. [CrossRef]
70. Thiel, M.; Romano, M.; Kurths, J.; Meucci, R.; Allaria, E.; Arecchi, F. Influence of observational noise on the recurrence quantification analysis. *Phys. D Nonlinear Phenom.* **2002**, *171*, 138–152. [CrossRef]
71. Rural and Environment Research and Analysis Directorate. *Agricultural Census (June); Series of Annual Census Data*; Scottish Government; 1939-Present. Available online: https://www.webarchive.org.uk/wayback/archive/20150218195914/http://www.gov.scot/Topics/Statistics/Browse/Agriculture-Fisheries/Publications/histagstats (accessed on 19 June 2007).
72. Department of Agriculture for Scotland. *Agricultural Statistics 1939 to 1944 for Scotland*; Department of Agriculture for Scotland: Edinburgh, UK, 1948; Volume 130.
73. Department of Agriculture and Fisheries for Scotland. *Agricultural Statistics. 1959 and 1960*; Department of Agriculture and Fisheries for Scotland: Edinburgh, UK, 1962.
74. Department of Agriculture and Fisheries for Scotland. *Agricultural Statistics. 1970*; Department of Agriculture and Fisheries for Scotland: Edinburgh, UK, 1971.
75. Ministry of Agriculture, Fisheries and Food and Department of Agriculture and Fisheries for Scotland. *A Century of Agricultural Statistics, Great Britain 1866–1966*; HMSO: London, UK, 1967.
76. Rural and Environment Research and Analysis Directorate. *Economic Report on Scottish Agriculture, Annual Series, 1940 to Present*; Scottish Government: Edinburgh, UK, 2010.
77. Department of Agriculture for Scotland. *Scottish Agricultural Economics*; 1950; Volume 39. Available online: https://www.webarchive.org.uk/wayback/archive/20150219044830/http://www.gov.scot/Topics/Statistics/Browse/Agriculture-Fisheries/Publications/histagstats/ScAgEconV1-X (accessed on 5 February 2021).
78. Scottish Government Agriculture and Rural Economy Directorate. *Economic Report on Scottish Agriculture*, 2020th ed.; Scottish Government: Edinburgh, UK, 2020.
79. Gibson, A.J.S.; Smout, T.C. *Prices, Food and Wages in Scotland, 1550–1780*; Cambridge University Press: Cambridge, UK, 1995.
80. Jansson, T.; Heckelei, T. Estimating a Primal Model of Regional Crop Supply in the European Union. *J. Agric. Econ.* **2010**, *62*, 137–152. [CrossRef]
81. Cai, R.; Mullen, J.D.; Wetzstein, M.E.; Bergstrom, J.C. The Impacts of Crop Yield and Price Volatility on Producer's Cropping Patterns: A Dynamic Optimal Crop Rotation Model. *Agric. Syst.* **2013**, *116*, 52–59. [CrossRef]

82. Royal Society of Edinburgh. *Committee of Inquiry in the the Future of Scotland's Hills and Islands*; Royal Society of Edinburgh: Edinburgh, UK, 2008; Volume 170.
83. Jacks, D.S. *From Boom to Bust: A Typology of Real Commodity Prices in the Long Run*; National Bureau of Economic Research: Cambridge, MA, USA, 2013.
84. Schumacher, D. The 1973 Oil Crisis and Its Aftermath. In *Energy: Crisis or Opportunity?* Palgrave: London, UK, 1985.
85. Beddington, J. Food security: Contributions from science to a new and greener revolution. *Phil. Trans. R. Soc. B Biol. Sci.* **2010**, *365*, 61–71. [CrossRef]
86. Coleman, W.D.; Tangermann, S. The 1992 CAP Reform, the Uruguay Round and the Commission: Conceptualizing Linked Policy Games. *J. Common Mark. Stud.* **1999**, *37*, 385–405. [CrossRef]
87. Royal Society of Edinburgh. *Inquiry into Foot and Mouth Disease in Scotland*; Royal Society of Edinburgh: Edinburgh, UK, 2002; Volume 77.
88. Conroy, M.J. Mapping Biodiversity for Conservation and Land Use Decisions. In *Spatial Information for Land Use Management*; Hill, M.J., Aspinall, R.J., Eds.; Gordon and Breach: Amsterdam, The Netherlands, 2000; pp. 145–158.
89. Billings, S.A. *Nonlinear System Identification: Narmax Methods in the Time, Frequency, and Spatio-Temporal Domains*; John Wiley & Sons: Hoboken, NJ, USA, 2013.
90. Aspinall, R.J.; Staiano, M.; Pearson, D.M. Data, Time, Change and Land System Dynamics. *J. Land Use Sci.* **2021**, *16*, 129–141. [CrossRef]

land

Article

Old Ways, New Ways—Scaling Up from Customary Use of Plant Products to Commercial Harvest Taking a Multifunctional, Landscape Approach

Julian Gorman [1,2,*], Diane Pearson [3] and Penelope Wurm [2]

1 School of People Environment and Planning, Massey University, Palmerston North 4442, New Zealand
2 Research Institute for the Environment and Livelihoods, Charles Darwin University, Darwin NT 0810, Australia; Penny.Wurm@cdu.edu.au
3 Farmed Landscape Research Centre, School of Agriculture and Environment, Massey University, Palmerston North 4442, New Zealand; d.pearson@massey.ac.nz
* Correspondence: julian.gorman@cdu.edu.au; Tel.: +64-212571719

Received: 21 April 2020; Accepted: 22 May 2020; Published: 25 May 2020

Abstract: Globally, the agricultural sector is facing many challenges in response to climate change, unsustainable farming practices and human population growth. Despite advances in technology and innovation in agriculture, governments around the world are recognizing a need for transformative agricultural systems that offer solutions to the interrelated issues of food security, climate change, and conservation of environmental and cultural values. Approaches to production are needed that are holistic and multisectoral. In planning for future agricultural models, it is worth exploring indigenous agricultural heritage systems that have demonstrated success in community food security without major environmental impacts. We demonstrate how indigenous practices of customary harvest, operating in multifunctional landscapes, can be scaled up to service new markets while still maintaining natural and cultural values. We do this through a case analysis of the wild harvest of Kakadu plum fruit by Aboriginal people across the tropical savannas of northern Australia. We conclude that this system would ideally operate at a landscape scale to ensure sustainability of harvest, maintenance of important patterns and processes for landscape health, and incorporate cultural and livelihood objectives. Applied to a variety of similar native products, such a production system has potential to make a substantial contribution to niche areas of global food and livelihood security.

Keywords: agricultural systems; indigenous economic development; production systems; landscape ecology; wild harvest

1. Introduction

Population growth, climate change, and unsustainable farming practices are some of the most pressing global challenges in the 21st century. Governments from around the globe have acknowledged these 'wicked' problems and are developing strategies to mitigate their impacts [1–3]. The agricultural sector is one area that is under considerable pressure to adjust and perform to meet the increased challenges of sustainable food production practices under increasingly uncertain climatic conditions. However, food production is currently occurring in a way which is having adverse impacts on climate, water, topsoil, biodiversity and marine environments [4]. If not addressed, these practices will undermine the world's ability to adequately feed future populations, and solutions are desperately needed to provide more sustainable options.

1.1. Food Security

The global population is predicted to reach 9.73 billion people by 2050 and will require more food, more stock feed and more biofuel to meet these demands [1]. Globally, food security is currently dependent on a small number of cultivated species with only 12 species contributing to 80% of total dietary intake [5]. Wheat, maize and rice account for over 50% of the world's daily requirement of protein and calories [6]. Despite there being limitations in crop–climate modeling, the impacts of climate change are almost certain to decrease global crop production [7]. This is a real concern for food security with a growing global population and there is now substantial research into global plant genetic resources being conducted, focused on improved cultivars, breeding lines, landraces and crop wild relatives diversity [8–10].

Global per capita consumption has increased considerably since the 1970s, with doubling of milk, dairy products and vegetables, while meats products have tripled (Alexandratos and Bruinsma in [1]). However, total productivity resulting from agricultural investment and technology is now thought to have slowed as a result of food loss and wastage, degradation of natural resources, biodiversity loss and spread of transboundary pests and disease, and resistance to antimicrobials [1].

There is now a marked global disparity of agricultural outputs among countries. Developed countries are currently facing chronic obesity rates, which have more than doubled since 1980s [11]. Issues added to these human health problems include an increased carbon footprint and an increase in area lost to landfill and biodiversity losses. Some 25–30% of food produced worldwide goes to waste, which costs about US$1 trillion per year and accounts for 10% of the greenhouse gas emission from food systems [12]. Third world countries continue to suffer from poverty and food shortages as well as chronic undernutrition [4]. Given the serious threat to food security and the far-reaching impacts of climate change on crops, livestock and fisheries production, agriculture needs to balance research and implementation strategies to be able to face these challenges [2].

1.2. Impacts of Climate Change

The impacts of climate change on agriculture are predicted to be significant. Climate change is a certainty [12] and the agriculture sector is, and will continue to be, impacted in various ways. The Intergovernmental Panel on Climate Change (IPCC) predicts a temperature rise of +1.5 °C above pre-industrial levels sometime between 2030 and 2052, if global warming continues at its current rate. The impacts of such changes, despite increasing confidence in prediction modeling, are not altogether certain. In 2019, the IPCC Special Report provided new evidence for the benefits of limiting global warming to the lowest possible level, in line with the goals set in the 2015 Paris Agreement [12,13].

There are numerous predictions of climate change that will impact on the agricultural sector. These include more extreme weather events with both drought and floods becoming more common in areas making it more expensive and difficult to grow and sustain crops and livestock. A change in weather will influence growing seasons and cause an impact upon productivity due to non-alignment of crop growth with soil moisture levels and pests. In some areas, seasonal weather patterns may cause an increase in the frequency of wildfires that will result in physical damage to infrastructure and pasture as well as a number of 'secondary impacts' from smoke [6]. Rising temperatures can alter exposure to pathogens and toxins, and rising levels of carbon dioxide in the atmosphere can decrease dietary iron, zinc, protein and other macro- and micro-nutrients [1,14].

1.3. Environmental Impacts

The rapid global expansion in food production and economic growth that the world has seen since the 1960s has come at a heavy cost to the natural environment. Adoption of high input and resource intense farming systems have caused massive deforestation, soil depletion, water scarcities and contributed to high levels of greenhouse gas emissions [1]. Klitgaard [15] suggests that systems in overshoot such as these require new economic theories to achieve sustainable futures.

As agriculture has expanded in recent decades, there has been greater competition for natural resources, increased carbon emissions, and land degradation. There has been a narrowing of cropping choice [9] and lack of diversity in crop rotation, which is coincident with an overuse of pesticides and other chemicals, which is damaging to human and ecosystem health [16].

Globally, agriculture is responsible for using 70% of all freshwater withdrawn from the natural system and 60% of biodiversity loss [4]. Half of the world's forests have been cleared, an ever-increasing volume of greenhouse gases is going into the atmosphere and ground water has been contaminated or depleted. However, agricultural land use has resulted in land values increasing as the system has become more capital-intensive, and requires greater vertical integration, which leads to big cooperatives dictating land-use. This is impacting on the social and cultural structure of rural towns, removing safety nets and increasing levels of rural poverty. This then drives migration into large cities which exacerbates welfare, food, employment and health issues.

Agricultural expansion is resulting in habitat loss [17–19], which in turn, has been identified as the primary contributor to what has been described as the 'Anthropocene' [20] or sixth mass extinction event [21].

The Australian agricultural and rural sectors are following the global trends outlined above and are currently facing extreme social and economic pressures, many of which are interrelated. These include depopulation of rural areas, a reduction in participation in agricultural education, low levels of uptake in the farming sector (especially by young women), low incomes for farm businesses and poor health outcomes for farmers and others in rural areas [3]. Thus, despite the many benefits of industrialized agriculture, these farming practices can be seen to be contributing to declining rural employment and rural depopulation [22]. There is a need to consider alternative agricultural paradigms and transformative agricultural systems. Current conventional agricultural practice may not always be the best way forward for all regions in Australia.

These undesirable direct and indirect impacts of agriculture on the environment are becoming less acceptable to the global community and pressure is being put on governments to find alternative paradigms for food production.

1.4. Response from Agricultural Sector

The agricultural sector can respond quickly to change when required and has been seen to triple agricultural production over a 50-year period (1961–2011) due to the new technologies available during the Green Revolution [1]. These increases in production were mirrored by improved transport and post-harvest techniques which contributed to substantially longer value chains (farm gate to plate) as well as increased consumption of processed, packaged and prepared food.

Globally, there has been recognition that there is a need for a shift away from high-input, resource intensive farming to more innovative systems that can continue to increase productivity but at the same time protect and enhance the natural resource base [1,23–25]. A few of the many such alternative production systems include agro-ecology; agro-ecosytsems; agroforestry; climate smart agriculture; diversified farming systems; 'socially-modified' crops; sustainable intensification; and conservation agriculture [1,26–30]. The commonality between these approaches is that they are often multi-use, more holistic and, in many cases, built on indigenous traditional knowledge. Of the 250,000 plant species globally, 4% (20,000) have edible products (many from trees), but only 0.3% of edible plants are cultivated in agriculture [31], making plants a highly underutilized resource.

The Food and Agriculture Organization of the United Nations (FAO) [32] proposed four dimensions of food security, which include: reducing greenhouse gas emissions to limit and adapt to change; reducing impacts of different types of agricultural production on the world's ecosystems; developing rural areas to improve livelihoods and create jobs for poor people; and maintaining ecosystem services [4]. In this research paper, we describe an agricultural paradigm based on the commercial use of native foods which is in line with FAO proposed criteria [32]. We advocate for the use of a landscape ecological approach in understanding, evaluating and developing such a paradigm.

2. Alternative Agricultural Systems

Much of the discourse around climate change, vulnerability and food security, emphasizes cultivated foods, new animal breeds and crop varieties, and climate–crop modeling as the solution [2,5,33]. This is likely to be the main approach to meeting the challenges of global food security in the future. However, there are additional pathways that could also contribute with less impacts on the natural and cultural environment. This vision is reflected in recent times in affluent, western societies where rural change has transitioned away from a dominance of production values towards a variable mix of production and environmental protection values [34].

2.1. Agricultural Heritage Systems

Agriculture is defined by the Merriam-Webster dictionary as 'the science, art, or practice of cultivating the soil, producing crops, and raising livestock and in varying degrees the preparation and marketing of the resulting products'. There are, however, many alternative production systems that do not fit neatly within this definition. For example, the harvest of forest products often involves a degree of forest custodianship and management which contributes to the growth, quality and abundance of a harvested product. It could be argued that such practices should be considered as agricultural practice. Therefore, we will refer to some of these alternative practices as 'agriculture' in this research paper.

There are globally important agricultural heritage systems that have been developed by indigenous cultures over millennia [35]. These are often very complex, diverse and specific to local areas, involving techniques and practices that have contributed to community food security often in conjunction with conservation of natural resources and biodiversity. Agricultural heritage systems can still be found globally, with about 5 million hectares providing a vital combination of social, cultural, ecological and economic services to humankind [34]. An estimated 1.4 billion people manage such agricultural systems and landscapes globally, mostly family farmers, peasants and indigenous communities [34]. Many scientists acknowledge that traditional agricultural systems have the potential to provide solutions to the predicted changes and transformations facing humanity in an era of climate change, biodiversity loss and sociocultural issues [34].

2.2. Wild Foods

'Wild foods' constitute a niche area of food production that involve production and harvest with minimal impacts and interventions on the surrounding environment while at the same time providing incentives not to clear natural habitats. A 'wild food' can be described as an animal or plant product which is found in an undomesticated state in nature. Many of the commonly used products that the world relies on today have wild origins including most staple foods (corn, potatoes, tea, spices), medicines (aspirin, codeine), fibers (cotton, hemp), dyes (indigo and saffron), intoxicants (tobacco, opium) [35]. There is still a high demand from western markets for wild genetic plant stocks, with 25% of prescription drugs currently in use today having plant origins. Between 1981 and 2006, approximately 75% of new anticancer drugs were derived from plant compounds [36,37]. Ensuring future biodiscovery will require the conservation and management of the world's remaining natural habitats.

Non-Timber Forest Products' (NTFPs) are an example of a type of wild food. NTFPs were defined by FAO in 1995 as consisting of 'goods of biological origin other than wood, derived from forests, other wooded land and trees outside forests' [38]. Wild foods and NTFPs are an area of agriculture which contributes to millions of livelihoods worldwide. Globally, there are 300 million people living in predominantly forest ecosystems, with a large percentage of these people dependent on forests and their products for their livelihoods [39]. As such, NTFPs make up a considerable component of the world's food economy and are an important safety net during extreme events. These products are sometimes termed the 'hidden harvest' because their direct and indirect values are often not measured nor included as part of official agricultural outputs [40]. NTFPs are collected for customary and commercial purposes, mostly managed sustainably by local people, communities and customary law.

Globally, many indigenous communities still have a high dependence on wild-collected plant products for their health, nutritional, cultural and spiritual wellbeing [39,41]. Agricultural and forager communities within 22 countries in Asia and Africa have been recorded as using an average of

90–100 species per location [5]. Much of the literature on food security emphasizes the production of cultivated foods, but clearly wild foods are making a substantial contribution to the global food basket [5]. Furthermore, many NTFPs are actively managed, which suggests there is a false dichotomy between agriculture and use of wild products.

Australian Aboriginal and Torres Strait Islander people (hereon Aboriginal people) are the custodians of the oldest culture on earth [42]. They continue to have extensive ecological knowledge and a deep, spiritual connection to their traditional lands [43]. Through customary care for and use of natural resources over tens of thousands of years, they have developed an intricate knowledge of the value of plant products [44,45]. A wide range of enterprises are emerging from this knowledge, including bushfood enterprises, native plant derived industries such as nurseries, seed harvesting, cut flowers, and a variety of botanical based medicinal and beauty products [46]. The resulting enterprises are largely based on wild harvest from traditionally managed estates, but also involve different models of cultivation such as enrichment planting and horticulture [47].

2.3. Niche Markets

Plant products play an important role in local, regional and international markets. At a local and regional level, they are often part of an indigenous customary harvest which trades, transports and sells products over vast distances along a diversity of value chains. In addition to market demand of wild plant material for specialized medical and pharmaceutical development, there is a rapidly growing consumer consciousness about links between health, diet and the environment, and an increased awareness of foods that are produced in safe, ethical and sustainable production systems [48]. This group of foods is referred to as "functional foods", which potentially have a positive effect on health above their nutritional values, in areas such as the prevention and management of health conditions [49]. Estimations of the revenue generated by the global functional food market vary considerably, however, it is estimated to have grown considerably over recent years. Market research estimates the global functional food market size as being 161.49 billion USD in 2018 and predicted to grow to 275.77 billion USD by 2025 [50].

Australia is well positioned to take advantage of this growing demand for functional foods. It has a very diverse endemic flora [51] with many species already having commercial applications in the fields of pharmacy, medicine, food, beverage, cosmetic, perfumery, and aromatherapy [52–56]. Coupled with this, Australian Aboriginal people have been using native foods for more than 40,000 years [57]. In recent years, there has been considerable interest among Aboriginal people in the commercialization of these products [46,58,59].

2.4. Sustainable Landscape Management

Aboriginal stakeholders are major landowners across northern Australia's tropical rangelands and have shown interest in a range of natural resource-based enterprise development opportunities [58,60–62]. The environments in which customary harvest practices take place are generally relatively intact ecosystems. Much of this area is under Aboriginal land tenure, is remote and has Aboriginal communities and Aboriginal Ranger groups actively involved in its management. If communities desire to scale up their customary use to commercial use, then they need knowledge of the impacts on the ecosystems in which they occur. This will require a greater understanding of the harvested species, the interconnectedness in the landscape and, more broadly, the ecology of the landscape relevant to the harvested species. Knowledge of important landscape patterns, processes and change will be fundamental in understanding and managing the dynamics of the systems in which the species occur. Traditional Aboriginal ecological knowledge, coupled with sound harvest and scientific monitoring data, will be important information sources that can help to determine a culturally appropriate basis for establishing good management practices.

There are also large areas across the Australian Rangelands which are not in pristine conditions, having been impacted from frequent, intense wildfire, high densities of feral animals and modification from other land uses such as pastoral use, cropping and mining. This has resulted in soil erosion and biodiversity loss and, in instances, involved high levels of tree removal, altered water flows and introduced pastures [63,64]. These areas may require different consideration to those landscapes that are more ecologically and culturally intact. For example, priority considerations may include cultural and environmental restoration, alongside commercial priorities.

The discipline of landscape ecology has an important role to play in helping understand and inform the sustainable harvest for traditional and commercial use under changing climatic conditions. The focus of landscape ecology has largely involved spatial heterogeneity and ecological values with recent recognition that cultural values are also important elements in a landscape [65]. In turn, ecological and cultural knowledge informs the selection of appropriate business models. These models include wild harvest of natural resources, cultural values and practices, and remote rural economic settings. This integration necessitates a sustainable approach, addressing the triple bottom line.

Applying a landscape ecological framework can bring together different knowledge systems, values and priorities to measure the impacts and develop strategies for sustainable use without destroying the ecological integrity of the landscapes. Sustainability must include consideration of the socio-economic context of the communities harvesting the species. Integrated approaches help to understand the characteristics of species that have value, the markets that are likely to be interested in these characteristics, and the communities that harvest the species. Integrated, landscape approaches can also help to maintain important socio-ecological systems whilst providing for increased livelihood opportunities and allowing multifunctionality of land uses (conservation and development) [66]. However, landscapes are dynamic and can progress in different directions [67], especially under changing climatic conditions.

We posit that Australian Aboriginal people are well positioned to scale up their customary harvest of native foods to service rapidly growing, niche, functional food markets. For reasons discussed later, we suggest wild harvest as a suitable initial production model for supply of native plant products to niche markets, while ensuring benefits from resource use are retained in the landscape. There will be issues around sustainable use that need to be considered and management plans will need to be developed if they have not been already [68]. However, a landscape-focused framework will be most appropriate to measure landscape system health and impacts, as there are many overarching cultural, social, ecological and political factors that need consideration [65].

3. Case Analysis—*Terminalia ferdinandiana* (Kakadu plum) Enterprise

To demonstrate the potential of wild foods as an important contributor to food production, we take a case analysis approach. To explore the scaling up of customary harvest to commercial use, a case analysis of the wild harvest of the fruit Kakadu plum on Aboriginal owned traditional lands is reviewed. This case involves the wild harvest of a native, endemic fruit, *Terminalia ferdinandiana* Exell., by Aboriginal people across northern Australia. We discuss the species, its customary use, markets and production options available for scaling up from customary use, by small remotely located populations situated in a multifunctional savanna landscape.

A participatory research methodology was used in conducting this analysis, along with an ethnographic account of factors that have influenced the progress of this enterprise over the last 15 years [48,61,62]. The main method used for qualitative data collection was participant observations, which is a tool in many disciplines for collecting data about people, processes and culture [69]. A literature review using published and unpublished papers and reports was also used in gathering data to describe the north Australia Kakadu plum industry.

3.1. Properties

T. ferdinandiana is best known by the common name 'Kakadu plum' in Northern Territory (NT); '*gubinge*' in the Kimberley, Western Australia (WA), and many other Australian Aboriginal language names across its range. It will be referred to as 'Kakadu plum' in this paper as this is one of its most widely used common names. It is a member of the Combretaceae family [70], is endemic to northern Australia and is one of 200 species in the genus *Terminalia*, of which 29 species or subspecies are native to Australia [71]. *T. ferdinandiana* is a small to medium sized semi-deciduous tree that is found in the woodlands of the upper rainfall band of the Australian wet/dry tropics (see Figure 1). Its density is very variable over its range, but this species can occur in very high densities on or near the coast [61].

Kakadu plum is well known for its phytochemical properties. It has the highest vitamin C (ascorbic acid) of any fruit in world [72]. These exceptionally high levels of vitamin C were first detected in 1982 through a study of the nutritional composition of bushfood used by Australian Aboriginal people [73,74]. The fruit and leaves also have very high levels of ellagic and gallic acid and other polyphenolic compounds. These, along with the vitamin C, provide high antioxidant values which are known to reduce risk of diseases such as cardiovascular disease, cancer, stroke, and rheumatoid arthritis [72,75–79]. These phytochemicals have also been proven to have high antimicrobial properties [80]. These phytochemical properties, and their demonstrated applications, have created a commercial demand from the food and beverage, pharmaceutical, cosmetic and nutraceutical industries [81].

3.2. Customary Use

Traditional foods continue to be an important part of the diets of Aboriginal people. A study of five Aboriginal communities in the NT, Australia, found that 89% of the people interviewed consumed a variety of traditional foods fortnightly [82]. Aboriginal people have a strong affiliation with the Kakadu plum and many Aboriginal language groups across its range have a close cultural connection and varying uses for this species and its products [45,83]. The fruit has been recorded as being consumed for quick energy and as a refreshment on hunting trips and is used for a variety of other medicinal purposes, such as treating colds and congestion [44,45,84,85]. The inner bark is used to treat skin disorders and as well as fungal infections such as ringworm [61].

3.3. Kakadu Plum Markets

The ongoing research and biodiscovery of different phytochemicals in Kakadu plum and the identification of potential commercial applications has stimulated several market sectors. However, until recently, market signals have been very inconsistent between years, which has contributed to the formation of a disjunct and poorly coordinated supply network. Response to these market signals for the supply of Kakadu plum has come from both Aboriginal and non-Aboriginal people, and involved three main production systems, namely horticulture, enrichment planting and wild harvest [47,48,86]. We describe these production systems below.

3.4. Kakadu Plum Production Systems

Wild harvest is the production system that most Aboriginal people (cooperatives, communities, family groups, and individuals) across northern Australia are involved in, through several different business structures. Kakadu plum has been commercially harvested from the wild in the Northern Territory (NT) since 2005 and was initially trialed through several Aboriginal Ranger groups [61]. One of these groups, the Thamarrurr Rangers from the Thamarrurr Region, NT, hosted a Kakadu plum enterprise and acted as the consolidator, by managing the fruit collected and linking with markets (Figure 1). The Thamarrurr Ranger's primary responsibility is natural resource management rather than commercial development, so after a few years, they handed over the consolidator role to a local Aboriginal owned and operated business, the Palngun Wurnangat Aboriginal Corporation (PWAC). In 2020, approximately 15 years since the Kakadu plum collection trial first started, the Thamarrurr Kakadu plum Enterprise still operates as a community owned and operated business. It now engages PWAC and the local Aboriginal development corporation, Thamarrurr Development Corporation, to assist in supporting operational activities. Annually, this community enterprise has purchased tons of wild harvested fruit from the community members who wild harvested it from their traditional estates [87]. This enterprise provides significant monetary and non-monetary benefits for the community.

Another Ranger Group in the NT, the Bawinanga Rangers, who are supported by the Bawinanga Aboriginal Corporation (BAC) in Maningrida, Central Arnhem Land (Figure 1), were also involved in a small, wild harvest of Kakadu plum between 2005–2008. BAC is currently trialing harvest of a range of native bushfoods for sale to restaurants and other markets around Australia [88] and hope to expand this activity to include some of the 32 Aboriginal clan estates in their jurisdiction.

The Kimberley area of WA is another geographical area where Aboriginal people wild harvest Kakadu plum commercially. Some examples of Aboriginal owned and operated businesses operating in this area include: Twin Lakes Cultural Park, Kimberley Wild Gubinge, Lombadina Community and Mayi Harvests. Twin Lakes Cultural Park is a family business located north of Broome on the Dampier Peninsula (Figure 1). The Aboriginal traditional owner, Bruno Dann, is a Nyulnyulan person who lives on his traditional land making a living from wild harvest of Kakadu plum and cultural tourism. He sells his fruit to a non-Aboriginal company, Living Earth Pty Ltd., which uses the Kakadu plum powder as an ingredient in chocolate [89]. Kimberley Wild Gubinge is another Aboriginal owned business which is situated north of Broome in the Dampier Peninsular, WA (Figure 1). It purchases fruit from local harvesters which it processes into powder in its solar-powered premises, providing local livelihoods and preservation of local knowledge though resource use and appropriate land management [90]. The Lombadina Community and Lombadina Aboriginal Corporation, established in 1985 on the Dampier Peninsular, in the Kimberley Region of WA, have been in partnership with traditional owners and communities to pick and sell wild harvested Kakadu plum and have recently started growing Kakadu plum in orchards [91]. Mayi Harvests was established in 2006 and is situated in West Kimberley on traditional family lands in Ngumbarl and Jabirr-Jabirr country. They harvest several native plant species, including Kakadu plum (locally called *gabiny*) which they sell as both dried and frozen products [92].

Enrichment planting is another production system in which Kakadu plum trees are grown in the Kimberley area of WA. Enrichment planting involves additional planting into wild populations using wild harvested seed stock of desirable properties to increase tree density [47]. The Kimberley Training Institute (KTI) in Broome, WA, has established an enrichment planting trial at its Balu Buru site, in partnership with WA Department of Conservation and Land Management [47]. KTI is a horticultural training provider which uses the Balu Buru trial site in their training. KTI also supports Aboriginal communities to establish other Kakadu plum enrichment plantings and orchards. The Balu Baru site has over 1000 Kakadu plum trees that have been enrichment planted over a 5 year period. They are developing a 'Savannah Enrichment' approach which combines traditional burning practices with modern horticultural techniques. These trials being conducted at Balu Buru provide valuable lessons in the propagation and growth of Kakadu plum in competition amongst the dense acacia thickets which have developed because of unmanaged fire regimes. They indicate that at a landscape scale, savanna enrichment practices could help change the structure of plant communities and reduce the fuel loads and occurrence of higher intensity wildfire. In turn, this would reduce the coinciding biodiversity loss while at the same time providing livelihood opportunities for local people and an economic incentive to manage the landscape differently [47].

Kakadu plum is also grown as a monoculture in horticultural settings. Kakadu Life Pty Ltd. is the main non-Aboriginal company that grows Kakadu plum at scale. Their production is based in NT, Australia, with distribution based in Perth, WA, and they sell a variety of Kakadu plum products, many with organic status [93]. There are also several Aboriginal owned communities that grow Kakadu plum in the Kimberley area of WA in a horticultural setting. These include WA's largest Aboriginal community, Bidyadanga Aboriginal Community, situated 180 km south of Broome, which started growing Kakadu plum almost 20 years ago [94] (Figure 1). Mamabulanjin Aboriginal Corporation is an Aboriginal Resource Centre based in Broome and is another group that is growing Kakadu plum in a horticultural setting [95]. More recently, GoGo Station Pty Ltd., an Aboriginal pastoral station situated near Fitzroy Crossing, WA, set up a trial plot of 200 Kakadu plum trees under drip irrigation [96].

The Kakadu plum industry across northern Australia is established and growing. In summary, the exceptional phytochemical properties of Kakadu plum, the commercial applications and market demand, and knowledge from generations of customary use, underpin an established and growing Kakadu plum supply chain. The main model of production of Kakadu plum is wild harvest from Aboriginal traditional estates, supported by local Aboriginal corporations. Value chains involve both Aboriginal and non-Aboriginal actors within or outside of these estates. In many cases, enterprises have been developed by local Aboriginal people who have customarily harvested this species but who have now expanded these practices and incorporated business practices for commercial use. Enrichment and horticultural plantings are emerging as new modes of supply for this species, with the support of regional training institutions. Inspection of the known distribution of *T. ferdinandiana* would suggest potential for wider uptake of the commercial use of Kakadu plum, particularly in the NT (Figure 1).

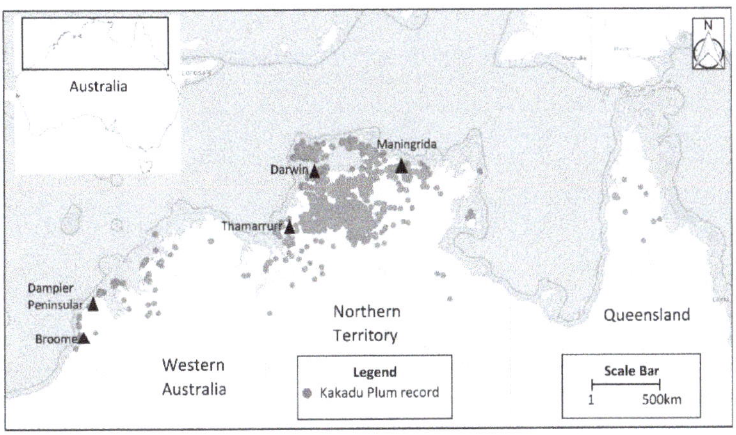

Figure 1. The distribution of the endemic *T. ferdinandiana* in Western Australia (WA), the Northern Territory (NT) and Queensland (QLD). Triangles represent place names and dots represent herbarium records of *T. ferdinandiana*, from Atlas of Living Australia [97].

4. Discussion

We now discuss the important considerations in wild harvest for customary purposes as well as commercial use and the scaling up to meet larger commercial demands for supply. We will also discuss the use of a landscape ecology approach for valuing, understanding and developing use of native foods for commercial purposes.

4.1. Customary Sector and Wild Harvest

Around Australia, Aboriginal people have used a variety of methods of food production prior to colonization. These range from fire being used to modify vegetation structure and composition for wild harvest purposes [98] through to more conventional types of agricultural practices including domesticating plants, sowing, harvesting, irrigating and storing crops, and implementing aquaculture and other farming practices [99–101].

The current Aboriginal economy has also been described as a 'hybrid' economy comprising of three sectors: customary, state and market [102]. The customary sector refers to subsistence harvest for food and cultural purposes; the state includes social security and government-funded programs such as 'work for the dole' schemes; and the market sector relates to the free market, most notably the fine arts and craft industry established in Aboriginal Australia. The customary sector constitutes a range of productive activities that are based on cultural continuities and cultural identity, such as hunting, fishing, gathering of bushfood, art and craft production, caring for kin and caring for country [103]. The Aboriginal economy has changed over time [104], but customary harvest is still a very important component of Aboriginal livelihoods. It is based on an intricate ecological knowledge and connection with their traditional lands. Aboriginal people relate well to the idea of a 'culture-based economy' which incorporates their knowledge, connection to their traditional estates and epistemology [105]. A culture-based economy involves a multifunctional approach and operates at a landscape scale.

Wild harvest of bushfoods for commercial purposes can be viewed as an extension of customary practice. It relies on Aboriginal ecological knowledge and practices, often still within cultural perimeters, but aims to service larger and external markets. As the scale of harvest increases, there will be concerns around sustainability and impacts of harvest that will need to be addressed, alongside the many other economic, ecological, social, and cultural benefits that come from this activity. Even though Kakadu plum is an abundant species and has an extensive natural distribution, the impacts of increasing harvest levels for commercial use, in conjunction with changing climatic conditions, will require careful land management practices. In response, the Northern Territory Government has developed a management program for Kakadu plum (*Terminalia ferdinandiana*) to ensure its sustainable use [68]. However, there are several case studies that demonstrate that Aboriginal people have established production systems which utilize wildlife for both customary (non-market) and commercial purposes, including use of saltwater crocodiles, long-necked turtles and raw materials for artworks [59,106–108].

Tropical savannas across northern Australia need active land management for their natural and cultural values to remain intact. This requires people to be living on country in the savannas, with knowledge of that country and how to manage it (such as traditional burning practices). People also need livelihoods to be able to thrive in these remote places across the savannas. However, except for the Aboriginal Ranger Program, there are very limited employment and enterprise opportunities in remote Aboriginal townships. The economic status of Aboriginal people is the lowest of any demographic group of Australians [109,110], with unemployment rates being as high as 90%, if various government welfare programs were not taken into consideration [111]. Commercial use of natural resources offers livelihood opportunities for Aboriginal people who have expertise in both customary use and land management as well as the right and a strong desire to be involved [61,62,112].

4.2. Scaling up from Customary Harvest

Extending harvesting from customary to commercial purposes is a manageable transition for Aboriginal harvesters because they have ecological knowledge about the resource (when it should be picked, where the best picking sites are, landscape management requirements, etc.,) and know the cultural protocols in which the resource must be harvested (access and harvest permissions, cultural sensitivities or prohibitions etc.). However, the component of commercialization that some Aboriginal communities might find difficult is building an appropriate business structure.

There are many complexities and challenges in developing an Aboriginal business, especially in remote and regional localities within the savanna landscapes [48,113–115]. Business development in remote Aboriginal communities is different to that in other Australian communities [116]. Enterprise development in Aboriginal communities is often funded through government programs and is likely to have originated without commercial intent and often involves subsidized community-based activities [117]. Many of these enterprises are largely focused on social goals in the absence of economic criteria for success. These enterprises often lack the business acumen required to make decisions that lead to viable long-term businesses in this distinctive landscape context [48]. This confusion between social and economic objectives has been cited as an important contributor to business failure in Aboriginal businesses [87,118]. There are, however, many examples of Aboriginal community-based enterprises (such as those described in Section 3.3 above) which have managed social, cultural and economic priorities successfully.

For Aboriginal people to have greater control of the Kakadu plum industry, there is a need for them to take a 'whole of industry' approach and become leaders in all aspects of the business, including research, harvest, processing and marketing. The Indigenous Land and Sea Corporation is working with Aboriginal people to achieve this goal. In 2018, it established the Northern Australian Kakadu Plum Alliance (NAAKPA). NAAKPA currently consists of a consortium of eight Aboriginal owned enterprises which ethically harvest and process Kakadu plum across northern Australia [119]. This Alliance provides support to its members as they grow their businesses while at the same time providing stability and reliability to the Kakadu plum supply chain. Such cooperative or collective ventures are likely to play an important role in the development of savanna enterprises, in so far as they support a focus on Aboriginal economic development as well as ecological, social and cultural priorities. A cooperative model across several savanna sites will also help manage the risks to supply, inherent in wild harvest.

As markets develop and demand for Kakadu plum increases in the future for Kakadu plum there may be a need for greater uptake of alternative production systems to complement wild harvest. This will place greater emphasis on domestication of Kakadu plum to meet demand. Enrichment planting is one alternative form of domestication which has been described earlier in this paper (Section 3.4) [47], which could contribute to more consistent yields and greater volumes of supply. More conventional horticultural production systems may also need be considered. Domestication may be desirable for some Aboriginal producers to meet the demands of larger markets. It is often seen as comprising of a spectrum of increasing levels of human intervention in the production of a species for human benefit [120]. There are some very relevant resources documenting the process of domesticating culturally important, indigenous food-tree species over a 25-year period in tropical/subtropical Africa [25]. This body of applied research is focused on domesticating trees and creating multifunctional landscapes which can reverse the cycle of land degradation and its associated social deprivation issues [26].

Leakey [121] describes food species as falling into four categories: i. Internationally important and widely cultivated staple foods, ii. Widely cultivated case crops, iii. Locally domesticated and cultivated species or 'orphan crops', which also have wider potential, iv. Culturally important species used for customary use and little known outside their natural range. He suggests that the first three categories have made the transition from ethnobotany to agriculture hundreds, if not thousands, of years ago, while the fourth category is currently making that step following recent research. Despite being focused more on agroforestry than wild harvest, this long-term participatory research project has demonstrated some interesting findings that could be incorporated into Australian Aboriginal agricultural development. This is particularly the case for a tree species like Kakadu plum with so many valuable attributes—edible fresh or processed fruit, a source of vitamin C and rich in antioxidants. The identification of 'ideotypes' to capture ideal phenotypic trait combinations for different end products and associated markets [122], can differentiate suppliers across the savanna landscape. Taking a geographically decentralized approach to reduce risk in narrowing the genetic base of species [123] is an approach that would be very applicable to Kakadu plum, as there are many varieties across its range on numerous traditional estates and there may be cultural reasons for land owners wanting to keep genetic strains represented by 'ideotypes' separate. Creating Rural Resource Centers to assist in technical training and business support [25] and making partnerships which link domestication and commercialization programs, is considered critical to the success of commercialization and demonstrates the long-term level of support and commitment required to progress the development of community-based enterprise development. Examples of successful Kakadu plum enterprises developed to date demonstrate this.

Given there are also many degraded landscapes across the Australian rangelands, there may be potential to establish an agroforestry domestication program. Processes to domesticate native species need to protect the interest of the traditional estate owners in Australia. Approaches, such as the *sculptured seedling technique* to revegetation, which relies on a knowledge and understanding of the natural vegetation in areas to match site capability with appropriate species, may prove useful [124]. Other processes, such as utilizing socially modified crops rather than genetically modified crops, are more likely to prevent the loss of genetic variability from the landscape and protect local interests and benefits [25,26]. This process could, in time, supplement yields from wild harvest and help provide the volume and consistency of supply to secure relationships with larger markets. This work will need to be done in close consultation with traditional landowners as there will be many issues (access, identifying desirable phenotypic traits, cultivar development, cultivation techniques) that will require their participation and customary authority.

4.3. An Intregrated Landscape Approach to Management of Country

Aboriginal people own vast areas of land across northern Australia. In the NT alone, Aboriginal people make up around a third of the population and own over half of the land, mostly under a communal title. Much of the natural range of Kakadu plum is found on Aboriginal lands across the northern Australian tropical savannas, which still consist of relatively intact landscapes. These lands are managed by a mixture of traditional Aboriginal and western land management practices, in collaboration with the Aboriginal Ranger groups and in conjunction with traditional land management practices or authority [125]. Landscapes managed by traditional land management practices in tropical northern Australia have been shown to have greater ecological integrity than those managed in the absence of Aboriginal land managers and their traditional ecological knowledge non-Aboriginal managed sites [126].

Access to native plant resources on Aboriginal land, their commercial use and sustainability, are regulated through customary lore and state legislation. Traditional owners, who are the designated decision makers for individual clan estates, have cultural obligations to look after their country and this relates to caring for both natural and cultural resources and maintaining their spiritual connections. People are integrally linked to place and place is integrally linked to people [127]. In the NT, the *Aboriginal Land Rights Act* (NT) 1976 states that any commercial activity must be approved by the traditional owner(s). Permits to access and commercially harvest on Aboriginal lands must be with the authority of the traditional owner(s) and must be captured in a Land Use Agreement between the Aboriginal Land Trust, on behalf of the traditional owner(s), and the proponent. Sustainable use of native plants and animals is regulated through the NT Department of Environment and Natural Resources through a permit system and has long been a focus and part of the NT Government's conservation strategy [128]. The sustainability of commercial use of Kakadu plum is of utmost importance to the NT Government and a 'Management Plan for *Terminalia ferdinandiana* in the Northern Territory 2018–2022' is in place to ensure wild populations and the species habitat are adequately maintained across the NT of Australia [68].

There is a risk that over time the benefits of commercial use of native plants could be realized off Aboriginal owned lands and to the exclusion of indigenous peoples. Kakadu plum is an industrial crop and there has already been an incident in 2004 where two multinational companies have tried to export Kakadu plum tissue culture out of Australia without permission or benefit-sharing agreements [85,129]. However, there are two key ways that the interests of Aboriginal peoples can be protected. Firstly, by targeting premium markets that value culturally identified and ethically sourced products. Secondly, there are legislative mechanisms in place to protect interests and ensure benefit sharing with landowners, including traditional landowners and traditional knowledge holders [130]. Australia is a signatory of the 'Convention on Biological Diversity 1992', under which the 'Nagoya Protocol' on 'Access to Genetic Resources' and the 'Fair and Equitable Sharing of Benefits Arising from their Utilization' has been framed to protect the interests of indigenous peoples and communities. In the Northern Territory, Australia, land held under Aboriginal Freehold title, and awarded under the *Aboriginal Land Rights (Northern Territory) Act* 1976, requires special land use agreements with traditional landowners before parties can access or use natural resources from this land. Finally, Australian states and territories, including the three jurisdictions in which Kakadu plum occurs, have biodiscovery acts and regulation to manage the accessing, collection and transfer of biological materials collected, and the benefit that flows from their use.

Australian Aboriginal people are the custodians of the oldest continuous culture on earth [42] and have a deep, spiritual connection to their 'country' [43]. Tens of thousands of years of Aboriginal land management can be described as 'sustainable' in that it has resulted in a productive and sustaining relationship between humans and their environment [100]. An integrated landscape approach seeks to understand the relationship between diverse values, which requires a transdisciplinary approach, that incorporates the ecological, economic, social, and cultural considerations with people from diverse cultural, educational and philosophical backgrounds [65].

5. Conclusions

The agricultural sector faces significant challenges now and into the future, with changing climates and a rapidly growing global population. Historically, indigenous agricultural systems, in their many forms, have accounted for the food security and livelihoods of many millions of people globally. In Australia, wild harvested foods continue to make an important contribution to Aboriginal livelihoods, health and wellbeing, and provide economic opportunity where remoteness, education and infrastructure allow few alternatives [131].

Aboriginal Australians are major landowners across northern Australia with strong cultural connections and intricate knowledge of their land and the plants and animals within. Many groups still rely on customary harvest for their livelihoods and use these products for a diverse range of nutritional, medicinal and cultural purposes. A north Australian native plant, *T. ferdinandiana*, which provides customary food and medicine, is currently being commercially wild harvested, enrichment cultivated and horticulturally grown by Aboriginal people. We conclude that a scaling up of customarily harvested products, such as Kakadu plum, is both desired by Aboriginal people and an appropriate alternative agricultural paradigm to meeting high value niche market demands, thus contributing to global food security by broadening the base of food species used and valued. As markets grow, domestication may become necessary. Models of domestication that promote community participation and local benefit and that protect the natural genetic diversity of a species across its landscape should be prioritized. Investment in regional training is required to support this process [25,121,132]. In parallel with the scaling-up of customary harvest, domestication may allow for greater consistency and volume of supply leading to greater market confidence and more competitive pricing structures.

Wild harvest can be conducted with minimal impact on the surrounding environment and the many opportunities that exist for scaling up the supply of native plant products that have market demand, must be explored. This will require further surveys and research to identify commercial potential through bioprospecting and incorporation of Aboriginal knowledge. There are already many species being harvested for customary purposes by Australian Aboriginal people that may have commercial potential [4,61,62]. We posit that globally there are many other similar customary harvested products found on other natural landscapes that could be harvested sustainably to meet market demands and contribute to food security.

This approach to food production would be best managed at a landscape scale. This would draw on concepts from landscape ecology to ensure sustainable production and protection of the multifunctionality of landscapes, through the conservation of all the important landscape values (ecological, cultural and social). North Australian landscapes are expansive and largely intact due to their poor soils, remoteness from markets and highly seasonal rainfall. There have been several failed attempts to 'develop the north' with the planning being largely influenced by external factors with little regard to Aboriginal aspirations [65] or local conditions [63]. Landscapes across northern Australia have been managed for thousands of years by Aboriginal custodians and are rich in both natural and cultural heritage values. Such holistic and integrated landscapes should not be compromised by inappropriate agricultural models. Aboriginal custodians have obligation to 'care for country' and in most cases, these obligations exclude large scale clearing and development. An alternative development practice is a 'cultural approach' to land use which is holistic in nature, building on customary lore and practice to service new and expanding markets that value sustainable practice and organic, ethically sourced foods.

This idea of holistic planning for multifunctional landscapes is not new and has been supported by several authors [65,131,133]. Valuing ecosystem services, Aboriginal knowledge, economic values of native plants, and managing for multiple values, are common threads of an integrated landscape approach. Many of the landscapes across northern Australia are "undeveloped", have relatively intact ecosystem services, along with people who have maintained cultural connectedness, albeit to varying degrees. This is an ideal time to prevent loss of important multifunctionality by taking an integrated and holistic landscape approach to land management and economic development in this region [134]. This can help to capture the ecological, economic, social and cultural values that are inherent in these landscapes and plan a shared vision for the future that provides sustainable livelihoods for current and future generations resident in these northern landscapes [135]. This vision of landscape as an agricultural system also prioritizes the retention of benefit locally, in an economically poor but culturally and biologically diverse landscape. We need to effectively conserve ecological and cultural integrity whilst creating novel future farming systems while we have them, rather than trying to apply these principles to landscapes which have been altered and no longer have Aboriginal cultural connections.

This paper has introduced the concept of an agriculture paradigm which is compatible with the concept of a 'customary economy' that Aboriginal people aspire to develop [94,136]. Future agribusiness models in these landscapes should build on the existing customary knowledge to create an innovative and appropriate agricultural paradigm which will contribute towards Aboriginal livelihoods and global food security issues in a changing world.

Author Contributions: Conceptualization, J.G., D.P. and P.W.; methodology, J.G.; investigation, J.G.; writing—original draft preparation, J.G.; writing—review and editing, D.P. and P.W.; supervision, P.W. All authors have read and agreed to the published version of the manuscript.

Funding: This research received no external funding and in part of a PhD.

Acknowledgments: Charles Darwin University for hosting this PhD research. The people and organizations in the Thamarrurr Region that have shared their experiences over many years. Chris Brady for his supervision and providing comments on a draft.

Conflicts of Interest: The authors declare no conflict of interest.

References

1. Food and Agriculture Organization of the United Nations (FAO). *The Future of Food and Agriculture—Trends and Challenges*; FAO: Rome, Italy, 2017.
2. Campbell, B.M.; Vermeulen, S.J.; Aggarwal, P.K.; Corner-Dolloff, C.; Girvetz, E.; Loboguerrero, A.M.; Ramirez-Villegas, J.; Rosenstock, T.; Sebastian, L.; Thornton, P.; et al. Reducing risks to food security from climate change. *Glob. Food Secur.* **2016**, *11*, 34–43. [CrossRef]
3. Lockie, S. Australia's Agricultural Future: The Social and Political Context. In *Report to SAF07—Australia's Agricultural Future Project*; Australian Council of Learned Academies: Melbourne, Australia, 2015.
4. TEEB (The Economics of Ecosystems and Biodiversity). *TEEB for Agriculture & Food: Scientific and Economic Foundations*; UN Environment: Geneva, Switzerland, 2018.
5. Bharucha, Z.; Pretty, J. The roles and values of wild foods in agricultural systems. *Philos. Trans. R. Soc.* **2010**, *365*, 2913–2926. [CrossRef] [PubMed]
6. Jaenicke, H.; Höschle-Zeledon, I. (Eds.) *Strategic Framework for Underutilized Plant Species Research and Development*; Colombo and Global Facilitation Unit for Underutilized Species; ICUC: Rome, Italy, 2006.
7. Porter, J.R.; Xie, L.; Challinor, A.J.; Cochrane, K.; Howden, M.; Iqbal, M.M.; Lobell, D.B.; Travasso, M.I. Chapter 7: Food security and Food Production Systems. In *Climate Change 2014: Impacts, Adaptation and Vulnerability; Working Group II Contribution to the IPCC Fifth Assessment Report of the Intergovernmental Panel on Climate Change*; Field, C.B., Barros, V.R., Dokken, D.J., Mach, K.J., Mastrandrea, M.D., Bilir, T.E., Chatterjee, M.,

Ebi, K.L., Estrada, Y.O., Genova, R.C., et al., Eds.; Cambridge University Press: Cambridge, UK; New York, NY, USA, 2014; pp. 485–533.

8. Byrne, P.F.; Volk, G.M.; Gardner, C.; Gore, M.A.; Simon, P.W.; Smith, S. Sustaining the future of plant breeding: The critical role of the USDA-ARS National Plant Germplasm System. *Crop Sci.* **2018**, *58*, 451–468. [CrossRef]
9. Khoury, C.K.; Bjorkman, A.D.; Dempewolf, H.; Ramirez-Villegas, J.; Guarino, L.; Jarvis, A.; Rieseberg, L.H.; Struik, P.C. Increasing homogeneity in global food supplies and the implications for food security. *Proc. Natl. Acad. Sci. USA* **2014**, *111*, 4001–4006. [CrossRef] [PubMed]
10. Hendre, P.S.; Muthemba, S.; Kariba, R.; Muchugi, A.; Fu, Y.; Chang, Y.; Song, B.; Liu, H.; Liu, M.; Liao, X.; et al. African Orphan Crops Consortium (AOCC): Status of developing genomic resources for African orphan crops. *Planta* **2019**, *250*, 989–1003. [CrossRef] [PubMed]
11. World Health Organization (WHO). Obesity and Overweight. Fact Sheet 311. 2012. Available online: http://www.who.int/mediacentre/factsheets/fs311/en/ (accessed on 24 February 2020).
12. IPCC. 2019. Available online: https://www.ipcc.ch/2019/ (accessed on 25 February 2020).
13. United Nations Paris Agreement. 2015. Available online: https://unfccc.int/process-and-meetings/the-paris-agreement/the-paris-agreement (accessed on 25 February 2020).
14. Cenacchi, N.; Sulser, T. Effects of Rising CO_2 on Protein, Iron and Zinc Availability in Global Diets. 18 July 2019. Available online: https://www.ifpri.org/blog/effects-rising-co2-protein-iron-and-zinc-availability-global-diets (accessed on 17 March 2020).
15. Klitgaard, K. Sustainability as an Economic Issue: A BioPhysical Economic Perspective. *Sustainability* **2020**, *12*, 364. [CrossRef]
16. Pimentel, D. Environmental and economic costs of the application of pesticides primarily in the United States. *Environ. Dev. Sustain.* **2005**, *7*, 229–252. [CrossRef]
17. Hansen, M.C.; Stehman, S.V.; Potapov, P.V. Quantification of global gross forest cover loss. *Proc. Natl. Acad. Sci. USA* **2010**, *107*, 8650–8655. [CrossRef]
18. Foley, J.A.; Ramankutty, N.; Brauman, K.A.; Cassidy, E.S.; Gerber, J.S.; Johnston, M.; Mueller, N.D.; O'Connell, C.; Ray, D.K.; West, P.C.; et al. Solutions for a cultivated planet. *Nature* **2011**, *478*, 337–342. [CrossRef]
19. Galluzzi, G.; Van Duijvendijk, C.; Collette, L.; Azzu, N.; Hodgkin, T. *Biodiversity for Food and Agriculture: Contributing to Food Security and Sustainability in a Changing World*; PAR Platform; FAO: Rome, Italy, 2011.
20. Lewis, S.L.; Maslin, M.A. Defining the anthropocene. *Nature* **2015**, *519*, 171–180. [CrossRef]
21. Ceballos, G.; Ehrlich, P.; Barnosky, A.D.; Garcıa, A.; Pringle, R.M.; Palmer, T.M. Accelerated modern human—Induced species losses: Entering the sixth mass extinction. *Sci. Adv.* **2015**, *1*, e1400253. [CrossRef] [PubMed]
22. Hazell, P.; Woods, S. Drivers of change in global agriculture. *Philos. Trans. R. Soc. Lond. B. Biol. Sci.* **2008**, *363*, 495–515. [CrossRef] [PubMed]
23. International Assessment of Agricultural Knowledge, Science and Technology for Development (IAASTD). Agriculture at a Crossroads. In *International Assessment of Agricultural Knowledge, Science and Technology for Development Global Report*; Island Press: Washington, DC, USA, 2009.
24. De Schutter, O.F. *Report Submitted by the Special Rapporteur on the Right to Food*; Human Rights Council Sixteenth Session; Agenda item 3 A/HRC/16/49; UN General Assembly, 2010; Available online: https://www2.ohchr.org/english/issues/food/docs/A-HRC-16-49.pdf (accessed on 23 February 2020).
25. Leakey, R.B. A holistic approach to sustainable agriculture: Trees, science and global society. In *Agroforestry for Sustainable Agriculture*; Mosquera-Losada, M.R., Prabhu, R., Eds.; Burleigh Dodds Science Publishing: Cambridge, UK, 2019.
26. Leakey, R.R.B. Greatly enhanced food security in Africa without high-level trade-offs: Multiple environmental, economic and social benefits from 'Socially Modified Crops'. *Food Secur.* **2018**, *10*, 505–524. [CrossRef]
27. Kremen, C.; Iles, A.; Bacon, C. Diversified farming systems: An agroecological, systems-based alternative to modern industrial agriculture. *Ecol. Soc.* **2012**, *7*, 44. [CrossRef]
28. Petersen, B.; Snapp, S.S. What is sustainable intensification: Views from experts. *Land Use Policy* **2015**, *46*, 1–10. [CrossRef]
29. Smith, A.; Snapp, S.; Chikowoa, R.; Thorneb, P.; Bekundac, M.; Gloverd, J. Measuring sustainable intensification in smallholder agroecosystems: A review. *Glob. Food Secur.* **2017**, *12*, 127–138. [CrossRef]

30. Wezel, A.; Bellon, S.; Doré, T.; Francis, T.; Vallod, T.; David, C. Agroecology as a science, a movement and a practice: A review. *Agron. Sustain. Dev.* **2009**, *29*, 503–515. [CrossRef]
31. Leakey, R.R.B.; Tomich, T.P. Domestication of tropical trees: From biology to economics and policy. In *Agroforestry in Sustainable Ecosystems*; CRC Press and Lewis Publishers: New York, NY, USA, 1999; pp. 319–338.
32. Food and Agriculture Organization of the United Nations (FAO). *Rome Declaration on World Food Security*; World Food Summit, 13–17 November 2016; FAO: Rome, Italy, 1996.
33. Eriksen, P.J.; Ingram, J.S.I.; Liverman, D.M. Food security and global environmental change. *Environ. Sci. Policy* **2009**, *12*, 373–377. [CrossRef]
34. Koohafkan, P.; Altieri, M.A. *Globally Important Agricultural Heritage Systems a Legacy for the Future*; Food and Agriculture Organization of the United Nations: Rome, Italy, 2011.
35. Shackleton, C.M. Multiple roles of non-timber forest products in ecologies, economies and livelihoods. In *Routledge Handbook of Forest Ecology*; Peh, K., Corlett, R., Bergeron, Y., Eds.; Routledge: London, UK, 2015; pp. 559–570.
36. Newman, D.J.; Cragg, G.M. Natural products as sources of new drugs over the last 25 years. *J. Nat. Prod.* **2007**, *70*, 461–477. [CrossRef]
37. Walsh, G. *Biopharmaceuticals: Biochemistry and Biotechnology*, 3rd ed.; Wiley: Chichester, UK, 2003.
38. Belcher, B.M. What isn't a NTFP? *Int. For. Rev.* **2003**, *5*, 161–168. [CrossRef]
39. WWF. Poverty and Conservation. 2018. Available online: http://wwf.panda.org/our_work/people/people_and_conservation/wwf_social_policies/poverty_and_conservation/ (accessed on 5 December 2019).
40. Scoones, I.; Nelnyk, M.; Pretty, J.N. *The Hidden Harvest: Wild Foods and Agricultural Systems. A Literature Review and Annotated Bibliography*; International Institute for the Environment and Development: London, UK, 1992; p. 245.
41. Cunningham, A.B. *Applied Ethnobotany: People, Wild Plant Use and Conservation*; Earthscan Publications Ltd.: London, UK, 2001; p. 300.
42. Bowler, J.M.; Johnston, H.; Olley, J.M.; Prescott, J.R.; Roberts, R.G.; Shawcross, W.; Spooner, N.A. New ages for human occupation and climatic change at Lake Mungo, Australia. *Nature* **2003**, *412*, 837–840. [CrossRef] [PubMed]
43. Rose, D.B. *Nourishing Terrains—Australian Aboriginal Views of Landscape and Wilderness*; Australian Heritage Commission: Canberra, Australia, 1996.
44. Lindsay, B.Y.; Waliwararra, K.; Milijat, F.; Kuwarda, H.; Pirak, R.; Muyung, A.; Pambany, E.; Marryridj, J.; Marrfurra, P.; Wightman, G. *MalakMalak and Matngala Plants and Animals—Aboriginal Flora and Fauna Knowledge from the Daly River Area, Northern Australia*; Conservation Commission of the Northern Territory: Darwin, Australia, 2001.
45. Puruntatameri, J.; Puruntatameri, R.; Pangiraminni, A.; Burak, L.; Tipuamantymirri, C.; Tipakalippa, M.; Puruntatameri, J.; Puruntatameri, P.; Pupangamirri, J.B.; Kerinaiua, R.; et al. *Tiwi Plants and Animals: Aboriginal Flora and Fauna Knowledge from Bathurst and Melville Islands, Northern Australia*; Parks and Wildlife Commission of the Northern Territory and Tiwi Land Council: Darwin, Australia, 2001.
46. Woodward, E.; Jarvis, D.; Maclean, K. *The Traditional Owner-Led Bush Products Sector: An Overview*; CSIRO: Canberra, Australia, 2019.
47. Lee, L.S.; Courtenay, K. Enrichment plantings as a means of enhanced bush food and bush medicine plant production in remote arid regions: Are view and status report. *Learn. Communities Int. J. Learn. Soc. Contexts Spec. Issue Synth. Integr.* **2016**, *19*, 64–75. [CrossRef]
48. Gorman, J.; Brady, C.; Courtenay, K. Commercialisation of Terminalia ferdinandiana Exell. in northern Australia. In *Australian Native Plants: Cultivation and Uses in Alternative Medicine and the Food Industry*; Sultanbawa, Y., Sultanbawa, F., Eds.; CRC Press: Boca Raton, FL, USA, 2016; pp. 89–104.
49. Tapsell, L.G.; Williams, P.; Droulez, V.; Southee, D.; Patch, C.; Lethbridge, A. *Functional foods for the Australian Industry*; National Centre for Excellence in Functional Food, University of Wollongong: New South Wales, Australia, 2006.
50. Grand View Research. Functional Foods Market Size, Share & Trends Analysis Report by Ingredient (Carotenoids, Prebiotics & Probiotics, Fatty Acids, Dietary Fibres), by Product, by Application, and Segment Forecasts. 2019–2025. Available online: https://www.grandviewresearch.com/industry-analysis/functional-food-market (accessed on 24 February 2019).

51. Chapman, A.D. *Numbers of Living Species in Australia and the World*, 2nd ed.; Report for the Australian Biological Resources Study; Canberra, Australia, September 2009; ISBN 978 642 56860 1. Available online: https://www.environment.gov.au/system/files/pages/2ee3f4a1-f130-465b-9c7a-79373680a067/files/nlsaw-2nd-complete.pdf (accessed on 25 March 2020).
52. Bandon, P. *Useful Bush Plants*; Western Australian Museum: Perth, Australia, 1996.
53. Clarke, M. *Australian Native Food Stocktake*; RIRDC No.12/066; RIRDC: Canberra, Australia, 2012.
54. Graham, C.; Hart, D. *Prospects for the Australian Native Bushfood Industry*; RIRDC No.97/022; RIRDC: Canberra, Australia, 1997.
55. Lands, M. *Mayi: Some Bush Fruits of Dampierland*; Magabala Books; Kimberley Aboriginal Law and Culture Centre: Broome, Australia, 1987.
56. Sultanbawa, Y.; Sultanbawa, F. *Australian Native Plants: Cultivation and Uses in Alternative Medicine and the Food Industry*; CRC Press: Boca Raton, FL, USA, 2016.
57. White, J. Indigenous Peoples' Livelihoods and Emerging Bush Produce Industries—Recent Experiences from Australia's Arid Zone. Ph.D. Thesis, University of South Australia, Adelaide, Australia, 2012.
58. Whitehead, P.J.; Gorman, J.; Griffiths, A.D.; Wightman, G.; Massarella, H.; Altman, J. *Small Scale Commercial Plant Harvests by Indigenous Communities*; A Report for the RIRDC/Land and Water Australia/FWPRDC/MDBC Joint Agroforestry Program; RIRDC: Barton, Australia, 2006.
59. Fordham, A.; Fogarty, W.; Fordham, D.A. *The Viability of Wildlife Enterprises in Remote Indigenous Communities of Australia: A Case Study*; CAEPR Working Paper No. 63; Australian National University: Canberra, Australia, 2010.
60. Fleming, A.E.; Petheram, L.; Stacey, N. Australian indigenous women's seafood harvesting practices and prospects for integrating aquaculture. *J. Enterp. Communities People Places Glob. Econ.* **2015**, *9*, 156–181. [CrossRef]
61. Gorman, J.; Whitehead, P.J.; Griffiths, A.D. An Analysis of the Use of Plant Products for Commerce in Remote Aboriginal Communities of Northern Australia. *Econ. Bot.* **2006**, *60*, 362–373. [CrossRef]
62. Gorman, J.; Griffiths, A.D.; Whitehead, P.J.; Petheram, L. Production from marginal lands: Indigenous commercial use of wild animals in northern Australia. *Int. J. Sustain. Dev. Wildlife Ecol.* **2008**, *15*, 240–250. [CrossRef]
63. Woinarski, J.; Dawson, F. Limitless Lands and Limited Knowledge: Coping with Uncertainty and Ignorance in Northern Australia. In *Ecology, Uncertainty and Policy: Managing Ecosystems for Sustainability*; Handmer, J.W., Norton, T.W., Dovers, S.R., Eds.; Prentice-Hall: New York, NY, USA, 2002; pp. 83–115.
64. Fitzhardinge, G. Australia's rangelands: A future vision. *Rangel. J.* **2012**, *34*, 33–45. [CrossRef]
65. Pearson, D.M.; Gorman, J. Exploring the relevance of a landscape ecological paradigm for sustainable landscapes and livelihoods: A case study from the Northern Territory Australia. *J. Landsc. Ecol.* **2010**, *25*, 1169–1183. [CrossRef]
66. Pretty, J.; Toulmin, C.; Williams, S. Sustainable intensification in African agriculture. *Int. J. Agric. Sustain.* **2011**, *9*, 5–24. [CrossRef]
67. Leakey, R.R.B.; Prabhu, R. Towards multifunctional agriculture—An African initiative. In *Leakey RRB, Multifunctional Agriculture: Achieving Sustainable Development in Africa*; Academic Press: San Diego, CA, USA, 2017; pp. 393–414.
68. Gorman, J.; Brady, C.; Clancy, T. *Management Program for Terminalia ferdinandiana in the Northern Territory of Australia 2018–2022*; Northern Territory Department of Environment and Natural Resources: Darwin, Australia, 2018.
69. Kawulich, B.B. Participant observation as a data collection method. *Forum Qual. Soc. Res.* **2005**, *6*. [CrossRef]
70. Wheeler, J.R. Family 81 Combretaceae. In *Flora of the Kimberley Region*; Wheeler, J.R., Ed.; Western Australian Herbarium, Department of Conservation and Land Management: Perth, Australia, 1992.
71. Dunlop, C.R.; Leach, G.J.; Cowie, I.D. *Flora of the Darwin Region*; Northern Territory Botanical Bulletin No. 20; Conservation Commission of the Northern Territory: Darwin, Australia, 1995; Volume 2.
72. Williams, D.J.; Edwards, D.; Pun, S.; Chaliha, M.; Sultanbawa, Y. Profiling ellagic acid content: The importance of form and ascorbic acid levels. *Food Res. Int.* **2014**, *66*, 100–106. [CrossRef]
73. Brand, J.C.; Cherikoff, V.; Lee, A. An Outstanding Food Source of Vitamin C. *Lancet* **1982**, *2*, 873. [CrossRef]
74. Miller, J.B.; James, K.W.; Maggiore, P.M. *Tables of Composition of Australian Aboriginal Foods*; Aboriginal Studies Press: Canberra, Australia, 1993; p. 256.

75. Mohanty, S.; Cock, I.E. The chemotherapeutic potential of *Terminalia ferdinandiana*: Phytochemistry and bioactivity. *Pharm. Rev.* **2012**, *6*, 29–36.
76. Ohno, Y.; Fukuda, K.; Takemura, G.; Toyota, M.; Watanabe, M.; Yasuda, N.; Xinbin, Q.; Maruyama, R.; Akao, S.; Gotou, K.; et al. Induction of apoptosis by gallic acid in lung cancer cells. *Anti-Cancer Drug* **1999**, *10*, 845–851. [CrossRef] [PubMed]
77. Sirdaarta, J.; Matthews, B.; Cock, I.E. Kakadu Plum fruit extracts inhibit growth of the bacterial triggers of rheumatoid arthritis: Identification of stilbene and tannin components. *J. Funct. Foods* **2015**, *17*, 610–620. [CrossRef]
78. Tan, A.C.; Konczak, I.; Ramzan, I.; Sze, D.M.-Y. Potential antioxidant, anti-inflammatory, and proapoptotic anticancer activities of Kakadu Plum and illawarra plum polyphenolic fractions. *Nutr. Cancer* **2011**, *63*, 1074–1084. [CrossRef] [PubMed]
79. Willet, W.C. Balancing life-style and genomics research for disease prevention. *Science* **2002**, *296*, 695–698. [CrossRef] [PubMed]
80. Cock, I.E.; Mohanty, S. Evaluation of the antibacterial activity and toxicity of Terminalia ferdinandia fruit extracts. *Pharmacogn. Mag* **2011**, *3*, 72–79. [CrossRef]
81. Sultanbawa, Y.; Chaliha, M.; Cusack, A.; Edwards, D.; Williams, D. *Monitoring Quality and Bioactivity of Kakadu Plum in the Northern Territory*; AgriFutures Australia Publication No. 18/024; AgriFutures: Wagga Wagga, Australia, 2018.
82. Ferguson, M.; Brown, C.; Georga, C.; Miles, E.; Wilson, A.; Brimblecombe, J. Traditional food availability and consumption in remote Aboriginal communities in the Northern Territory, Australia. Australian and New Zealand. *J. Public Health* **2017**, *41*, 294–298.
83. Raymond, E.; Blutja, J.; Gingina, L.; Raymond, M.; Raymond, O.; Raymond, L.; Brown, J.; Morgan, Q.; Jackson, D.; Smith, N.; et al. *Wardaman Ethnobiology*; Government Printer of the Northern Territory: Darwin, Australia, 1999.
84. Brock, J. *Top End Native Plants: A Comprehensive Guide to the Trees and Shrubs of the Top End of the Northern Territory*; Reed New Holland: Sydney, Australia, 2001.
85. Isaacs, J. *Bush Food, Aboriginal Food and Herbal Medicine*; Ure Smith Press: Sydney, Australia, 1987.
86. Cunningham, A.B.; Courtenay, K.; Gorman, J.T.; Garnett, S. Eco-enterprises and Kakadu Plum (*Terminalia ferdinandiana*): "best laid plans" and Australian policy lessons. *Econ. Bot.* **2009**, *63*, 16–23. [CrossRef]
87. Gorman, J.; Bentivoglio, M.; Brady, C.; Wurm, P.; Vemuri, S.; Sultanbawa, Y. Complexities in developing Australian Aboriginal enterprises based on natural resources. *Rangel. J.* **2020**. (under review).
88. Bawinanga Aboriginal Corporation. Available online: https://www.bawinanga.com/ (accessed on 25 February 2020).
89. Twin Lakes Cultural Park. Available online: http://www.twinlakes.net.au/ (accessed on 25 February 2020).
90. Kimberley Wild Gubinge. Available online: https://www.kimberleywildgubinge.com.au/about-us (accessed on 25 February 2020).
91. Lombadina. Available online: http://www.lombadina.com/ (accessed on 24 February 2020).
92. Mayi Harvests. Available online: https://www.mayiharvests.com.au/ (accessed on 24 February 2020).
93. Kakadu Life. Available online: https://kakadulife.com/ (accessed on 24 February 2020).
94. Bidyadanga. Available online: http://www.bidyadanga.org.au/ (accessed on 24 February 2020).
95. Mamabulanjin Aboriginal Corporation. Available online: http://mama.org.au/ (accessed on 24 February 2020).
96. Fowler, C.; Stanley, M.; Hudson, R. Centre-Pivot Irrigation Leading Pastoral Industry Attempts to Drought-Proof WA's North. Available online: https://www.abc.net.au/news/rural/2019-11-17/wa-pastoralists-attempt-to-drought-proof-using-centre-pivots/11663698 (accessed on 17 March 2020).
97. Atlas of Living Australia. Mapping and Analysis. Available online: https://www.ala.org.au/mapping-and-analysis/ (accessed on 5 January 2020).
98. Jones, R. Firestick farming. *Aust. Nat. Hist.* **1969**, *16*, 224–231.
99. Gerritsen, R. Evidence for Indigenous Agriculture. *Aust. Sci.* **2010**, *31*, 35–37.
100. Gammage, B. *The Biggest Estate on Earth: How Aborigines Made Australia*; Allen and Unwin: Sydney, Australia, 2011.
101. Pascoe, B. *Dark emu: Black seeds: Agriculture or Accident?* Magabala Books: Broome, Australia, 2014.
102. Altman, J.C. *Sustainable Development Options on Aboriginal land: The Hybrid Economy in the Twenty-First Century*; Discussion Paper No. 226; Centre for Aboriginal Economic Policy Research, Australian National University: Canberra, Australia, 2001.

103. Buchanan, G. Hybrid economy research in remote Indigenous Australia: Seeing and supporting the customary in community food economies. *Local Environ.* **2014**, *19*, 10–32. [CrossRef]
104. Vemuri, S.; Gorman, J. Poverty Alleviation in Indigenous Australia. *Int. J. Environ. Cult. Econ. Soc. Sustain.* **2012**, *8*, 19–36. [CrossRef]
105. Morrison, J. *Managing for the Future: A Culture-Based Economy for Northern Australia*; Charles Darwin Symposium Series; Charles Darwin University: Darwin, Australia, 11 May 2006.
106. Altman, J.C. *Hunter-Gatherers Today: An Aboriginal Economy in Northern Australia*; Australian Institute of Applied Sciences: Canberra, Australia, 1987.
107. Griffiths, A.D.; Philips, A.; Godjuwa, C. Harvest of Bombax ceiba for the Aboriginal arts industry, central Arnhem Land, Australia. *Biol. Conserv.* **2003**, *113*, 295–305. [CrossRef]
108. Vardon, M. Aboriginal wildlife use in the Top End of the Northern Territory. In *Veterinary Conservation Biology: Wildlife Health and Management in Australasia Conference*; Taronga Park Zoo: Sydney, Australia, 2001.
109. Altman, J.C. Economic development and Indigenous Australia: Contestations over property, institutions and ideology. *J. Agric. Resour. Econ.* **2004**, *48*, 513–534. [CrossRef]
110. Australian Institute of Health and Welfare (AIHW). *The Health and Welfare of Australia's Aboriginal and Torres Strait Islander Peoples*; Cat. No. IHW 147; AIHW: Canberra, Australia, 2015. Available online: http://www.aihw.gov.au/publication-detail/?id=60129550168 (accessed on 5 December 2018).
111. Abbott, T. Grassroots Capitalism. In Proceedings of the Indigenous Employment Conference, Canberra, Australia, 25 September 2002.
112. Gorman, J.; Brady, C.; Sultanbawa, Y. 'Looking after country' through wildlife utilisation. In *Advances in Conservation through Sustainable Use of Wildlife: Proceedings of the Sustainable Use Conference, Brisbane, Australia, 30 August–1 September 2016*; Baxter, G., Finch, N., Murray, P., Eds.; Wildlife Science Unit, School of Agriculture and Food Sciences, University of Queensland: Gatton, Australia, 2018; pp. 235–239.
113. Bodle, K.; Brimble, M.; Weaven, S.; Frazer, L.; Blue, L. Critical success factors in managing sustainable indigenous businesses in Australia. *Pac. Account. Rev.* **2018**, *30*, 35–51. [CrossRef]
114. Cahn, M. Indigenous entrepreneurship, culture and microenterprise in the Pacific Islands: Case studies from Samoa. *Entrep. Reg. Dev.* **2008**, *20*, 1–18. [CrossRef]
115. Nikolakis, W. Barriers to Indigenous enterprise development on community owned land. *Glob. Bus. Econ. Rev.* **2010**, *12*, 85–99. [CrossRef]
116. Flamsteed, K.; Golding, B. *Learning through Indigenous Business: The Role of Vocational Education and Training in Indigenous Enterprise and Community Development*; NCVER: Adelaide, Australia, 2005.
117. Evans, L. Opportunities, impediments and capacity building for enterprise development by Australian Aboriginal communities. *JAED* **2007**, *5*, 44–55.
118. Arthur, W. *What's new? The 1997 Parliamentary Inquiry into Indigenous Business*; CAEPR Discussion Paper No 177/1999; Centre for Aboriginal Economic Policy Research; Australian National University: Canberra, Australia, 1999.
119. NAAKPA. North Australian Aboriginal Kakadu Plum Alliance. Available online: https://naakpa.com.au/ (accessed on 25 February 2020).
120. Zeder, M.A. Central questions in the domestication of plants and animals. *Evol. Anthropol.* **2006**, *15*, 105–117. [CrossRef]
121. Leakey, R.R.B. *Living with the Trees of Life—Towards the Transformation of Tropical Agriculture*; CABI: Wallingford, UK, 2012; p. 200.
122. Leakey, R.R.B.; Page, T. The 'ideotype concept' and its application to the selection of 'AFTP' cultivars. *For. Trees Livelihoods* **2006**, *16*, 5–16. [CrossRef]
123. Leakey, R.R.B.; Akinnifesi, F.K. Towards a domestication strategy for indigenous fruit trees in the tropics. In *Indigenous Fruit Trees in the Tropics: Domestication, Utilization and Commercialization*; Akinnifesi, F.K., Leakey, R.R.B., Ajayi, O.C., Sileshi, G., Tchoundjeu, Z., Matakala, P., Kwesiga, F.R., Eds.; CAB International: Wallingford, UK, 2008; pp. 28–49.
124. Jacobson, E.T.; Wark, D.B.; Arnott, R.G.; Haas, R.J.; Tober, D.A. Sculptured Seeding, an ecological approach to revegetation. *Restor. Manag. Notes* **1994**, *12*, 46–50.
125. Country Needs People. Available online: https://www.countryneedspeople.org.au/ (accessed on 3 March 2020).

126. Yibarbuk, D.; Whitehead, P.J.; Russell-Smith, J.; Jackson, D.; Godjuwa1, C.; Fisher, A.; Cooke, P.; Choquenot, D.; Bowman, D.M.J.S. Fire ecology and Aboriginal land management in central Arnhem Land, northern Australia: A tradition of ecosystem management. *J. Biogeogr.* **2001**, *28*, 325–343. [CrossRef]
127. Bradley, J. Landscapes of the mind, landscapes of the spirit: Negotiating a sentient landscape. In *Working on Country: Contemporary Indigenous Management of Australia's Lands and Coastal Regions*; Baker, R., Davies, J., Young, E., Eds.; Oxford University Press: Melbourne, Australia, 2001.
128. Parks and Wildlife Commission for the Northern Territory (PWCNT). *A Strategy for Conservation through the Sustainable Use of Wildlife in the Northern Territory of Australia*; Parks and Wildlife Commission of the Northern Territory: Darwin, Australia, 1997.
129. Gorman, J.; Wurm, P.A.S.; Vemuri, S.; Brady, C.; Sultanbawa, Y. Kakadu Plum (*Terminalia ferdinandiana*) as a Sustainable Indigenous Agribusiness. *Econ. Bot.* **2019**. in print. [CrossRef]
130. Robinson, D.F.; Abdel-Latif, A.; Roffe, P. (Eds.) *Protecting Traditional Knowledge: The WIPO Intergovernmental Committee on Intellectual Property and Genetic Resources, Traditional Knowledge and Folklore*; Taylor & Francis: London, UK, 2017.
131. Burgess, C.; Johnston, F.; Berry, H.; McDonnell, J.; Yibarbuk, D.; Gunabarra, C.; Mileran, A.; Bailie, R. Healthy country, healthy people: The relationship between Indigenous health status and "caring for country". *Med. J. Aust.* **2009**, *190*, 567–572. [CrossRef] [PubMed]
132. Leakey, R.R.B. The role of trees in agroecology and sustainable agriculture in the tropics. *Annu. Rev. Phytopathol.* **2014**, *52*, 113–133. [CrossRef]
133. Holmes, J. The Multifunctional Transition in Australia's Tropical Savannas: The Emergence of Consumption, Protection and Indigenous Values. *Geogr. Res.* **2010**, *48*, 265–280. [CrossRef]
134. Pearson, D.M. Landscape ecology: Its role as a trans-disciplinary science for rangeland sustainability. *Rangel. J.* **2013**, *35*, 363–371. [CrossRef]
135. Pearson, D.; Gorman, J.T. Managing landscapes in Australian Northern Territory for sustainability: Visions, issues and solutions. *Futures* **2010**, *42*, 711–742. [CrossRef]
136. Russell-Smith, J.; James, G.; Pedersen, H.; Sangha, K.K. (Eds.) *Sustainable Land Sector Development in Northern Australia: Indigenous Rights, Aspirations and Cultural Responsibilities*; CRC Press: Boca Raton, FL, USA, 2018.

land

MDPI

Article

Leveraging Traditional Agroforestry Practices to Support Sustainable and Agrobiodiverse Landscapes in Southern Brazil

André Eduardo Biscaia Lacerda [1], Ana Lúcia Hanisch [2] and Evelyn Roberta Nimmo [3,*]

1 EMBRAPA Forestry, Colombo 83411-000, Brazil; andre.biscaia@embrapa.br
2 EPAGRI Empresa de Pesquisa Agropecuária e Extensão Rural de Santa Catarina, Canoinhas 89466-500, Brazil; analucia@epagri.sc.gov.br
3 Department of History, Universidade Estadual de Ponta Grossa, Ponta Grossa 84030-900, Brazil
* Correspondence: evelynnimmo@uepg.br; Tel.: +55-41-99211-0645

Received: 11 March 2020; Accepted: 28 May 2020; Published: 1 June 2020

Abstract: Integrated landscape approaches have been identified as key to addressing competing social, ecological, economic, and political contexts and needs in landscapes as a means to improve and preserve agrobiodiversity. Despite the consistent calls to integrate traditional and local knowledge and a range of stakeholders in the process of developing integrated landscape approaches, there continues to be a disconnect between international agreements, national policies, and local grassroots initiatives. This case study explores an approach to address such challenges through true transdisciplinary and multi-stakeholder research and outreach to develop solutions for integrated landscapes that value and include the experience and knowledge of local communities and farmers. Working collaboratively with small-scale agroforestry farmers in Southern Brazil who continue to use traditional agroecological practices to produce erva-mate (*Ilex paraguariensis*), our transdisciplinary team is working to collect oral histories, document local ecological knowledge, and support farmer-led initiatives to address a range of issues, including profitability, productivity, and legal restrictions on forest use. By leveraging the knowledge across our network, we are developing and testing models to optimize and scale-out agroforestry and silvopastoral systems based on our partners' traditional practices, while also supporting the implementation of approaches that expand forest cover, increase biodiversity, protect and improve ecosystem services, and diversify the agricultural landscape. In so doing, we are developing a strong evidence base that can begin to challenge current environmental policies and commonly held misconceptions that threaten the continuation of traditional agroforestry practices, while also offering locally adapted and realistic models that can be used to diversify the agricultural landscape in Southern Brazil.

Keywords: yerba mate; agroforestry; integrated landscape; agrobiodiversity; silvopastoral systems

---

1. Introduction

Several high-level reports from a range of international agencies highlight the need to rethink conventional agricultural systems through innovative and sustainable approaches, including agroecology, forest landscape restoration, and agroforestry, among many others [1–6]. These reports emphasize that business as usual in terms of conventional agriculture is continuing to have lasting negative impacts on agricultural biodiversity, soil health, water and landscape management, greenhouse gas (GHG) emissions, and human health and food security. The United Nation Food and Agricultural Organization's (FAO) recent report on the state of the world's biodiversity for food and agriculture (BFA) argues that "many of the drivers that have negative impacts on BFA, including overexploitation, overharvesting, pollution, overuse of external inputs, and changes in land and water management, are at least partially caused by inappropriate agricultural practices" [2] (p. xxxviii). Meanwhile, Padoch and Sunderland [7] argue that using conventional practices and technologies for sustainable intensification may not necessarily have the desired effects on forest and biodiversity conservation, but rather lead to greater loss of forests and associated ecosystem services, with little or no benefits for some agricultural regions in which small-scale farming is predominant. They highlight that "producing food in diverse, multifunctional landscapes challenges dominant agricultural development paradigms, but it also presents issues and difficulties. For example, many types of integrated landscape approach have not been studied by scientists, and the existing research and policy framework may be insufficiently integrated to improve either agricultural production or environmental protection in such diverse landscapes" [7] (p. 6).

From a landscape ecology perspective, taking a holistic approach to land management planning and modeling is a key aspect of the discipline [8,9]. Understanding the mechanisms and impacts of land use and land cover change (LULCC) over time in farming regions, including forest fragmentation, habitat loss, and human–environment interactions, is crucial to determining the likely impacts of the continuation of conventional agriculture not only at the local scale, but also how this will affect rural and urban human and non-human populations in the long term. In order to support sustainable management and changes to land use and land cover (LULC) that focus on biodiversity conservation, increased ecosystem services and connectivity, as well as human food security and livelihoods, debates have focused on land sharing vs. land sparing as methods to address the competing needs in landscapes [7]. This debate either calls for land to be set aside for conservation with intensive agriculture conducted separately, or land to be shared across a range of goals from food production to biodiversity conservation through less intensive practices such as agroecology [10]. Although many involved in the debate acknowledge that both land set aside for conservation and alternative approaches to agriculture can occur simultaneously, there are few examples of how this might work on a practical level or how to scale up what works on individual farms to address issues of managing sustainable, biodiverse productive landscapes.

One of the methods used in landscape management planning that bridges the divide across the various competing social, ecological, economic, and political contexts and needs in landscapes are integrated landscape approaches. As defined by the Consultative Group on International Agricultural Research (CGIAR) [11], such approaches consider not only multiple land uses (including agriculture and forests) but also the livelihoods dependent on such land uses, moving beyond conventional perceptions of management and governance. It seeks "to provide tools and concepts to identify, understand and address a complex set of environmental, social and political challenges, and to enable evidence-based and inclusive prioritization, decision-making and implementation" [6] (p. 1). Importantly, such an approach highlights stakeholder engagement as key to managing conflicting perceptions of the value and function of land use types in a landscape across a range of scales, from the local to the national [6]. What is important here is that analyzing and developing solutions for integrated landscapes requires a truly transdisciplinary lens, in which a range of researchers and other stakeholders, including local communities and farmers, are actively engaged in the research design, data collection and analysis, implementation, and assessment.

Agroforestry systems have been identified as one of the key approaches that can be implemented in integrated landscape management as they offer a range of ecological and social benefits. As noted in the recent High Level Panel of Experts (HLPE) report, forests contribute extensively to food and nutritional security (FSN) through the "direct provision of food; provision of energy, especially for cooking; income generation and employment; and provision of ecosystem services that are essential for FSN, human health and well-being" [6]. Specifically, the implementation of agroforestry systems offers multifunctional landscapes that support the development of regenerative agricultural systems that offer not only a diverse, multi-layer food production system, but also land use that can restore or conserve ecological resources [12]. The International Union for Conservation of Nature (IUCN) together with the World Resources Institute (WRI) have highlighted agroforestry methods as an important strategy in forest landscape restoration (FLR) to address climate change mitigation and food security issues worldwide [4]. Research on the benefits of agroforestry systems and their ecological, social, economic, and cultural importance have been conducted in a range of contexts around the world (see for example [13,14]).

In Southern Brazil, previous research has shown that forest fragments, including those managed in agroforestry systems, are important havens of biodiversity on the landscape scale, particularly in terms of tree species [15–18], but also act as crucial connectivity corridors that enable genetic flows and buffer the impacts of anthropogenic activities along waterways [19]. Traditional agroforestry systems have continued in the central-south of Paraná state and Northern Santa Catarina state mainly due to the extraction of erva-mate (*Ilex paraguariensis*, also known as yerba mate), a tree species that grows well in the shaded understory of the region's iconic Araucaria Forest. These systems have been important in maintaining ecosystem services and biodiversity corridors, but they are also important to the maintenance of cultural and traditional agroecological practices on small-scale family farms that include a heterogenic mosaic of crops, livestock, vegetable gardens, and productive forest areas, all of which are essential to family and local food security [20].

We understand these traditional systems as outcomes of generations of adaptive practices based on local ecological knowledge (LEK), resource management techniques, and cultural and historic subjectivities [21]. We leverage the premise outlined by Berkes et al. [22] and Fonseca-Cepeda et al. [21] of traditional ecological knowledge (TEK), which Berkes [23] (p. 3) defined as "a cumulative body of knowledge and beliefs, handed down through generations by cultural transmission, about the relationship of living beings (including humans) with one another and with their environment. Further, TEK is an attribute of societies with historical continuity in resource use practices." Although such a concept is often associated with indigenous knowledge paradigms, we consider how such an approach can apply to settler communities that have continual, historical resource use practices, as this knowledge is "cumulative and dynamic, builds on the experience of generations", but also adapts to new technological and socioeconomic realities [24] (p. 281). Nevertheless, these systems have received little research attention, particularly from federal and state agricultural research and outreach institutions, as they tend to be viewed as remnants of outdated, subsistence agricultural practices that require modernization, rather than being valued as systems developed and adapted to the forest environment in which they have continued for generations.

In this paper, we discuss some preliminary results of our ongoing participatory research and outreach project with small-scale traditional agroforestry producers in Southern Brazil and present models being developed to optimize and scale-out agroforestry systems based on our partners' traditional practices. Through the implementation of these models and systems, we are beginning to address some of the main concerns of small-scale farmers, mainly profitability, productivity, and legal restrictions on forest use, while also developing strategies that can be used in landscape management to expand forest cover, increase biodiversity, protect and improve ecosystem services, and diversify the agricultural landscape. Our collaborative approach ensures that the research addresses the needs of communities and is applicable to local realities. In so doing, we are developing a strong evidence base that can begin to challenge current environmental policies and commonly held misconceptions that threaten the continuation of traditional agroforestry practices.

2. Traditional Land Use and Conventional Agriculture

The land use legacy of Paraná state helps to clarify the current LULC in the region and how the continuation of forest resources in some regions has been more pronounced than others, as shown in Figure 1. At the beginning of the nineteenth century, most of Southern Brazil was covered by forests, from the coastal Atlantic Forest through to the sub-tropical Araucaria Forest biome on the central plateaus, and the tropical semi-deciduous forests of the Paraná River basin in the west. Although the forests had been managed by indigenous groups for thousands of years [25] and while there was continued indigenous and settler occupation of these forest landscapes from the sixteenth century [26], forest cover was relatively uniform, interspersed with the naturally occurring grasslands on the higher elevation plains. By the end of the eighteenth century, a process of westward colonization began in Paraná, and to some extent Santa Catarina state, originating from the coastal/eastern region and moving through the region's highlands. This colonization process was characterized by an economy based on cattle husbandry, erva-mate harvesting (a resource that had only recently begun to be exploited in the regional economy, despite its economic importance since early Spanish colonization in the seventeenth century [27]) and the logging of araucaria or Paraná pine (*Araucaria angustifolia*) [28]. More intense colonization did not happen until later, with a migratory influx of Germans, Italians, Poles, and Ukrainians, among others, in the nineteenth and early twentieth centuries, as part of a government policy to occupy the 'unoccupied' hinterlands of the country [29]. This second wave of migration, along with the expansion of the railroad and extensive exploitation of araucaria, led to much more intensive land occupation. Yet, in our study region, many settler communities maintained forest cover on their lands for animal husbandry and erva-mate harvesting, developing farming systems that integrated European and indigenous practices with local crops (corn, manioc, beans, etc.) and forests. Today, forests still occupy significant portions of the landscape, as shown in Figure 1c, in which small-scale farmers continue to use traditional practices that have been passed on for generations.

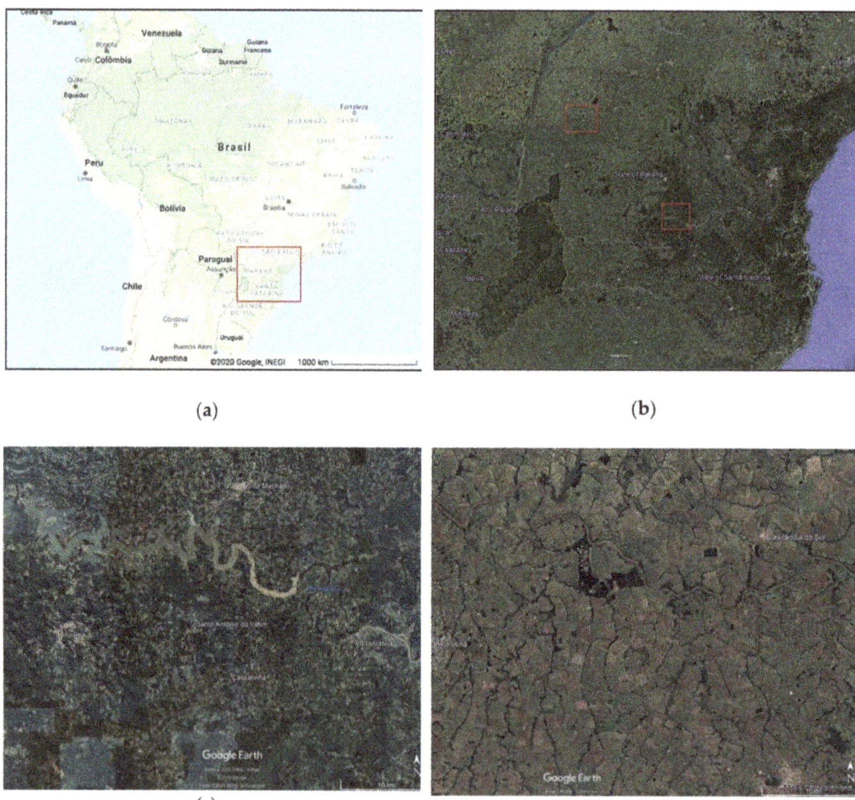

(a) (b) (c) (d)

Figure 1. Forest and land use and land cover (LULC) in Paraná state as a result of differing land use legacies: (**a**) location of the region in Brazil; (**b**) Paraná state with northern and southern regions highlighted; (**c**) land use in southern region with a significant incidence of forest cover; (**d**) land use in northern region mostly covered by soy plantations.

In other areas of Paraná, where forest cover has been almost completely decimated, the process of colonization was much later and with quite a different focus. Beginning in the early to mid-twentieth century, the Southwest was occupied by colonization companies expanding from the southernmost state of Rio Grande do Sul; their economies were linked to a subsistence agriculture based on grains and pork husbandry. Finally, a colonization wave coming from the north (São Paulo state) in search of land for coffee production occupied the north of Paraná [29]. The vast majority of the area originally occupied by the colonization from São Paulo and Rio Grande do Sul is currently covered by large-scale soy farms, as shown in Figure 1d.

3. Methods to Leverage Traditional Agroforestry Practices

Our approach to implementing participatory methods to develop land management planning systems that integrate multiple uses of natural resources in a socially and politically complex context was conceptualized as Locally Adapted Participatory Sustainable Forest Management Systems—lapSFM, outlined by Lacerda et. Al [30]. This approach provides a roadmap for managing rural properties, focusing on forested lands, while the decision-making process includes local ecological knowledge (LEK) as a key input necessary for establishing the goals and objectives and is based on the demands and interests of landowners. As such, in 2011, we began by leveraging long-term agroecological initiatives in place for more than 30 years that involved key partners (Federation of Family Farmers' Unions—FETRAF; Agronomic Institute of Paraná—IAPAR; and the Brazilian Agricultural Research Company—EMBRAPA), and started to conceptualize these productive systems by consolidating communication, information, and activities among farmers, outreach technicians, researchers, and environmental agencies through workshops, field-days, social media, farmers' union meetings, and scientific conferences.

Beginning in 2017, we began a new phase of the project which focused on conducting oral history interviews with traditional erva-mate producers as a means of gaining a better understanding of the historical, social, and cultural aspects of traditional practices, how landscape and environmental changes are perceived by farmers, as well as the socio-political implications of their lived experiences [31]. As Williams and Riley [32] argue, oral history interviews provide an understanding of the ways in which people produce meaning of the places they inhabit, and how they perceive and value the natural world around them. As such, they offer unique perspectives on issues of the environment, forests, and conservation as narratives are situated within the environment in question, grounded in the everyday challenges of rural life. To date, we have conducted interviews with 39 erva-mate producers and members of their families across seven different municipalities in Paraná and Santa Catarina [33].

Across the range of participatory methods employed, participants were encouraged to discuss the challenges faced in conducting agroecological and traditional agroforestry practices on small-scale farms (economic, technical, social, and political) and possible solutions that included creation of co-ops, cooperative/participatory research to deal with gaps in scientific knowledge, youth-focused farming, and increased participation of women, among others. As a result, we have established action plans and models that integrate a range of perspectives and issues in terms of the technical, social, cultural, and economic.

Herein, we discuss two particular outcomes of project to date: optimized LEK-based agroforestry systems; and models of Productive Agroforestry Restoration. These optimized productive systems have been implemented at EMBRAPA Research Station in Caçador (ERSC) and in over 50 properties across more than 20 municipalities in Paraná, Santa Catarina, and Rio Grande do Sul states, for a total of 3000 hectares under management. Both lines of research seek to add value to these systems and the knowledge behind them, while also testing alternatives that do not require high rates of investment or debt, and may offer small-scale farmers the opportunity to transition some of their land from high-input commodities (i.e., corn, tobacco, soy) to other more sustainable products. The dissemination of results and co-creation of knowledge include technical and scientific documents [19,20,34–40], conferences e.g., [41], monthly technical visits to ERSC, and bi-monthly visits to farms across the region.

3.1. *Traditional Agroforestry Optimization*

The diversity of forest management systems based on the traditional use of erva-mate reflect the variations in the natural environmental that led to an extensive accumulation of LEK over generations. Forest structure, tree diversity, presence of dominant or invasive species, and land use history and legacy are all integral factors that play a role in the decision-making process related to how forests are managed for erva-mate production and have been described in depth by Mattos [42], Chaimsohn and Souza [18], Marques [43], and Hanisch [44]. In most cases, forest structure and diversity are gradually managed, aiming at a spatial distribution that favors an understory with a homogeneous light availability considered empirically as ideal for erva-mate development. The intensity and frequency of forest interventions depend on various factors that include forest development (successional stage), historic and contemporary use, presence of dominant or invasive species, and natural occurrence of erva-mate trees, among others. Despite the fact that traditional agroforestry systems in Southern Brazil in most cases have erva-mate as one of the key products, practices vary widely between properties and municipalities, with the presence of cattle husbandry as one of the most significant characteristics that differentiates the systems in the region [20].

3.1.1. Agrisilvicultural Systems

Erva-mate production occurs across a range of forest successional stages, from well-developed (late successional) forest stands relying mainly on native, naturally regenerating trees, to younger, secondary forests that rely more heavily on planting and silvicultural management. Well-developed forest stands typically have lower density (number of trees), higher diversity with long-living species (i.e., *Araucaria angustifolia* and *Ocotea porosa*), with a more complex structure (trees with various sizes distributed in forest layers), while younger forests commonly have a much simpler structure (homogeneous sizes), lower levels of diversity with short-lived species (i.e., *Mimosa scabrella*) but with much greater density. Research has shown that agroforestry systems with erva-mate in Southern Brazil present significant levels of tree diversity, with 107 tree species identified across 39 botanical families [18], which represent a significant proportion of the region's diversity [15]. They also show high levels of nutrient cycling through litter that far exceed the nutrient exportation that takes place during erva-mate harvesting [45].

In well-developed forests, erva-mate occurs naturally as large trees and is harvested through radical pruning in 2- to 4-year production cycles. Erva-mate can be also be planted and managed at shrub size for ease of harvesting, but production is often limited due to low levels of luminosity for plant development. Although well-developed forests are mostly found in an "open" state [43], in which historic management has reduced tree density, as shown in Figure 2a, some management is required to ensure optimal conditions for erva-mate growth. However, canopy management as a means to increase light in the understory is limited by very restrictive legislation governing the use of such forests (see the Brazilian Forest Code [46,47] and Atlantic Forest Law [48]). Consequently, production in well-developed forests tends to be restricted to animal husbandry with low stock density and naturally growing erva-mate plants. Contrastingly, younger forests tend to be more actively managed in Southern Brazil for erva-mate production. In most cases, the understory is maintained mostly clear in order to make space for erva-mate plants, while thinning is applied if insufficient light permeates the forest canopy. The intensity of thinning is established empirically and ranges widely from intense tree reduction where producers aim at production levels similar to monoculture stands, to agroecological practices that try to maintain sustainable multispecies environments.

(**a**) (**b**) (**c**) (**d**)

Figure 2. Traditional agroforestry optimization: (**a**) traditional erva-mate production where the lack of adequate forest management caused canopy decline leading to monoculture-like erva-mate production; (**b**) understory management, with initial removal of invasive native bamboos, for production and species diversification; (**c**) induced forest regeneration for canopy restoration (white trunks), followed by erva-mate plantation (increased production) and diversified tree plantation (species and forest structure diversification); (**d**) sheep used for weed control in a newly implemented erva-mate intensification stand.

Our research on the management of young forests has led to the establishment of best practices that aim to achieve a more stable income while maintaining or increasing ecosystem diversity and complexity using innovative silvicultural treatments focused both on erva-mate plants and management of other forest elements [39]. In this method, erva-mate is harvested annually at much lower intensities (~50%) that maintain plant vigor as opposed to the 2- to 4-year cycles during which almost all plant foliage is removed, causing significant plant stress. Site-specific silviculture is used to define the plant's shape, height, and pruning, and regeneration tending. We introduced the use of spreaders (strings, bamboo) to widen erva-mate tree crowns into a goblet shape that allows for a greater production per plant compared to traditional growth based on a cluster of vertical branches growing from a main trunk (lower leaf-to-branch ratio). Furthermore, branches are trained away from the main vertical axis of the tree creating a sharp to near horizontal angle from which new sprouts can be harvested. Leaf harvesting from secondary branches instead of a radical pruning from the main trunk dramatically reduces the damage caused by water seepage that rots the trunk interior and often occurs after consecutive pruning, ultimately reducing the plant's life-span.

Additionally, areas where light is more available at the understory level will have plants pruned and harvested at 1–3m in height, whereas a darker understory will require taller trees (3–8m) to improve production. Similarly, spacing between trees will determine pruning methods: sparser plantations (e.g., 3 × 2 m) allows for shaping a much wider crown which in turn produces abundant foliage that can be harvested annually at less stressful levels (~50%).

We also introduced practices to ensure forest regeneration for maintaining forest cover and diversity in the long term, avoiding the expensive and impractical need to reintroduce trees through seedling planting. Two simple methods were successfully tested [39] and currently applied in farms: (i) identification and marking of seedlings from natural regeneration using bamboo/wood sticks prior to weed trimming; (ii) defining areas in which weed trimming is not conducted—plots of 1 m^2 are marked using sticks where natural regeneration is protected and encouraged. Our results show that after one year, at least one native tree seedling was successfully recruited in 75% of these plots (reaching up to more than 20 recruits in a single plot). We recommend that such areas of regeneration are introduced together with the planting of erva-mate, replacing one erva-mate seedling every 5–9 m to support forest succession.

One of the opportunities identified through this research and by others [43] is the management of young forests in the region that are dominated by native invasive bamboo species, as shown in Figure 2b. Bamboos are a determinant factor in forest dynamics as they tend to impede the development of seedlings and young trees [37,49,50]. Our previous research has shown that these bamboo species, when dominant, create a simplified plant community in which succession is arrested [37,51] and an impractical environment for the development of productive systems. In terms of landscape management planning, the areas in which bamboo are dominant require practical management solutions that kick-start forest succession to improve biodiversity and maintain forest connectivity, but also offer benefits to property owners. In these areas, farmers can manage the understory, eliminating bamboo cover and establishing erva-mate plantations in densities varying from 1500 to 10,000 seedlings per hectare, as shown in Figure 2c. Again, farmer's objectives define specific practices: production maximization tends to lead to very dense plantations with the use of chemical fertilizers and pesticides (an illegal process as no pesticide is regulated for use with erva-mate) along with the removal of forest regeneration and continuous canopy thinning; whereas traditional and agroecological producers use organic (or no) fertilizers with forest management aiming at maintaining forest structure and diversity in the long term. Typically, the former produces more leaves per area, while the latter often obtains a premium for the quality of the product.

New techniques to improve traditional agroforestry practices are also incorporating farmer-led initiatives that have found innovative ways to optimize production and minimize environmental impacts. One very promising solution is the use of sheep husbandry for weed control in erva-mate plantations, as shown in Figure 2d. The introduction of the sheep breed "Texel" for controlling weeds reduces substantially the demand for labor, which is one of the most pressing limitations in farming today. One important characteristic of the Texel breed is the fact that they do not graze on erva-mate plants and have minimal impact on soil compaction. Finally, sheep farming can play an important role in food security for farmers and can be a smart solution to halt the current trend of using glyphosate herbicides for weed control. Such low-tech, practical approaches to optimizing production and reducing the use of chemical inputs can be scaled out from individual small-scale farms to create regional approaches and best practices that recognize and support the knowledge and participation of farmers, and in turn can have lasting impact on the forested landscape.

3.1.2. Silvopastoral Systems

The use of animal husbandry in traditional agroforestry in Southern Brazil has two main systems: *caívas* in North Santa Catarina state and *faxinais* in South Paraná state. *Faxinais* were once a common feature in the landscape in Paraná, where communities use a large forest area as a commons for animal husbandry, as shown in Figure 3a, and erva-mate harvesting, with a wide diversity of food crops (including corn, beans, manioc, and rice) in fenced fields protected from animal grazing [43]. These multifunctional land use systems were typical of the indigenous descendants that occupied the region and later assimilated by settler communities, especially Ukrainian and Polish immigrants. On the other hand, *caívas* are generally found in the Northern Plateau region of Santa Catarina on individual properties in which dairy cattle husbandry is carried out in forest patches usually combined with erva-mate production, as shown in Figure 3b [20].

Figure 3. Traditional silvopastoral systems: (**a**) animal husbandry within forest stands in a *faxinal*; (**b**) *caíva* with low productivity pasture and senile erva-mate trees (small trees in the background, to the right); (**c**) view of a *caíva* with improved pasture.

The natural pasture in *caívas* typically has low levels of productivity, particularly in the winter when plant regrowth cannot keep up with grazing demands [44,52]. Thus, animal productivity is low, leading to food insecurity and low economic income and resilience. Undernourished animals have knock-on economic impacts on erva-mate production as they look for grazing alternatives and damage erva-mate trees or consume the leaves. Furthermore, environmental sustainability may also be affected because grazing on forest regeneration can compromise forest renovation and physical damage caused by bark consumption, which compromises tree health, ultimately reducing tree lifespan.

As farmers look for more profitable economic alternatives, *caívas* have been replaced by monoculture production based on forest plantations and commodity crops with direct loss of forest cover and biodiversity and traditional practices. As a response, a participatory research project carried out by EPAGRI, the Agricultural Research and Rural Outreach Company of Santa Catarina, and EMBRAPA has focused on developing strategies to increase household income by improving animal and erva-mate productivity while also maintaining or restoring forest structure and diversity [20,44,52]. The project framework includes testing innovative practices related to pasture improvement, renovation of forest stands and erva-mate trees, and defining ideal levels of canopy cover and regeneration management.

In 2010, EPAGRI initiated the implementation of improved practices for traditional *caívas* that include pasture overseeding during the winter that evolved into the development of the genetically improved *Axonopus catharinensis* SCS 315 (referred to locally as *missioneira-gigante* or giant missionary grass) [53], a pasture species that is better suited for *caíva* environments due to its tolerance to shade [44,54,55]. The process to implement this technique includes the removal of native grasses and the introduction of giant missionary grass (detailed information about the process can be found in [44]; Figure 3c). The improved *caíva* system has been implemented successfully in eight farms across four municipalities in Santa Catarina [44] and another ten properties will adopt the technology by the end of 2020; those farms will be used as reference properties for outreach agencies to disseminate the results.

Along with the goal of increasing animal productivity, the project also developed practices to improve erva-mate production. These include a set of activities aimed at creating a highly diverse, healthy forest with multi-aged and multi-strata elements. Due to years of neglect and prohibitive laws that severely restrict forest management, as noted above, most traditional *caívas* have inconsistent forest structure ranging from large, frequent gaps to very dense clusters of trees. Thus, we developed forest management guidelines to support forest restoration, canopy refinement, and erva-mate intensification. Forest restoration seeks to restore gaps in forest cover and reintroduce a multi-aged tree population which can be achieved by designating areas for restoration where animal grazing is temporarily restricted (usually by using electrical fencing) for 3–5 years, after which foraging is again allowed and a new area is fenced. Monitoring of these areas showed that regeneration was highly effective in restoring species diversity and structure as 59 different tree species were recorded in the fenced areas [56,57], which was greater than the diversity of the adult tree population. Additionally, we recommend thinning of abundant species and tree clusters in order to increase species diversity and establish a more even forest canopy, respectively. Simultaneously, in fenced areas, erva-mate can be planted at densities between 1000 and 3000 seedlings per hectare to increase productivity.

Through the implementation of the higher-productivity, shade-adapted perennial grass in *caívas*, farmers are producing five times more pasture per area, enabling a triplication in the stocking rate and consequent increases in milk production [43]. Furthermore, erva-mate production can increase tenfold depending of the level of degradation of the erva-mate trees. Finally, in order to monitor changes over time and help evaluate the impacts of new practices as they are implemented, we adapted the Sustainability Assessment of Food and Agriculture Systems (SAFA) developed by FAO [58] to assess the sustainability of farms considering 77 indicators across themes of environmental integrity, economic resilience, quality of life (social), and good governance [20]. The overall results showed that the implementation of the improved practices outlined above enabled farmers to obtain a ranking of good (based on a classification as unacceptable, limited, or good) for 87% of indicators, in comparison to 65% ranked as good for *caívas* that had not implemented improved practices.

As is the case in many other countries [7], current agricultural and environmental policies in Southern Brazil do not recognize traditional practices, as they are often excluded from scientific analyses or assessments and intensification through conventional agriculture is still seen as the way forward. As such, landscape policies and management strategies do not consider how traditional systems might be leveraged to mitigate the increasing homogeneity of the landscape through monoculture farming and the consequential impacts on human health and nutrition, biodiversity, gene flow, and ecosystem services. Without a recognition of, or support for, traditional systems, the tendency is for this local knowledge to be lost, along with the associated cultural identities and environmental subjectivities. Thus, analyses, participatory approaches, and farmer-led initiatives such as those we have outlined in this section that optimize traditional systems, provide the evidence base necessary for these practices to be integrated into policy and governance structures, which in turn can provide landscape managers with practical approaches that are culturally relevant and can be implemented across the landscape.

3.2. Productive Agroforestry Restoration

As noted above, Southern Brazil has been subjected to devastating rates of deforestation and forest fragmentation over the last century, as shown in Figure 1. While current policies have been essential in reducing rates of deforestation, they have been ineffective in reforesting already degraded ecosystems as legal restrictions severely limit the use of forested areas (Legal Reserves and Areas of Permanent Protection) on all rural properties [47]. These regulations have been difficult to implement as reforesting lands is viewed negatively by landowners as the assumption is that once the forest recovers, the land becomes worthless or untouchable. The question remains as to how to incentivize transformative changes to the landscape when the predominant perception is that monoculture crops, with inevitable forest loss and detriments to the agroecological landscape, produce higher yields?

Aiming at reintegrating degraded agroecosystems into ecologically and economically functional areas, we developed and implemented Productive Agroforestry Restoration models focusing on restoring degraded or underused agricultural land into a multispecies productive system maintained as a forested environment [40]. The goal is to allow for a variety of outputs that take advantage of the inherent spatial and temporal variations of the system and produce direct (e.g., crops) and indirect (e.g., ecosystem services) benefits. We designed Productive Agroforestry Restoration as a response to the need for innovative productive systems that can generate income and restore ecosystems for the benefit of rural communities and society in general, offering land management solutions that can be implemented across the region.

Thus, the process of developing Productive Agroforestry Restoration models began with the premise that any land to be restored or (re)integrated into a sustainable agroecosystem (namely agroforestry) should not only focus on the restoration of ecological attributes but also be integral to the socioeconomic reality of the farm; the system must be both ecologically and economically sustainable. In 2011, with financial support from the Brazilian National Council for Scientific and Technological Development (CNPq) and EMBRAPA, we began implementing a project to leverage LEK in order to optimize traditional management practices and create agroforestry models for restoration. In collaboration with farmers, we developed a comprehensive list of potential agroforestry systems and species that farmers would be interested in cultivating. From this, we collectively chose models that were deemed more feasible and implemented these models at the EMBRAPA Research Station in Caçador (ERSC) replicated across an area of 40 ha. The results [35,36,39,40] allowed us to expand the implementation of such models to more than 50 small-scale farms in the region. Importantly, we also integrated the needs and expectations of farmers as four key requirements of the models. Specifically, the models must be easy to understand and implement; fast, through the rapid (re-)establishment of a forest canopy by using fast-growing pioneer species; profitable, as investment should result in economic return; and flexible to regional characteristics, property, and goals. Implementation can take place at different places and scales in a property and be integrated into a landscape restoration program.

The species selection for Productive Agroforestry Restoration varies depending on the region in which it is implemented. Initially, a few key species should be identified in order to fulfill the need for rapid forest cover establishment and acting as a cash crop. In many tropical and subtropical regions, the use of native, fast-growing species from the legume family is highly recommended for their multiple benefits, which include use for firewood and lumber, rapid deposition of soil litter, and nitrogen fixation, among others [59,60]. As some fast-growing trees are short-lived, sometimes with cycles of less than 20 years, their replacement should be considered early in the management planning. On the other hand, cash crops should be based on long-lived species in order to create financial predictability. There is a myriad of species and combinations but the design of a system with initial complex arrangements should consider possible constraints, such as seed/seedling availability, potential market outlets, local labor capacity due to systems with higher management requirements, and lack of technical knowledge about the species and their interactions, among others. Thus, biological complexity and product variety is often better achieved after the establishment of an initial simple system.

Among the Productive Agroforestry Restoration systems implemented at the ERSC and currently being used in farms across Southern Brazil, the most successful was designed to be implemented in any region of the subtropical highlands of Southern Brazil and is based on two species: the fast-growing bracatinga (*Mimosa scabrella*) and erva-mate. While bracatinga is a legume pioneer tree that regenerates spontaneously in the region and is used for firewood and slab props, erva-mate is a shade tolerant low maintenance tree with a consolidated market.

This system was first used to restore a degraded agroecosystem that after several decades of high intensity commodity monoculture had extremely impoverished and compacted soil. Following the Productive Agroforestry Restoration model, we planted the fast-growing bracatinga in rows 6 m apart (1.5 m distance within rows; Figure 4a) that rapidly formed a forest canopy, as shown in Figure 4b. As a pioneer species, bracatinga is expected to very rapidly form a forest canopy as it reaches average heights of 2.8 m after one year, 6.8m by year two, and 9.5 by year three (at which point a forest environment is established), and 13.3 m by year four (diameter at breast height of 2.2, 7.6, 11, 13.3 cm, respectively) [36,61,62]. With ideal shaded conditions for erva-mate in place with the establishment of forest cover by year three, erva-mate seedlings are planted in double rows between bracatingas (~3000 plants per hectare), as shown in Figure 4c.

(a) (b) (c) (d)

Figure 4. Productive Agroforestry Restoration carried out in a degraded agroecosystem: (**a**) initial planting of the fast-growing bracatinga for rapid canopy development (year 1); (**b**) developed canopy allowed for the plantation of erva-mate (year 3); (**c**) commercial maturity of erva-mate generates profitability for the system (year 8); (**d**) different methods of bracatinga harvesting: left—alternate full row removal for optimized financial return from lumber; right—in row-alternate tree girdling (trees with brown crowns) for subsequent harvesting to minimize environmental change.

Bracatinga demands yearly pruning until at least year three to obtain trees with higher market value (timber without knots) and high branching that does not disturb erva-mate cultivation [40], while erva-mate should be cultivated following the silvicultural practices developed described above. Combined with the planting of bracatinga, in the first two years we cultivated soy between rows which helped to ameliorate revenue/investment financial ratio during this period. Later, by year five, income was again generated through bracatinga thinning (at 50%) and finally at year eight, erva-mate harvesting reached commercial levels when the system become profitable, as shown in Figure 4c, with cost–benefit ratio varying between 2.0 and 2.6 with higher levels obtained when bracatinga is harvested for lumber [36,40].

The system presented offers an alternative for land restoration, but its elements are open to variation and diversification. As environmental conditions gradually improve over time, especially soil structure and fertility, with much lower levels of humidity and temperature fluctuations, other species can be integrated to the model, taking advantage of the horizontal and vertical space available, which includes vines, herbs, and shrubs for various uses (food, medicinal, handcraft, etc.). Importantly, landscape managers can support the use of these systems as a means to reforest Legal Reserves on rural properties, while remaining productive.

We are currently undergoing a comprehensive monitoring and modeling effort in order to quantify and qualify the socioeconomic benefits and improvement of environmental services provided by the practices discussed herein. Our initial results show extensive improvement in terms of forest species diversity, which increased from the initially two tree species planted to 40 species successfully recruited [39] in five years. Furthermore, the occurrence of vines, shrubs, herbs, and an abundant natural regeneration in a multi-strata forest seems to confirm the widely held assumption that species diversity is expected to increase with habitat complexity [63]. Moreover, those results contrast sharply with our control area (no intervention since 2010; unpublished data) in which no trees to date have managed to recruit. In terms of the land sparing approach, our results suggest that merely setting aside land for forest restoration and conservation may be insufficient. Our model does not have some of the common pitfalls of other restoration systems that focus solely on maximizing substrate stability or primary productivity, often resulting in arrested succession and demanding additional efforts to encourage successional change [64]. While detailed carbon monitoring is underway, we have already been able to estimate large-scale benefits of restoring riparian forests in the region [65], which can be used as a proxy to estimate the benefits of restoring degraded lands and forests.

4. Stakeholder Engagement and Current Challenges

The introduction of new technologies in traditional agroforestry systems help initiate a process of social, economic, and environmental transformation in family farming involved in the participatory research. Firstly, our focus on participatory research that values the knowledge of small-scale farming families has led to self-reflection and a growing self-awareness of the value of this knowledge and the associated environmental identities in wider socioenvironmental discourses, and also their rights as farmers and food producers. Our focus on documenting and sharing knowledge across institutional, class, and gendered divisions has enabled the creation of a knowledge network that has led to many new initiatives, ideas, and solidarities. We are in the process, for example, of supporting the development of a network of women farmers that will identify the needs and challenges women face in participating as active members of decision-making and knowledge sharing circles, not only on the farm, but in the wider context of agroecological production. Farm visits that bring together a variety of farmers, practitioners, and researchers have also shown to be fruitful in knowledge exchange across various spheres, leading to innovative management strategies for pruning and pest management, among many others. Our initiatives are also helping to consolidate our partner communities into a collective network with a stronger political voice. One major advancement has been the creation of a strategic council (*Observatório dos Sistemas Tradicionais e Agroecologicos da Erva-mate do Paraná*) for traditional and agroecological erva-mate production systems, spearheaded by the Public Prosecutor of Labor of

Paraná, which supports the continuation and expansion of these systems. This council brings together 27 organizations that are working together to bring greater awareness to the ecological and cultural value of traditional erva-mate production, while also incentivizing new products, markets, and other economic benefits.

In terms of economic impact, optimization strategies, such as those tested and implemented in the *caíva* systems, are showing promise in terms of improved incomes for farmers [20], while the models being tested at the ERSC have shown possibilities of economic returns in relatively short periods of time [40]. The restoration model is also being rolled out through partnerships with industry to better test these productive systems at a larger scale and gain more concrete insights into the economic capacities of these models. Part of our ongoing research is to determine the indirect environmental benefits of these agroforestry systems, particularly in terms of carbon capture, water quality, soil health, and biodiversity, as support for the development of payment for ecosystem services models. As noted above, traditional erva-mate agroforestry systems have been shown to have significant levels of tree biodiversity [18,40], and despite consistent cattle grazing for several generations, the tree regeneration potential of *caíva* systems has been shown to be quite strong, with significant levels of diversity in terms of regeneration in comparison to those found in the Santa Catarina Forest Inventory [44]. Clearly, these productive forest systems have substantial environmental resilience and offer compelling strategies that can be implemented across the region. Preliminary results on ecosystem services, as noted above, are also showing greater potential for total carbon and nitrogen capture in erva-mate agroforestry systems than in monoculture erva-mate production areas. Nevertheless, more detailed data is required to continue to inform policy and regulatory frameworks.

Despite the benefits, several challenges still face the continuation and expansion of traditional and agroecological systems in the region. Current legislation related to forest management, for example, severely restricts silvicultural practices on private properties, with relatively arbitrary quotas placed on the number of trees that can be removed from native forests, while regulations related to livestock grazing and production in silvopastoral systems remain unclear. Although current legislation has been important in stemming the devastating loss of forests that occurred throughout the twentieth century, small-scale farmers feel disproportionately affected by the regulations, which has led to mistrust on both sides of the issue [38]. The oral history interviews conducted as part of our research have clearly underscored how tensions between small-scale farmers and environmental agencies have led many farmers to question the continuation of these systems as the current impasse seems insurmountable [33]. Yet, through research and advocacy in collaboration with farmers, changes are taking place, with environmental agencies such as Paraná State Environmental Institute (IAP) and the federal Brazilian Institute of the Environment and Natural Resources (IBAMA) participating in recent events and outreach activities organized as part of this project, and in the *Observatório*. Promisingly, these agencies are looking to update regulations and change legal restrictions based on the current state of forests in Brazil, supported by the data and experiences projects such as ours are sharing.

Although changes are taking place within policy circles, one of the biggest challenges we face is the inherent bias not only against forests, as they continue to be viewed as useless, which is directly related to current legislation that prohibits use, but also the culture of agricultural research and outreach agencies that are very much focused on conventional agriculture and mechanization and/or modernization at the expense of traditional knowledge and practices. Agroecological or traditional farming practices are generally excluded from agronomy courses in universities and technical colleges, and as such, the majority of outreach workers have little experience engaging with these alternative approaches. Despite the myriad policies that have been enacted to support small-scale family farming and agroecology/organic production in Brazil (see [66,67]), these policies have not necessarily trickled down to have clear impacts on small-scale farming communities, while others have reinforced the dominant model of intensification based on monoculture commodity crops. Nevertheless, recent developments occurring through ongoing engagement with environmental agencies and other institutions are starting to show promising shifts in the top-down approaches to governance and agricultural outreach.

Despite national policy initiatives implemented in the last 10 years that have focused on drastically increasing the amount of data related to land use in Brazil (for example the Environmental Rural Registry (CAR) [47] and the National Forest Inventory [68]), federal and state land management planning is still in its infancy. One of the major challenges land management attempts face, however, stems from conflicting perceptions of productive land use. On one side, agriculture research and outreach agencies and large agrobusiness are pushing to modernize production through intensive, high-input monocultures, such as soy and corn. While these systems are seen as more productive, offering greater yields and thus higher income than traditional systems, there is a range of negative consequences including human health and food security, loss of farmer autonomy and increased debt, deforestation and loss of biodiversity and LEK, and impacts on water and nutrient cycles. On the other hand, environmental policies focus on protecting the remaining old growth forests and attempt to increase forest cover through restrictive laws that prohibit the use of forest resources, for example through Legal Reserves in which 20% of the property must be forested [47]. Furthermore, there remains an underlying assumption across both agriculture and environmental policy areas that native forests are not productive land, meaning either that the land must be deforested to become 'productive', or the burden of maintaining the land as forest (i.e., untouchable) falls on the landowner. This conflict is disproportionately felt by small-scale producers who use traditional agroforestry systems because it is exactly these farms that continue to maintain forest cover, which often extends well beyond the required 20%. Yet the law prohibits most types of forest management and agencies that monitor and inspect farms tend to administer fines with the onus on the landowner to prove they are within their legal limits, an unrealistic requirement for most small-scale farmers (for a full discussion, see [30]).

Our research is demonstrating that there is middle ground between these two competing land use perceptions that offer economic, cultural, and environmental benefits at both the local and regional scales. Implementing such models and practical approaches at the landscape scale can have dramatic impacts on the amount of land under forest cover, with the inherent returns of improved ecosystem services, biodiversity, and carbon sequestration. It can also bring about significant changes to the economic, cultural, and social value associated with forests and the products derived from them. By supporting and disseminating the knowledge, environmental subjectivities, and intangible heritage associated with traditional agroforestry systems, small-scale producers have an opportunity to create and capture niche markets that value ecologically and culturally significant products and processes. Thus, agroforestry systems using native species can be productive and economically, culturally, and environmentally viable, and through a transdisciplinary approach, land managers can work with traditional producers to develop best practices that can be implemented on farms across the region to have a real impact on sustainable land use on a larger scale.

5. Conclusions

The research presented herein demonstrates innovative approaches to documenting, valuing, and leveraging traditional agroforestry practices as a means to support diverse, resilient agroecological landscapes. While the focus of our research is on small-scale farms, the models we are testing show potential for scaling-out, offering promising alternatives that landscape managers can use to support sustainable land use and land cover change. The collaborative approach to research has been fundamental in this project as we are integrating and valuing different perceptions of agroecosystems and ensuring that communities are central to all aspects of the work, from defining research questions, to developing monitoring systems, and disseminating the results. Grassroots initiatives and locally adapted agroecological practices, such as those used in erva-mate agroforestry, are often ignored by research and government agencies, resulting in serious disconnect between overarching policy frameworks, such as the United Nations Sustainable Development Goals, and actual strategies that have the potential for transformational change at the landscape scale. Our work is attempting to bridge this divide by leveraging the knowledge and practices small-scale farmers have been developing for generations before they are lost to the dominant paradigm of conventional agriculture.

Author Contributions: The authors contributed to the development, implementation, analysis, and writing of this study as follows: conceptualization and methodology, A.E.B.L. and E.R.N.; formal analysis, A.L.H., A.E.B.L., and E.R.N; resources and funding, A.E.B.L. and E.R.N.; writing—original draft preparation, A.E.B.L. and E.R.N; writing—review and editing, E.R.N., A.L.H., and A.E.B.L.; project administration, A.E.B.L. and E.R.N. All authors have read and agreed to the published version of the manuscript.

Funding: This research was funded by EMBRAPA, grant number (16.16.05.002.00.00), EPAGRI, Programa SC Rural of the Government of the State of Santa Catarina, Social Sciences and Humanities Research Council of Canada, FLEdGE Partnership Grant, and Wilfrid Laurier University's Centre for Sustainable Food Systems. ERN was supported PNPD/CAPES.

Acknowledgments: The authors are sincerely grateful to the farmers and their families who have shared their stories with us. Particularly we thank Maria Izabel Radomski, Ednilson Pereira Gomes, Demerval Pessin de Farias, Bernardo Vergopolen, Carlos Urio, Arnaldo Soares, João F.M.M. Nogueira, Ricardo Gomes Luiz, Alessandra I. de Carvalho, and Robson Laverdi.

Conflicts of Interest: The authors declare no conflict of interest.

References

1. Food and Agriculture Organization; Information Intergovernmental Technical Panel on Soils. *Status of the World's Soil Resources—Main Report*; Food and Agriculture Organization of the United Nations; Intergovernmental Technical Panel on Soils: Rome, Italy, 2015.
2. Food and Agriculture Organization. *The State of the World's Biodiversity for Food and Agriculture: A Call to Action?* Bélanger, J., Pilling, D., Eds.; FAO Commission on Genetic Resources for Food and Agriculture Assessments: Rome, Italy, 2019.
3. Food and Agriculture Organization. *Catalysing Dialogue and Cooperation to Scale up Agroecology: Outcomes of the FAO Regional Seminars on Agroecology*; FAO: Rome, Italy, 2018.
4. The International Union for Conservation of Nature; World Resources Institute. *A Guide to the Restoration Opportunities Assessment Methodology (ROAM): Assessing Forest Landscape Restoration Opportunities at the National or Sub-National Level*; IUCN: Gland, Switzerland, 2014.
5. Intergovernmental Panel on Climate Change; Summary for Policymakers. *Climate Change and Land: An IPCC Special Report on Climate Change, Desertification, Land Degradation, Sustainable Land Management, Food Security, and Greenhouse Gas Fluxes in Terrestrial Ecosystems*; Shukla, P.R., Skea, J., Calvo Buendia, E., Masson-Delmotte, V., Pörtner, H.-O., Roberts, D.C., Zhai, P., Slade, R., Connors, S., van Diemen, R., et al., Eds.; IPCC: Geneva, Switzerland, 2019.
6. High Level Panel of Experts. *Sustainabile Forestry for Food Security and Nutrition*; A Report by the High Level Panel of Experts on Food Security and Nutrition of the Committee on World Food Security; HLPE: Rome, Italy, 2017.
7. Padoch, C.; Sunderland, T. Managing landscapes for greater food security and improved livelihoods. *Unasylva* **2013**, *64*, 3–13.
8. Hobbs, R.; Wu, J. Perspectives and prospects of landscape ecology. In *Key Topics in Landscape Ecology*; Wu, J., Hobbs, R., Eds.; Cambridge University Press: Cambridge, UK, 2006.
9. Naveh, Z.; Lieberman, A.S. *Landscape Ecology: Theory and Application*, 2nd ed.; Springer Science Business Media: New York, NY, USA, 1994.
10. Fischer, J.; Abson, D.J.; Butsic, V.; Chappell, M.J.; Ekroos, J.; Hanspach, J.; Kuemmerle, T.; Smith, H.G.; von Wehrden, H. Land sparing versus land sharing: Moving forward. *Conserv. Lett.* **2014**, *7*, 149–157. [CrossRef]
11. Free Trade Agreement; Consultative Group for International Agricultural. *Research Program on Forests, Trees and Agroforestry: Landscape Approaches to Tackle Climate Change, and Achieve Sustainable Development and Food Security*; CGIAR: Bogor, Indonesia, 2017.
12. Elevitch, C.R.; Mazaroli, N.D.; Ragone, D. Agroforestry standards for regenerative agriculture. *Sustainability* **2018**, *10*, 3337. [CrossRef]
13. Food and Agriculture Organization; Convention on Biological Diversity. *Mainstreaming Ecosystem Services and Biodiversity into Agricultural Production and Management in East Africa*; Food and Agriculture Organization of the United Nations (FAO); Secretariat of the Convention of Biological Diversity: Rome, Italy, 2016.

14. Kumar, C.; Begeladze, S.; Calmon, M.; Saint-Laurant, C. *Enhancing Food Security Through Forest Landscape Restoration: Lessons from Burkina Faso, Brazil, Guatemala, Viet Nam, Ghana, Ethiopia and Philippines*; Kumar, C., Begeladze, S., Calmon, M., Saint-Laurant, C., Eds.; IUCN: Gland, Switzerland, 2015.
15. Vibrans, A.C.; Sevegnani, L.; de Gasper, A.L.; Lingner, D.V. *Diversidade e Conservação dos Remanescentes Florestais*; Edifurb: Blumenau, Brazil, 2012.
16. Vibrans, A.C.; Sevegnani, L.; de Gasper, A.L.; Lingner, D.V. *Floresta Ombrófila Mista*; Edifurb: Blumenau, Brazil, 2013.
17. Hanisch, A.L.; Vogt, G.A.; Marques, A.D.C.; Bona, L.C.; Bosse, D.D. Estrutura e composição florística de cinco áreas de caíva no planalto norte de Santa Catarina. *Pesqui. Florest. Bras.* **2010**, *30*, 303–310. [CrossRef]
18. Chaimsohn, F.P.; de Souza, A.M. *Sistemas de Produção Tradicionais e Agroflorestais de Erva-Mate no Centro-Sul do Paraná e Norte Catarinense*; Chaimsohn, F.P., de Souza, A.M., Eds.; IAPAR: Ponta Grossa, Brazil, 2013.
19. Lacerda, A.E.B. Conservation strategies for Araucaria Forests in Southern Brazil: Assessing current and alternative approaches. *Biotropica* **2016**, *48*, 537–544. [CrossRef]
20. Hanisch, A.L.; Negrelle, R.R.B.; Bonatto, R.A.; Nimmo, E.R.; Lacerda, A.E.B. Evaluating Sustainability in Traditional Silvopastoral Systems (caívas): Looking Beyond the Impact of Animals on Biodiversity. *Sustainability* **2019**, *11*, 3098. [CrossRef]
21. Fonseca-Cepeda, V.; Julián Idrobo, C.; Restrepo, S. The changing chagras: Traditional ecological knowledge transformations in the Colombian amazon. *Ecol. Soc.* **2019**, *24*, 8. [CrossRef]
22. Berkes, F.; Colding, J.; Folke, C. Rediscovery of Traditional Ecological Knowledge as Adaptive Management. *Ecol. Appl.* **2000**, *10*, 1251–1262. [CrossRef]
23. Berkes, F. Traditional Ecological Knowledge in Perspective. In *Traditional Ecological Knowledge: Concepts and Cases*; Inglis, J.T., Ed.; International Program on Traditional Ecological Knowledge; International Development Research Centre: Ottawa, ON, Canada, 1993; pp. 1–10.
24. Stevenson, M.G. Indigenous knowledge in environmental assessment. *Arctic* **1996**, *49*, 278–291. [CrossRef]
25. Bonomo, M.; Costa Angrizani, R.; Apolinaire, E.; Noelli, F.S. A model for the guaraní expansion in the La Plata Basin and littoral zone of southern Brazil. *Quat. Int.* **2015**, *356*, 54–73. [CrossRef]
26. Noelli, F.S. A Ocupação Humana na Região Sul do Brasil: Debates e perspectivas 1872–2000. *Rev. USP* **2000**, *44*, 218–269. [CrossRef]
27. Nimmo, E.R.; Nogueira, J.F.M.M. Creating hybrid scientific knowledge and practice: The Jesuit and Guaraní cultivation of yerba mate. *Can. J. Lat. Am. Caribb. Stud.* **2019**, *44*, 347–367. [CrossRef]
28. Westphalen, C.M.; Machado, B.P.; Balhana, A.P. *Nota Prévia ao Estudo da Ocupação da Terra no Paraná Moderno*; Boletim (Universidad Federal do Paraná. Departamento de História); Universidade Federal do Paraná, Conselho de Pesquisas de Pesquisas: Curitiba, Brazil, 1968.
29. Mota, L.T. *As Guerras dos Índios Kaingang: A História Épica dos Índio Kaingang no Paraná (1769–1924)*, 2nd ed.; Universidade Estadual de Maringá: Maringá, Brazil, 2009.
30. Lacerda, A.E.B.; Rosot, M.A.D.; Filho, A.F.; Garrastazú, M.C.; Nimmo, E.R.; Kellermann, B.; Radomski, M.I.; Beimgraben, T.; Mattos, P.P.; Oliveira, Y.M.M. Sustainable Forest Management in Rural Southern Brazil: Exploring Participatory Forest Management Planning. In *Sustainable Forest Management—Case Studies*; Martin-Garcia, J., Diez, J.J., Eds.; IntechOpen: Rijeka, Croatia, 2012; pp. 97–118. [CrossRef]
31. Persson, J.; Johansson, E.L.; Olsson, L. Harnessing local knowledge for scientific knowledge production: Challenges and pitfalls within evidence-Based sustainability studies. *Ecol. Soc.* **2018**, *23*, 38. [CrossRef]
32. Williams, B.; Riley, M. The Challenge of Oral History to Environmental History. *Environ. Hist. Camb.* **2020**, *26*, 207–231. [CrossRef]
33. Nimmo, E.R.; de Carvalho, A.I.; Laverdi, R.; Lacerda, A.E.B. Oral history and traditional ecological knowledge in social innovation and smallholder sovereignty: A case study of erva-Mate in Southern Brazil. *Ecol. Soc.* **2020**. Accepted.
34. Maran, J.C.; Rosot, M.A.D.; Rosot, N.C.; Radomski, I.; Cardoso, J.D.; Lacerda, A.E.B.; Kellermann, B. Análise de sobrevivência em plantios de enriquecimento com Araucaria angustifolia usando mudas de grande e pequeno porte. In *Congresso Florestal Paranaense*; Associação Paranaense de Empresas de Base Florestal: Curitiba, Brazil, 2015; p. 11.

35. Radomski, M.I.; Lacerda, A.E.B.; Kellermann, B. *Sistemas agroflorestais: Restauração ambiental e produção no âmbito da Floresta Ombrófila Mista*; Documentos 276; Empresa Brasileira de Pesquisa Agropecuária; Embrapa Florestas; Ministério da Agricultura, Pecuária e Abastecimento: Colombo, Brazil, 2014. Available online: http://ainfo.cnptia.embrapa.br/digital/bitstream/item/121202/1/Doc.-276-Izabel.pdf (accessed on 1 June 2020).
36. Radomski, M.I.; Lacerda, A.E.B. *Plantios de bracatinga na Estação Experimental da Embrapa em Caçador, SC*; Doc. Embrapa Florestas 328; Embrapa Florestas: Colombo, Brazil, 2019. Available online: https://www.embrapa.br/busca-de-publicacoes/-/publicacao/1111377/plantios-de-bracatinga-na-estacao-experimental-da-embrapa-em-cacador-sc (accessed on 20 February 2020).
37. Kellermann, B.; Lacerda, A.E.B. Arrested development? Investigating the role of bamboo in Araucaria Forest succession in Southern Brazil. *J. Plant Ecol.* **2019**, *12*, 1034–1046. [CrossRef]
38. Blay-Palmer, A.; Spring, A.; Nimmo, E.R.; Lacerda, A.E.B. Traditional food, the Right to Food and Sustainable Food Systems. In *The Routledge Handbook of Sustainable and Regenerative Food Systems*; Duncan, J., Carolan, M., Wiskerke, H., Eds.; Routledge: Abingdon, UK, 2020.
39. Lacerda, A.E.B. Série erva-Mate sombreada (1): Sistema de produção de erva-mate baseado no manejo tradicional: Bracatingais dominados por bambus (taquarais). *Comun. Técnico Embrapa Florestas* **2019**, *439*. Available online: http://ainfo.cnptia.embrapa.br/digital/bitstream/item/205467/1/CT-439-1682-final-1.pdf (accessed on 18 January 2020).
40. Lacerda, A.E.B. Série erva-mate sombreada (2): Sistema de restauração produtivo agroflorestal. *Comun. Técnico Embrapa Florestas* **2019**, *400*. Available online: http://ainfo.cnptia.embrapa.br/digital/bitstream/item/205685/1/CT-440-1677-final.pdf (accessed on 18 January 2020).
41. Chaimsohn, F.P.; Radomski, M.I. Anais do III Seminário sobre caracterização de sistemas de produção tradicionais e agroecológicos de erva-mate de agricultores familiares nas regiões Centro Sul e Norte Catarinense. *Doc. Embrapa* **2016**, *292*. Available online: https://ainfo.cnptia.embrapa.br/digital/bitstream/item/155047/1/Doc-292-Anais-do-III-Seminario-sobre-caracterizacao.pdf (accessed on 6 January 2020).
42. Mattos, A.G. Caracterização das Práticas de Manejo e das Populações de Erva-Mate (Ilex paraguariensis A. Sant. Hil) Nativa em Exploraçao no Planalto Norte Catarinense. Master's Thesis, Universidade Federal de Santa Catarina, Florianópolis, Brazil, 2011.
43. Da Marques, A.C. As Paisagens do Mate e a Conservação Socioambiental: Um Estudo Junto aos Agricultores Familiares do Planalto Norte Catarinense. Ph.D. Thesis, Universidade Federal do Paraná, Curitiba, Brazil, 2014.
44. Hanisch, A.L. Intensificação do uso da Pastagem em Sistema Silvipastoril Tradicional (Caívas) Como Estratégia de Uso Sustentável de Remanescentes de Floresta Ombrófila Mista. Ph.D. Thesis, Universidade Federal do Paraná, Curitiba, Brazil, 2018.
45. Machado, N.C. Produção de Serapilheira, Aportes de Nutrientes e Atributos de Fertilidade do Solo em Sistemas Agroflorestais de Erva-Mate. Master's Thesis, Universidade Estadual de Ponta Grossa, Ponta Grossa, Brazil, 2014.
46. Brasil Lei N° 12.651, de 25 de Maio de 2012; Dispõe Sobre a Proteção da Vegetação Nativa. 2012. Available online: https://www2.camara.leg.br/legin/fed/lei/2012/lei-12651-25-maio-2012-613076-normaatualizada-pl.pdf (accessed on 6 January 2020).
47. Machado, F. *Brazil's New Forest Code: A Guide for Decision-Makers in Supply Chains and Governments 2015 GUIA*; WWF Brazil: Brasilia, Brazil, 2016.
48. Brasil Lei N° 11428/2006—"Dispõe Sobre a Utilização e Proteção da Vegetação Nativa do Bioma Mata Atlântica, e dá Outras Providências". 2006. Available online: http://www.planalto.gov.br/ccivil_03/_Ato2004-2006/2006/Lei/L11428.htm (accessed on 2 February 2020).
49. Kellermann, B.; Lacerda, A.E.B. Bambus nativos como espécies invasoras no sul do Brasil. In *Bambus no Brasil: Da Biologia à Tecnologia*; Drumond, P.M., Wiedman, G., Eds.; Embrapa: Brasilia, Brazil, 2017; pp. 179–196.
50. Budke, J.C.; Alberti, M.S.; Zanardi, C.; Baratto, C.; Zanin, E.M. Bamboo dieback and tree regeneration responses in a subtropical forest of South America. *For. Ecol. Manag.* **2010**, *260*, 1345–1349. [CrossRef]
51. Lacerda, A.E.B.; Kellermann, B. What is the Long-Term Effect of Bamboo Dominance on Adult Trees in the Araucaria Forest? A Comparative Analysis between Two Successional Stages in Southern Brazil. *Diversity* **2019**, *11*, 165. [CrossRef]

52. Hanisch, A.L.; Radomski, M.I.; Bona, L.C.; Marques, A.D.C. Melhoria da produção animal em áreas de caíva e sua contribuição para a viabilização de corrredores ecológicos. *Desenvolv. Reg. Debate* **2016**, *6*, 170–188. [CrossRef]
53. EPAGRI SCS315 Catarina Gigante Pastagem Perene de Verão. Available online: www.epagri.sc.gov.br/index.php/solucoes/tecnologias/tecnologias-em-destaque-catarina-gigante/ (accessed on 10 January 2020).
54. Hanisch, A.L.; Negrelle, R.R.B.; Balbinot Junior, A.A.; de Almeida, E.X. Produção, composição botânica e composição química de missioneira-gigante consorciada com leguminosas perenes. *Rev. Bras. Ciências Agrárias* **2016**, *11*, 60–66. [CrossRef]
55. Hanisch, A.L.; Dalgallo, D.; Almeida, E.X.; Negrelle, R.R.B. Desempenho e composição química de missioneira-gigante cultivada em sistema silvipastoril tradicional em duas alturas de pastejo. *Rev. Ciências Agrárias* **2016**, *59*, 345–351. [CrossRef]
56. Pinotti, L.C.A.; Hanisch, A.L.; Negrelle, R.R.B. The impact of traditional silvopastoral system on the Mixed Ombrophilous Forest remnants. *Floresta Ambient.* **2018**, *25*, e20170192. [CrossRef]
57. Pinotti, L.C.A. Subsídios à Conservação e Manejo Sustentável de Remanescentes de Floresta Ombrófila Mista. Master's Thesis, Universidade Federal do Paraná, Curitiba, Brazil, 2017.
58. Food and Agriculture Organization; Sustainability Assessment of Food and Agriculture Systems. Smallholders App; Version: 2.0.0. Available online: https://apkdownloadforandroid.com/org.fao.mobile.safa (accessed on 15 February 2017).
59. Chaer, G.M.; Resende, A.S.; Campello, E.F.C.; De Faria, S.M.; Boddey, R.M.; Schmidt, S. Nitrogen-Fixing legume tree species for the reclamation of severely degraded lands in Brazil. *Tree Physiol.* **2011**, *31*, 139–149. [CrossRef]
60. Sprent, J.I. *Legume Nodulation: A Global Perspective*; Wiley-Blackwell: Chichester, UK, 2009.
61. Carpanezzi, A.A.; Laurent, J.-M.E.; Carvalho, P.E.R.; Pegoraro, A.; Baggio, A.J.; Zanon, A.; Oliveira, E.B.; Tadeu Iede, E.; Rotta, E.; Sturion, J.A.; et al. Manual técnico da bracatinga (*Mimosa scabrella* Benth). *Documentos* **1988**, *20*, 1–70.
62. Carvalho, P.E.R. Composição e crescimento da bracatinga em povoamento natural. In *Proceedings of the Seminário Sobre Atualidades E Perspectivas Florestais*; Embrapa—URPFCS: Curitiba, Brazil, 1991; pp. 67–75.
63. Falk, D.A.; Palmer, M.A.; Zedler, J. *Foundations of Restoration Ecology*; Island Press: Washington, DC, USA, 1997.
64. Mansourian, S.; Vallauri, D.; Vallauri, N. *Forest Restoration in Landscapes*; Springer: New York, NY, USA, 2005.
65. Garrastazú, M.C.; Mendonça, S.D.; Horokoski, T.T.; Cardoso, D.J.; Rosot, M.A.D.; Nimmo, E.R.; Lacerda, A.E.B. Carbon sequestration and riparian zones: Assessing the impacts of changing regulatory practices in Southern Brazil. *Land Use Policy* **2015**, *42*, 329–339. [CrossRef]
66. Sambuichi, R.H.R.; de Moura, I.F.; de Mattos, L.M.; de Ávilda, M.L.; Spínola, P.A.C.; da Silva, A.P.M. (Eds.) *A Política Nacional de Agroecologia e Produção Orgânica No Brasil: Uma Trajetória de Luta Pelo Desenvolvimento Rural Sustentável*; Ipea: Brasilia, Brazil, 2017.
67. Food and Agriculture Organization. *Experiencias Exitosas de Políticas Agroambientales en Brasil*; De Andrade, R.B., Cabral, P.G., Eds.; FAO: Santiago, Chile, 2018.
68. Serviço Florestal Brasileiro Inventário Florestal Nacional. Available online: www.florestal.gov.br/inventario-florestal-nacional (accessed on 15 April 2020).

land

MDPI

Article

Pollination Potential in Portugal: Leveraging an Ecosystem Service for Sustainable Agricultural Productivity

Caroline Wentling, Felipe S. Campos, João David and Pedro Cabral *

Campus de Campolide, NOVA Information Management School (NOVA IMS), Universidade Nova de Lisboa, 1070-312 Lisbon, Portugal; cwentling@novaims.unl.pt (C.W.); fcampos@novaims.unl.pt (F.S.C.); jcdavid@novaims.unl.pt (J.D.)
* Correspondence: pcabral@novaims.unl.pt

Abstract: As urbanization and agriculture increase worldwide, habitats and food sources for wild pollinators are often fragmented or destroyed. As wild pollinators contribute both resilience and variety to agricultural fields, it is desirable to implement land management practices that preserve their well-being and ability to contribute to food production systems. This study evaluates continental Portugal for its change in suitability to host bee's pollinator species (*Apis mellifera*) from 1990 to 2018. It uses the InVEST crop pollination modeling tool and CORINE Land Cover, as well as parameterization to produce pollinator abundance and supply maps. These are generalized to municipality boundaries to provide actionable insights to farmers and policymakers and strengthen land management practices. It finds that the potential for pollination services is growing, with averages of both pollinator abundance and supply indices improving by 8.76% across the continental territory in 28 years. The study results are validated using another pollination index derived from a study that is based on expert opinion and field sampling in a sub-region of Portugal. This method of aggregation of model results and comparison of the percent difference by administrative boundary has the potential to better inform both policymakers and farmers about the pollination potential on a local level, as well as inspire interventions for future productivity.

Keywords: land use changes; wild bees; land management practices; validation; InVEST model

Citation: Wentling, C.; Campos, F.S.; David, J.; Cabral, P. Pollination Potential in Portugal: Leveraging an Ecosystem Service for Sustainable Agricultural Productivity. *Land* **2021**, *10*, 431. https://doi.org/10.3390/land10040431

Academic Editor: Diane Pearson

Received: 12 March 2021
Accepted: 15 April 2021
Published: 17 April 2021

1. Introduction

Ecosystem services are natural processes from which human benefit, whether directly or indirectly [1]. Because natural "capital" (i.e., trees, atmosphere, carbon, information, nourishment, etc.) and human reliance on it is difficult to quantify economically [2] its value is often discounted in policy development. However, these services have tremendous effects on our wellbeing, resilience, and markets [3], making them a valuable addition to discussions about sustainable land management practices [4].

Pollination is one of these services from which humans reap significant benefit [5]. Though wild bees provide essential pollination services to both wild plants and crops alike [6], "agricultural intensification jeopardizes wild bee communities and their stabilizing effect on pollination services at the landscape scale" [7,8]. This type of loss has impacts on national economies. The estimated annual value of ecosystem services provided by wild insects and other animal pollinators (including pollination, dung burial, pest control, and wildlife nutrition) equates to more than USD 57 billion [9]. Other estimates project losses of USD 1.4 billion of the gross domestic product (GDP) between 2011 and 2050 in the US alone due to pollinator loss [4]. Insect pollination accounted for 35% of global food production in 2004 as well as 75% of crop types, [8,10]. Losses in crop pollinators are expected to affect the world supply of fruits, vegetables, oilseeds, and cotton, leading to direct and indirect effects on global commodity supplies and prices [4]. Worldwide declines of pollinators can catalyze similar trends in wild plant species [7]. These implications on both human well-being and environmental vibrancy necessitate the utilization of models that can characterize the effects of current trends, as well as predict and evaluate potential future scenarios [11].

Though there is a multitude of species responsible for the pollination of human consumable crops [7], many farmers employ domestic bees for managed pollination. However, the utilization of wild bees increases temporal stability as well as additional efficiency for certain crop species [6]. Though not strictly necessary, some vegetable species yield higher quality and more pest-resilient crops, in addition to improved seed production [10] after being visited by wild pollinators. Heterogeneous and organic fields are usually more suitable, both in terms of habitat appeal as well as in food resources, which may attract pollinators within their foraging ranges [4,12,13]. Understanding these types of behavior and interdependencies can improve the way farmers and policymakers adjust their practices to improve yields, as well as maintain sustainable supplies of pollination services into the future [7].

Previous studies on pollinator suitability have been performed in sub-national regions around the world [13] at more general continental or global levels [2], are limited to specific crop types [14], or landscape types [10]. Some of the previous literature provide frameworks for incorporation into future studies [5], some leverage or derive theoretical monetization models [15] and some do not incorporate spatial dependence into their models [9]. Derivation and validation of these studies range from labor intensive field sampling [16] or leveraging of primary sources [7], to expert opinion [17], to predictions of models derived from environmental inputs (such as Land Use Land Cover (LULC), climate, or topology [8]).

This study demonstrates the viability of applying a spatially dependent model of pollinator suitability to an entire continental area (corresponding to the mainland area of Portugal, designated by the Nomenclature of Territorial Units for Statistics (NUTS) level 1 code PT1: "Continente" in Portuguese, or "continental" in English) and then aggregating the interim results to subregions (NUTS subdivision 3) to evaluate both overall trends and local changes over time. This aggregation provides new opportunities for land management and innovation practices applicable to various levels of local administration. This type of investigation has not yet been applied to Portugal.

To this end, the study evaluates the changes in pollination suitability in continental Portugal from 1990 to 2018 of a representative guild characterizing the behavior of the European honeybee (*Apis mellifera*). It derives pollinator abundance and supply indices from input LULC raster maps for the area as well as parameterized pollinator guilds. These are processed by the Integrated Valuation of Ecosystem Services and Trade-offs (InVEST) crop pollination model, which incorporates spatial dependency of nearby floral resources and nesting sites in relation to pollinator foraging ranges. The resulting raster maps are aggregated to administrative municipalities, and the percent variation (PV) is calculated. The results are evaluated for their trajectories of change and validated via the extrapolation of a local pollinator index based on in-field sample collected data and expert opinion.

2. Materials and Methods

2.1. Study Area

Portugal is a European country of about 92,212 square kilometers on the southwestern corner of the Iberian Peninsula [18]. It contains the most western point in continental Europe and shares a land border with Spain. Portugal experiences Mediterranean climate of dry, hot summers and wet, cool winters [19], though this varies throughout the territory's microclimates (generally categorized as cooler and rainier in the north while drier and hotter in the south). According to CORINE Land Cover (CLC) of 2018, approximately 3.83% of the country's land cover is artificial surfaces, 47.81% is agricultural land, and 46.48% forests, with the remainder, made up of wetlands and water bodies (Figure 1) [20]. 28 years prior, artificial surfaces only covered 1.9% of the land, with 47.80% and 47.92% of the area dedicated to agricultural and forest land, respectively. This sizeable increase in artificial cover is consistent with the high rates of urbanization seen in and around Portugal's major cities. Desertification, the degradation of dryland also affects the changing classifications of land cover especially in the interior of the country [21].

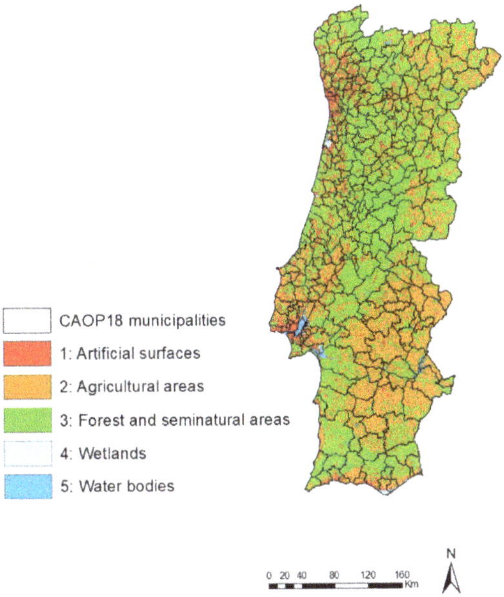

Figure 1. CORINE Land Cover of 2018 (level 1) distribution in Portugal, with municipal boundary definition (Carta Administrativa Oficial de Portugal: CAOP 2018).

Portugal is composed of 308 "concelhos" (municipalities, NUTS 3), with 278 of these located on the mainland [18]. Of Portugal's 240.7 billion USD GDP in 2018, 2.05% was produced by the agriculture, forestry and fishing industry [22]. With almost half of continental Portugal's land surface devoted to agriculture, there is tremendous value in ensuring the successful production of cultivated crops. It is estimated that Portugal is home to more than one thousand pollinating insect species, including a variety of bee, hoverfly, butterfly, and flower beetle species [16,23,24]. As of 2018, 680 distinct bee species have been collected and catalogued in Portugal [25]. Within the River Minho area alone, 200 distinct species were catalogued for a smaller scale study on pollination services [16]. According to the Joint Research Centre (JRC) Technical Report, Portugal demonstrated the "highest increase of pollination potential" from 2000 to 2006 in the European Union (EU) [26]. The main pollination season in Portugal can range from March to September, which includes the season of most active airborne pollen particles in the country as well as the general period for crop pollinator foraging in the north half of the globe [19,26].

2.2. *Software and Data Management*

This study was carried out using the free and open source InVEST crop pollination model, available under the open data license [27,28]. It also leverages the proprietary ArcGIS software (ArcMap 10.6) to perform the spatial temporal variation model, visualize the results, and for validation. All data included in the study is open data freely available to the public through the portals described in Section 2.3.

2.3. *InVEST Crop Pollination Model*

InVEST is a software platform of the Natural Capital Project and a suite of models to evaluate and chart a variety of ecosystem services, ultimately to inform decisions on how to manage these natural resources by quantifying their economic impact. It has been used in previous academic studies to marry macroeconomic scales with local environmental processes to predict multiple future scenarios of varying degrees of environmental action to "resonate with political economy audiences" [4].

The InVEST crop pollination model produces pollinator abundance and supply indices, which are scaled from 0 (least suitable) to 1 (most suitable) [27]. Abundance index represents the likely location of their activity, while supply index describes the likelihood, based on proximal nesting sites and food resources of the location and foraging ranges of the species, for pollinators to nest in a space. The results characterize wild bee pollinator guilds (groups of bee pollinators demonstrating similar nesting and foraging preferences as well as foraging distances and relative abundance).

The model utilizes land cover raster maps as well as persistent bio and guild tables as inputs. It incorporates habitat parameters (estimated nesting site and floral resource availabilities, and relative abundance per guild) for each cell of the input raster, considering the floral parameters of its neighbors [20,26]. One of the key features of the model is the incorporation of foraging distance, which allows the model to bridge the possible spatial separations of nesting and foraging habitats [8,12,29]. This model was selected for its accommodation of the spatial dependency required in such geographically explicit studies.

To make meaningful comparisons between time frames, pollinator abundance and supply indices for each pixel were generalized via zonal statistics into the 278 municipalities under study. The resulting statistical means of each municipality were utilized to calculate the variation from 2018 to 1990, according to Equation (1):

$$PV_c = \frac{\Delta Abundance_{2018c} - \Delta Abundance_{1990c}}{\Delta Abundance_{1990c}} \quad (1)$$

where PV_c is the percentage variation index for delivering pollination abundance for year 2018 in comparison to the baseline year 1990 for each concelho©. The general flow is depicted in Figure 2, with a comprehensive modeling workflow for the percent variation of abundance index. This process was executed in ArcGIS software (ArcMap 10.6).

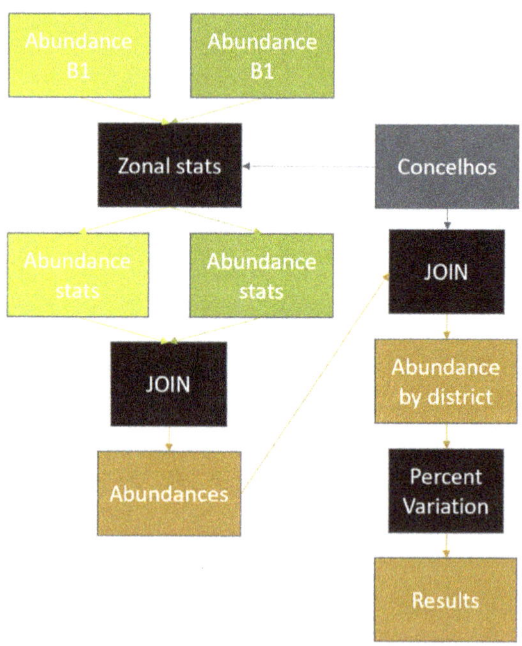

Figure 2. Schematic representation for analyzing the statistical comparison model flow.

2.4. Spatial Data

Land cover raster maps utilized are available from the Copernicus project, provided by the European Environmental Agency (EEA) [20]. The CLC classification includes 44 distinct subcategories that fall within five major areas: artificial surfaces, agricultural areas, forest and semi-natural areas, wetlands, and water bodies. As this study seeks to understand the change of ecosystem services over time, the earliest and latest available years (1990, 2018) are used. The raster data have a spatial resolution of 100 m and a minimum mapping unit of 25 ha [20]. All "slivers", landmasses associated with continental Portugal but removed by water, have been excluded from the study area. All data in the study is in the common coordinate system ETRS_1989_Portugal_TM06.

The biophysical table (required for InVEST crop pollination) corresponds to the LULC classifications to establish suitability for nesting and floral resources of each raster input (see [27] for additional details). Of particularly high suitability are certain agricultural areas and forest edges, both of which tend to provide heterogeneity of habitat in the form of diverse nesting space and floral resources within a small area, often promoting insect activity [8]. Nesting and floral resources parameter values are provided in the supplementary material of [8]. The values are derived from expert opinion and leveraged in their European continent level of pollination, also utilizing CLC input raster maps.

This study utilizes the CLC classification conversion provided in a study on pollination services across the European continent [8], as it is directly applicable to the study area (Appendix A). The conversion parameterizes all 44 classifications of CLC, generalized as a single season (versus representation of seasonal pattern variations spanning a calendar year) and a single nesting substrate (no distinction between cavity or ground preference), thus the results are representative of these generalizations.

Guild parameters assign values to represent the different behavior patterns of various bee species. These patterns include preferences for different nesting sites and floral resources, as well as relative prevalence and foraging ranges (Appendix A).

Though InVEST has the capability to model multiple nesting types, seasons, and bee guilds, insufficient data exists in previous literature to leverage the full potential of the tool (let alone the variation of parameter values due to environmental conditions [12]). Therefore, values for individual parameters were aggregated from a variety of sources [8,14,30] to describe one pollinator guild. *Apis mellifera*, better known as the European honeybee, is considered "the most economically valuable [pollinator] of crop monocultures worldwide" [8] and is widely employed in managed crop pollination and honey production and is native to mainland Portugal. This species is well studied and can be easily characterized as per requirements of the InVEST model.

The Carta Administrative Official de Portugal 2018 (CAOP 2018) is originally available from DGT (Direção-Geral do Território), a portal providing geodesic and geographic information services by the Portuguese ministry of agriculture, sea, and environment, and territorial management [2,11,13]. It includes 278 continental Portugal administrative territories, ranging in size from 7.94 to 1720 km^2 (São João da Madeira and Odemira municipalities, respectively).

2.5. Validation

The InVEST ecosystem service modeling toolset is well established and widely used in academic study [15,27], which supports its reliability. However, it has some recognized shortcomings and is subject to the quality of input data [16]. Validation of the results is required prior to their influence on future decisions on the management of ecosystem services.

So that the study results can be meaningfully compared to the validation methods, the pollination indices are normalized such that they are distributed between their reported minimum index (adjusted to 0) and their reported maximum value (normalized to 1). Another study performed in a subsection of continental Portugal is leveraged as validation. The study developed a Pollination Suitability Index for Riverine Landscapes (PSIRL) in the River Minho (norther border of Portugal with Spain) in 2018 [27]. Though this approach has its own limitations (the study considers insect pollinators in general, not just *Apis mellifera* and it derives specifically for riparian areas, requiring generalization and translation to the input CLC LULC), its index is derived from expert judgment, floral diversity, and actual field surveys, increasing the overall confidence in the results. No other spatially comprehensive yet reliable data exist in Portugal for validation.

The validation process is depicted in Figure 3. The original PSIRL index is translated to land use codes used by the CLC LULC (see Appendix A), The "unclassified" features representing water areas are selected then intersected to create water edge lines. These are buffered by 10 m and merged with the LULC polygons. These are then converted back to rasters and zonal statistics are applied. These are joined to the CAOP municipal boundaries, normalized, and then compared to the normalized index values of the results.

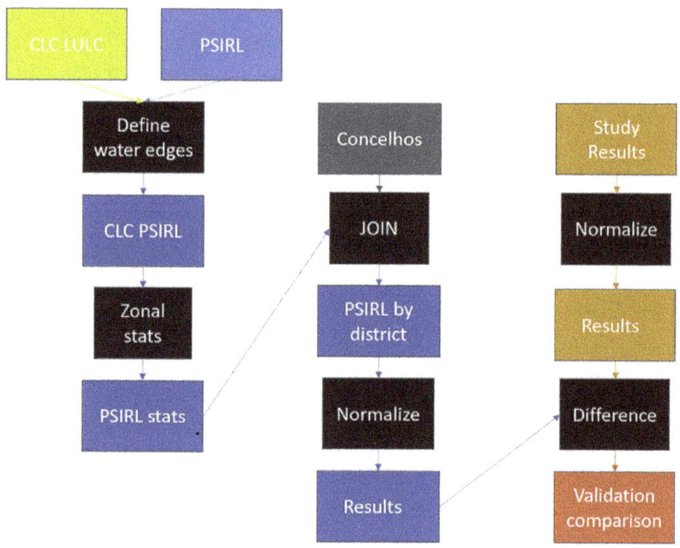

Figure 3. Pollination Suitability Index for Riverine Landscapes (PSIRL) validation flow.

3. Results

3.1. Land Use Land Cover Evolution from 1990 to 2018

Table 1 demonstrates the land surface utilization (LULC areal data) within Portugal in 1990, and then the percent variation from this baseline to 2018. These have been generalized to the broadest category (CORINE Land Cover designation level 1). Note that each L1 category is not associated with homogeneous biophysical parameters. The table demonstrates a doubling of artificial surfaces between 2018 and 1990 (largely inhospitable to bees), as well as a 1.8% and 3.0% drop in largely appealing habitats (agricultural and forest cover, respectively) over time. This would suggest an overall decrease in pollination services over time.

Table 1. Land Use Land Cover (LULC) percentage by 1990 and its percent variation (PV) by 2018.

LULC	1990 LULC (%)	2018 PV (%)
Artificial surfaces	1.90	101.27%
Agricultural areas	48.70	−1.83%
Forests and seminatural areas	47.92	−3.00%
Wetlands	0.32	7.08%
Water bodies	1.15	33.59%

3.2. Pollinator Abundance and Supply Indices

Figure 4 depicts the spatial variation of pollinator abundance and supply indicators for 2018. As one would expect, both indices follow similar spatial patterns: less hospitable in the urban and water areas (red), more hospitable in forested areas (green). Coastal areas are particularly unfriendly, both in the West and the South, with much of eastern Alentejo region exhibiting low suitability as well. On the other hand, much of north eastern Portugal and around the border of Alentejo and the Algarve regions appear to be quite suitable along both indices.

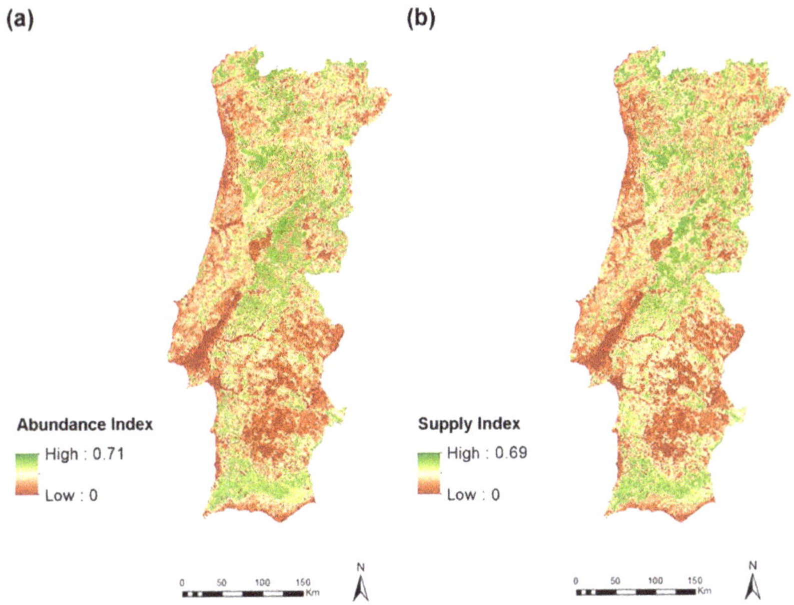

Figure 4. Spatial distribution of InVEST crop pollination model outputs (**a**) abundance index and (**b**) supply index in 2018.

Table 2 displays the general statistics of the raster results. The minimum index value remains zero for those areas that are unsuitable for pollinator activity (both inhospitable to bee nests as well as outside of the range of foraging). Overall, the means and maximum values for each index have increased between 1990 and 2018, indicating that the overall suitability for pollination services in Portugal is growing. This yields a slightly larger standard deviation, which reflects a greater range of index values distributed across continental Portugal.

Table 2. Statistics of InVEST crop pollination model results in 1990 and 2018.

Year	Index	Min	Max	Mean	Std
1990	Abundance	0.000	0.700	0.274	0.163
1990	Supply	0.000	0.709	0.274	0.163
2018	Abundance	0.000	0.711	0.298	0.174
2018	Supply	0.000	0.694	0.298	0.174

3.3. Pollination Service Changes from 1990 to 2018

Once the results are associated by municipality, inferences about trends for each administrative boundary are more easily understood. Ideally, this will contribute to better policy making at the district level. Figure 5 displays both the percent variation of abundance (color of polygon area) and supply (color and size of overlaying triangle,) from a baseline of 1990 to 2018. Red colors indicate negative changes in suitability indices, while green and blue indicate positive trends. Yellow indicates no significant change over the 28-year study period. The triangle size also indicates the degree of deviation of supply from the 1990 baseline. Both changes in supply and abundance tend to fall into the same categorizations. Those areas that experience differences are usually (but not exclusively) characterized by a supply index that is slightly more extreme than that of abundance. This suggests that the pollinators may be more selective about their habitats (origins) than in the areas they are willing to traverse in search of food.

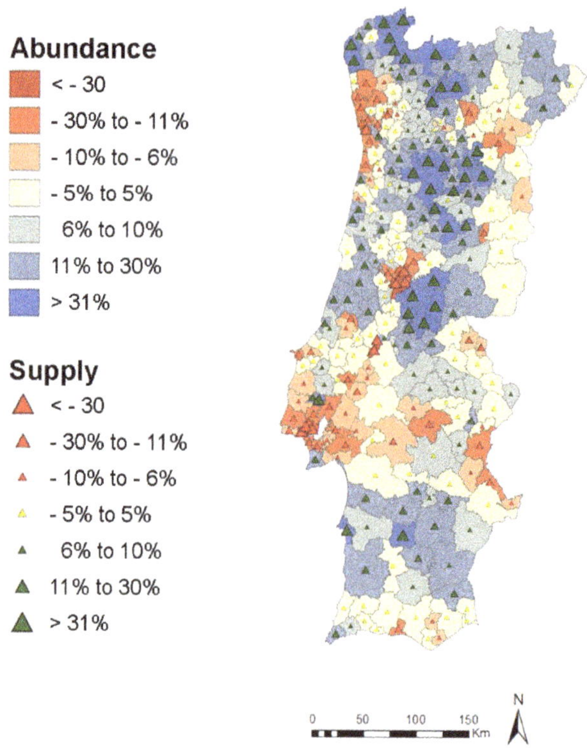

Figure 5. Percent variation (PV) from 1990 to 2018 of abundance and supply indices by municipality.

3.4. Validation

The PSIRL validation technique required translation of the given PSIRL index values to the original input LULC raster map (Appendix A). The resulting difference map (Figure 6) includes an overlay of the riparian areas (including a buffer of 300 m from water areas), as well as an indication of the area of which the PSIRL index was originally derived (the River Minho area in north western Portugal, identified with a red box). The yellow zones indicate those in which the validation demonstrates good coincidence with the results of the study (within a 5% tolerance), whereas the stronger purple and red colors indicate larger discrepancies between the two methods (a maximum discrepancy of 38%). Clearly the results incorporate some amount of spatial autocorrelation, though surprisingly these are not necessarily correlated with riparian adjacent areas as one might expect. The PSIRL tends to slightly over-predict the supply of pollinators as compared to the results of the study. The difference map indicates a strong spatial similarity between the two models, strengthening the confidence in the study results.

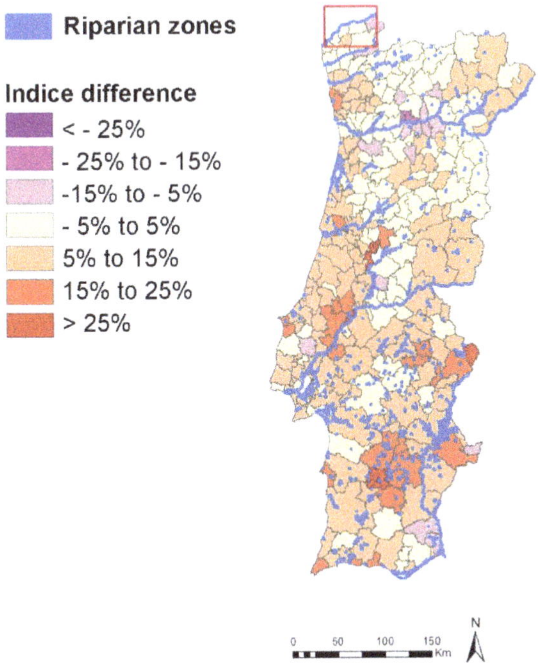

Figure 6. Difference between study results and Pollination Suitability Index for Riverine Landscapes (PSIRL) with riparian zone overlay. Red box indicates River Minho area, in north western Portugal.

4. Discussion

4.1. Study Significance

Bees require suitable places to nest and sufficient food sources near nesting sites to sustain them [1]. These and other factors have been applied to a model that produces maps of projected pollinator activity within Portugal. Too often, stakeholders (farmers, policy makers, economists, etc.) ignore the subtle interactions between ecosystem services and production, which can be to their own detriment when those nebulous costs outstrip their values [4]. It is estimated that, at the current rate of land use transformation, the United States gross domestic product (GDP) will suffer a loss of 0.02% (or 15 billion USD) due to reduced wild pollinator habitats near agricultural sites [8], which can have ripple effects in other industries to compensate for the deficit.

Results of this study and other such investigation will ideally support farmers, land developers, and policy makers alike with better information from which to make decisions about how to better manage these resources as well as improve economics systems that depend on them by maximizing their sustainability. Agriculture and thriving pollinator communities are not mutually exclusive. In fact, well-managed cropland can be economically and ecologically productive [6,7,10,11]. For instance, farmers could identify locations for crops based on maximizing exposure to wild pollinators, adjust their management towards organic practices, or maintain heterogeneous nesting substrates that would attract diverse and productive pollinator populations [13]. Likewise, configuring farms towards a variety of pollinators (instead of just the domesticated varieties) can produce better yields, as different pollinators are associated with varying levels of productivity for certain crops [29]. Even the understanding of the tendency of larger bees to populate new fields and smaller bees to prefer older fields [8], or the observed abundance and variety of pollinators in forest edges and grasslands [6] can assist with the development of management strategies. Further, the understanding of the relationships between space, crops, and pollinators may provide an incentive to better care for areas beyond crop fields [31] or mitigate the appeal of monoculture practices [7]. In fact, there is potential for the benefits of ecosystem services management practices to positively impact other areas within foraging distances of the appealing habitat sites. On the flip side, poor planning regarding conversion patterns of forest to agriculture can have devastating impacts on wild bee populations that will also undercut the productivity of the new agricultural land [7].

The association to municipalities provides a simplification of the detailed information to ease comprehension of the big picture, such that areas requiring intervention (those tending towards lower suitability) can be triaged and evaluated more efficiently. Though pollination may not be strictly required to achieve sufficient caloric intake, indeed many staple foods do not require this type of sexual reproduction, the production of many valuable nutrients require pollination [27], and pollination services haves been linked to qualitative (nutritional content, appearance) and quantitative (production yields) factors that boost economic value of agricultural production [32]. Ideally, stakeholders will be able to model and evaluate different policies and their effects on farm productivity, optimizing both resilient biodiversity as well as economic yields [1,6,10]. Methods to achieve this could be coordinating reserved land areas that provide pollination services through integration of natural areas throughout agricultural areas [31]. Further, incentive programs that promote healthy management practices or payment schemes could be organized, in addition to the inherent benefits experienced by the implementing farmers [26]. As this is a relatively new concept, there is much room for novel methods of accounting for ecosystem services within the economic structure of farmers and other land managers.

4.2. Critical Analysis

This study provides a valuable baseline indicator of pollinator services within continental Portugal. Overall, since 1990 there have been significant, polarizing tendencies of municipalities across Portugal. The percent variation of likelihood for pollinators to be active between 1990 and 2018 swings from −69% (Pedrógão Grande) to 107% (Ponte da Barca and Vila de Rei). As one would expect, there are concentrations of negative trend areas that are associated with major city areas and likely rapid urbanization (such as Lisbon, and Porto areas), as well as some areas of vast agriculture swaths (the south west portion of Alentejo Central), which are less hospitable. On the other hand, it is promising to see the constant improvement through central northern Portugal. Interestingly the areas of greatest percent increase and decrease are located adjacent to each other: the municipalities along the border of Médio Tejo and Beira Baixa both exhibit extremely positive trends since 1990, yet just across the border, several municipalities in Região de Leiria include some of the most negative changes in the same time frame. This is due to recent fires resulting in large swatch of burnt areas in these regions, making them inhospitable to pollinators, though their neighboring forested areas demonstrate favorable habitats. Some areas have significantly changed from the 1990 baseline. Ponte da Barca continues to improve its tendency towards pollinator likelihood in both abundance and supply, rising to a high of 107 and 109 percent variation increases, respectively.

The positive trend of pollinator abundance and supply indicators are consistent with the findings of another study that Portugal demonstrated impressive improvements in pollination potential [7]. From the maps of Figure 5, policy makers and farmers alike may better understand the existing trends of pollinator suitability since 1990, using this information to support new interventions that may increase pollinator suitability within each municipality. Of course, pollinator suitability may not be a priority to certain urban areas such as Lisbon or Porto, or municipalities specifically cultivating crops that do not require pollination such as the Douro region. Other areas of agricultural swatches that demonstrate reduced or unchanging suitability may benefit from a re-organization of agricultural land to better suit natural pollinator activity. These include areas such as Alentejo Central and Algarve areas, though many other persistent or worsening regions are distributed throughout Portugal.

4.3. Additional Findings

The results also demonstrate the stark impacts of forest fires on pollinator suitability, such as the dramatic changes in the Região de Leiria. Portugal is prone to fires in the hot, dry summers. Though these are often uncontrollable natural phenomena, understanding their effects on pollinator activity among other ecosystem services in conjunction with social loss may strengthen the attempts to better manage forest areas and inspire more radical interventions to recover the areas in the wake of such devastation. More granular studies may consider excluding burnt areas from their studies, though they were retained here as they contribute to the overall trends (encompassing both natural and human influence) in mainland Portugal.

4.4. Research Limitations

Though the results of this study are promising, there are several limitations of note and opportunities for future improvement. The results of this study are limited the to the available data and certainly leave much opportunity for further evolution. Because pollinators can differ significantly from ecosystem to ecosystem [16], leveraging the parameters of similar studies in other regions is often inappropriate. Better characterization of local bee species throughout the study area may yield more accurate characterization of the potential of this ecosystem service. For example: the pollination potential characterized in this study is relative only to *Apis mellifera*, which has different habitat and foraging preferences and activities (such as potential foraging distances) than other smaller, wild species. However, due to scarcity of data on the behaviors and preferences of other wild bee guilds in Portugal, only a single bee specie was characterized. The application of the methodology undertaken by [26] in the Minho river area (counting the number and characterization of pollinators active in a particular area) to the entire country was outside of the scope of this project but this and the inclusion of expert based models (EBM) could enhance future research [27]. In addition to better characterization of pollinators, more detailed parameterization of the biophysical table of LULC designations (such as the inclusion of multiple nesting substrates and seasons) could more accurately reflect the actual pollination activity throughout the year.

InVEST models measure the potential of the study area to provide pollination for bee pollinators. Additional considerations outside of the model purview will affect the actual pollination supply, such as the lack of accounting for pollinator persistence over time. Likewise, many other non-bee pollinators (such as butterflies, bats, moths, and birds) that are active in Portugal are not accommodated in the model. These and other such inherent limitations are described in the InVEST documentation in greater detail [11]. Notably, the model does not distinguish between natural or artificially initiated changes in pollination potential-discerning the source requires savvy technicians and good understanding of the local context to presume.

Regarding the input raster maps, the minimum mapping unit of 25 hectares of the CLC LULC data does not accommodate the impact of potential pollinator habitats or foraging supplies smaller than this area (such as in green spaces in urban areas). Similarly, the study does not accommodate the implementation of agricultural practices that may alter the desirability of the area for pollinators, such as the accommodation of nesting sites or use of pesticides. Though pollination is sensitive to both aggregation and spatial resolution as an ecological service that involves stocks and dynamics, it is expected that the CLC mapping units are appropriate for the scale of study [27]. Likewise, the study is subject to the accuracy of the CLC classifications. Any assumptions or misclassifications will propagate through this study.

4.5. Future Opportunities

The InVEST crop pollination model, in additional to providing suitability indices of pollinator habitat and foraging supply, can model a yield index for pollination impacts on existing agriculture [8]. This requires vector data detailing the geospatial location of farms along with their crop types, dependence on pollinators, abundance of managed pollinators, farm nesting sites and floral resources. This study did not have access to national farm data, and the establishment of the required attributes would require additional investigation outside of the scope of this project, likely including the need of expert opinion to properly assign values to these parameters. This is an area of potential study in future endeavors.

Suitability of edge environments has been noted to differ from that of non-edge environments. For example, forest edges are particularly suitable as pollinator habitat [13,16], but are not accommodated in the InVEST model. Other studies have included additional characterization of these areas, which could improve the approach with this additional nuance. Roadsides and riparian areas are further examples of opportunities for model adjustment and finer characterization. The study could also benefit from the accommodation of more granular parameterization, including that of urban areas hosting managed bee colonies or integrating green and biodiverse areas within the built environment, distinction between forest compositions, or refuge areas that may experience different pollinator assemblages.

5. Conclusions

The land use land cover impact on pollination services distributed over time and space was studied in the context of continental Portugal. The InVEST crop pollination service modeling tool was leveraged to understand the spatial relationship between pollinator abundance relative to a landscape's available habitat and food resources in accordance with their behavior and preferences. The results demonstrated an overall improvement in wild pollinator hospitality across the country, though several municipalities are becoming increasingly weaker in their suitability for such services. The relative distribution of pollinator hospitality indices was validated via a local pollination index based on field sampling and expert opinion.

In Portugal the measured distribution of tons of crop production in the country is almost equally distributed between known dependency (34.2%), non-dependency (33.4%) and unknown dependency on pollinators [32] With more than a third (and potentially up to two thirds) of the production weight relying on pollinators, there is a large economic incentive for agro-farmers, the primary beneficiary of pollinator services [33], to incorporate pollination ecosystem services into their practices and protect these resources. Further, secondary beneficiaries—such as consumers of more nutrient dense crops or governments receiving greater tax revenues—will experience positive effects from the products of these measures as well.

Author Contributions: Conceptualization, C.W., F.S.C., J.D. and P.C.; methodology, C.W., F.S.C., J.D. and P.C.; formal analysis, C.W.; investigation, C.W., F.S.C., J.D. and P.C.; writing—original draft preparation, C.W.; writing—review and editing, C.W., F.S.C. and P.C.; visualization, C.W. and F.S.C.; supervision, F.S.C. and P.C.; project administration, P.C. All authors have read and agreed to the published version of the manuscript.

Funding: This study was supported through the FCT (Fundação para a Ciência e a Tecnologia) under the projects PTDC/CTA-AMB/28438/2017—ASEBIO and UIDB/04152/2020—Centro de Investigação em Gestão de Informação (MagIC).

Institutional Review Board Statement: Not applicable.

Informed Consent Statement: Not applicable.

Data Availability Statement: The data that support the findings of this study are available on reasonable request from the corresponding author.

Acknowledgments: The authors would like to thank the anonymous Reviewers for their comments which helped improving the quality of this manuscript.

Conflicts of Interest: The authors declare no conflict of interest.

Appendix A. Model Parameterization

Table A1. Land Use Land Cover (LULC) classes, parameterization of the biophisical table for the InVEST model (nesting and floral resources), and PSIRL final scores used in validation.

Classes	Nesting Resourses	Floral Resourses	PSIRL Final Scores
Riparian scrubland	0.8	0.9	0.83
Broad-leaved forest	0.8	0.9	0.83
Natural grassland	0.8	1	0.81
Moors and heathland	0.8	1	0.81
Sclerophyllous vegetations	0.8	1	0.81
Transitional woodland scrub	0.8	1	0.81
Riparian forest	0.8	0.5	0.78
Fruit trees and berry plantations	0.4	0.9	0.6
Olive groves	0.5	0.4	0.6
Mixed forest	0.8	0.6	0.55
Sparsely vegetated areas	0.7	0.35	0.52
Inland marshes	0.3	0.75	0.52
Salt marshes	0.3	0.55	0.52
Coniferous forest	0.8	0.3	0.49
Annual crops associated with permanent crops	0.4	0.5	0.47
Complex cultivation patterns	0.4	0.4	0.47
Land principally occupied by agriculture	0.7	0.75	0.47
Agro-forestry areas	1	0.5	0.47
Non-irrigated arable land	0.2	0.2	0.39
Permanently irrigated land	0.2	0.05	0.39
Rice fields	0.2	0.05	0.39
Pastures	0.3	0.2	0.39
Continuous urban fabric	0.1	0.05	0.23
Discontinuous urban fabric	0.3	0.3	0.23
Industrial or commercial units	0.1	0.05	0.23
Road and rail networks	0.3	0.25	0.23
Port areas	0.3	0	0.23
Airports	0.3	0	0.23
Mineral extraction sites	0.3	0.05	0.23
Dump sites	0.05	0	0.23
Green urban areas	0.3	0.25	0.23
Sport and leisure facilities	0.3	0.05	0.23
Vineyards	0.4	0.6	0.2
Burnt areas	0.3	0.2	0.13

References

1. Costanza, R.; D'Arge, R.; de Groot, R.; Farber, S.; Grasso, M.; Hannon, B.; Limburg, K.; Naeem, S.; O'Neill, R.V.; Paruelo, J.; et al. The Value of the World's Ecosystem Services and Natural Capital. *Nature* **1997**, *387*, 253–260. [CrossRef]
2. Chaplin-Kramer, R.; Sharp, R.P.; Weil, C.; Bennett, E.M.; Pascual, U.; Arkema, K.K.; Brauman, K.A.; Bryant, B.P.; Guerry, A.D.; Haddad, N.M.; et al. Global Modeling of Nature's Contributions to People. *Science* **2019**, *366*, 255–258. [CrossRef]
3. Grafius, D.R.; Corstanje, R.; Harris, J.A. Linking Ecosystem Services, Urban Form and Green Space Configuration Using Multivariate Landscape Metric Analysis. *Landsc. Ecol.* **2018**, *33*, 557–573. [CrossRef]
4. Johnson, J.A.; Baldos, U.; Hertel, T.; Liu, J.; Nootenboom, C.; Polasky, S.; Roxburgh, T. *Global Futures: Modelling the Global Economic Impacts of Environmental Change to Support Policy-Making*; WWF UK: Woking, UK, 2020.
5. Patrício-Roberto, G.B.; Campos, M.J.O. Aspects of Landscape and Pollinators-What Is Important to Bee Conservation? *Diversity* **2014**, *6*, 158–175. [CrossRef]
6. Kennedy, C.M.; Lonsdorf, E.; Neel, M.C.; Williams, N.M.; Ricketts, T.H.; Winfree, R.; Bommarco, R.; Brittain, C.; Burley, A.L.; Cariveau, D.; et al. A Global Quantitative Synthesis of Local and Landscape Effects on Wild Bee Pollinators in Agroecosystems. *Ecol. Lett.* **2013**, *16*, 584–599. [CrossRef]
7. Klein, A.M.; Vaissière, B.E.; Cane, J.H.; Steffan-Dewenter, I.; Cunningham, S.A.; Kremen, C.; Tscharntke, T. Importance of Pollinators in Changing Landscapes for World Crops. *Proc. R. Soc. B* **2007**, *274*, 303–313. [CrossRef] [PubMed]
8. Zulian, G.; Maes, J.; Paracchini, M.L. Linking Land Cover Data and Crop Yields for Mapping and Assessment of Pollination Services in Europe. *Land* **2013**, *2*, 472–492. [CrossRef]
9. Losey, J.E.; Vaughan, M. The Economic Value of Ecological Services Provided by Insects. *BioScience* **2006**, *56*, 311–323. [CrossRef]

10. Picanço, A.; Gil, A.; Rigal, F.; Borges, P.A.V. Pollination Services Mapping and Economic Valuation from Insect Communities: A Case Study in the Azores (Terceira Island). *Nat. Conserv.* **2017**, *18*, 1–25. [CrossRef]
11. Grafius, D.R.; Corstanje, R.; Warren, P.H.; Evans, K.L.; Hancock, S.; Harris, J.A. The Impact of Land Use/Land Cover Scale on Modelling Urban Ecosystem Services. *Landsc. Ecol.* **2016**, *31*, 1509–1522. [CrossRef]
12. Greenleaf, S.S.; Williams, N.M.; Winfree, R.; Kremen, C. Bee Foraging Ranges and Their Relationship to Body Size. *Oecologia* **2007**, *153*, 589–596. [CrossRef]
13. Groff, S.C.; Loftin, C.S.; Drummond, F.; Bushmann, S.; McGill, B. Parameterization of the InVEST Crop Pollination Model to Spatially Predict Abundance of Wild Blueberry (*Vaccinium angustifolium*) Native Bee Pollinators in Maine, USA. *Environ. Model. Softw.* **2016**, *79*, 1–9. [CrossRef]
14. Lonsdorf, E.; Kremen, C.; Ricketts, T.; Winfree, R.; Williams, N.; Greenleaf, S. Modelling Pollination Services across Agricultural Landscapes. *Ann. Bot.* **2009**, *103*, 1589–1600. [CrossRef] [PubMed]
15. Roxburgh, T.; Ellis, K.; Johnson, J.A.; Baldos, U.L.; Hertel, T.; Nootenboom, C.; Polasky, S. *Global Futures: Assessing the Global Economic Impacts of Environmental Change to Support Policy-Making*; WWF UK: Woking, UK, 2020.
16. Santos, A.; Fernandes, M.R.; Aguiar, F.C.; Branco, M.R.; Ferreira, M.T. Effects of Riverine Landscape Changes on Pollination Services: A Case Study on the River Minho, Portugal. *Ecol. Indic.* **2018**, *89*, 656–666. [CrossRef]
17. Koh, I.; Lonsdorf, E.V.; Williams, N.M.; Brittain, C.; Isaacs, R.; Gibbs, J.; Ricketts, T.H. Modeling the Status, Trends, and Impacts of Wild Bee Abundance in the United States. *Proc. Natl. Acad. Sci. USA* **2016**, *113*, 140–145. [CrossRef] [PubMed]
18. *Direção Geral do Território (DGT)*; Carta Administrativa Oficial de Portugal: Lisboa, Portugal, 2018.
19. Camacho, I.; Caeiro, E.; Nunes, C.; Morais-Almeida, M. Airborne Pollen Calendar of Portugal: A 15-Year Survey (2002–2017). *Allergol. Immunopathol.* **2020**, *48*, 194–201. [CrossRef]
20. Copernicus Program, CORINE Land Cover. Available online: https://land.copernicus.eu/paneuropean/corine-land-cover (accessed on 30 March 2020).
21. ECA. *Desertification in the EU.*; Background Paper; European Court of Auditors (ECA)—Guardians of the EU Finances: Luxembourg, 2018.
22. World Bank Open Data—Indicators. Available online: https://data.worldbank.org/indicator/SP.POP.TOT?locations=PT&most_recent_value_desc=true (accessed on 3 April 2020).
23. Ciência Viva. Polinizadores em ação. Available online: http://www.cienciaviva.pt/aprenderforadasalaeaula/index.asp?accao=showobj&id_obj=1532 (accessed on 31 March 2020).
24. BioDiversity4All. Observations, *Apis Mellifera*. Available online: https://www.biodiversity4all.org/observations?place_id=7122&q=abelha&view=species&iconic_taxa=Insecta (accessed on 30 March 2020).
25. Baldock, D.W.; Wood, T.J.; Smit, J.T. The Bees of Portugal (Hymenoptera: Apoidea: Anthophila). *Entomofauna Suppl.* **2018**, *22*, 164.
26. Vallecillo, S.; la Notte, A.; Polce, C.; Zulian, G.; Alexandris, N.; Ferrini, S.; Maes, J. *Ecosystem Services Accounting: Part I—Outdoor Recreation and Crop Pollination*; Publications Office of the European Union: Luxembourg, 2018; ISBN 9789279773334.
27. Sharp, R.; Douglas, J.; Wolnie, S.; Arkema, K.; Bernhardt, J.; Bierbower, W.; Chaumont, N.; Denu, D.; Fisher, D.; Glowinski, K.; et al. *InVEST 3.8.9 User's Guide*; The Natural Capital Project, Standford University, University of Minnesota, The Nature Conservancy, and World Wildlife Fund: Standford, CA, USA, 2020; Available online: https://invest-userguide.readthedocs.io/en/latest/ (accessed on 9 April 2021).
28. BSD License Definition. Available online: http://www.linfo.org/bsdlicense.html (accessed on 23 September 2020).
29. Gathmann, A.; Tscharntke, T. Foraging Ranges of Solitary Bees. *J. Anim. Ecol.* **2002**, *71*, 757–764. [CrossRef]
30. Hammond, G.; Blankenship, M. Apis Mellifera. Available online: https://animaldiversity.org/accounts/Apis_mellifera/ (accessed on 17 September 2020).
31. Chaplin-Kramer, R.; Sharp, R.P.; Mandle, L.; Sim, S.; Johnson, J.; Butnar, I.; Milà, I.; Canals, L.; Eichelberger, B.A.; Ramler, I.; et al. Spatial Patterns of Agricultural Expansion Determine Impacts on Biodiversity and Carbon Storage. *Proc. Natl. Acad. Sci. USA* **2015**, *112*, 7402–7407. [CrossRef]
32. Porto, R.G.; de Almeida, R.F.; Cruz-Neto, O.; Tabarelli, M.; Viana, B.F.; Peres, C.A.; Lopes, A.V. Pollination Ecosystem Services: A Comprehensive Review of Economic Values, Research Funding and Policy Actions. *Food Secur.* **2020**, *12*, 1425–1442. [CrossRef]
33. Tashie, A.; Ringold, P.A. Critical Assessment of Available Ecosystem Services Data According to the Final Ecosystem Goods and Services Framework. *Ecosphere* **2019**, *10*, e02665. [CrossRef] [PubMed]

land

Article

Modelling the Impacts of Habitat Changes on the Population Density of Eurasian Skylark (*Alauda arvensis*) Based on Its Landscape Preferences

Nándor Csikós and Péter Szilassi *

Department of Geoinformatics, Physical and Environmental Geography, University of Szeged, Egyetem utca 2-6, H-6722 Szeged, Hungary; csikos@geo.u-szeged.hu
* Correspondence: toto@geo.u-szeged.hu

Abstract: The dramatic decline of the abundance of farmland bird species can be related to the level of land-use intensity or the land-cover heterogeneity of rural landscapes. Our study area in central Europe (Hungary) included 3049 skylark observation points and their 600 m buffer zones. We used a very detailed map (20 × 20 m minimum mapping unit), the Hungarian Ecosystem Basemap, as a land-cover dataset for the calculation of three landscape indices: mean patch size (MPS), mean fractal dimension (MFRACT), and Shannon diversity index (SDI) to describe the landscape structure of the study areas. Generalized linear models were used to analyze the effect of land-cover types and landscape patterns on the abundance of the Eurasian skylark (*Alauda arvensis*). According to our findings, the proportions of arable land, open sand steppes, closed grassland patches, and shape complexity and size characteristics of these land cover patches have a positive effect on skylark abundance, while the SDI was negatively associated with the skylark population. On the basis of the used statistical model, the abundance density (individuals/km*) of skylarks could be estimated with 37.77% absolute percentage error and 2.12 mean absolute error. We predicted the skylark population density inside the Natura 2000 Special Protected Area of Hungary which is 0–6 individuals/km* and 23746 ± 8968 skylarks. The results can be implemented for the landscape management of rural landscapes, and the method used are adaptable for the density estimation of other farmland bird species in rural landscapes. According to our findings, inside the protected areas should increase the proportion, the average size and shape complexity of arable land, salt steppes and meadows, and closed grassland land cover patches.

Keywords: land cover; land use; landscape structure; Eurasian skylark; farmland birds; prediction; Natura 2000

Citation: Csikós, N.; Szilassi, P. Modelling the Impacts of Habitat Changes on the Population Density of Eurasian Skylark (*Alauda arvensis*) Based on Its Landscape Preferences. *Land* **2021**, *10*, 306. https://doi.org/10.3390/land10030306

Academic Editor: Diane Pearson

Received: 15 February 2021
Accepted: 16 March 2021
Published: 17 March 2021

Publisher's Note: MDPI stays neutral with regard to jurisdictional claims in published maps and institutional affiliations.

1. Introduction

In the terrestrial ecosystems of the world, the dominant land-cover category is agriculture (38%), including the arable-land use type [1]. In Europe, this value is much higher, at 45% (EBCC, 2015). The agricultural land-cover category contains various land-use types with different levels of human impact. The heterogeneity and spatial structure of these land-use/land-cover (LULC) patches vary greatly across rural areas, which has strong impact on farmland-bird diversity in Europe [2,3]. Many articles have determined that the decreasing trend of farmland birds is strongly connected with the intensity of agricultural management (level of use of fertilizers etc.) [4–7]. Very few studies have investigated the dramatic decline of the abundance of farmland birds, and its connection with change in landscape structure and land-cover heterogeneity [7–9]. There are some regional (country)-scale studies that analyze the connection between land-cover types and farmland-bird population data [10–15]. These studies have indicated that the abundance of farmland birds is significantly connected with the intensity of agricultural cultivation, crop heterogeneity, and land-use change. Most articles focus on small, local study areas and analyzing the connection between Eurasian skylark (*Alauda arvensis*) abundance, and the proportions of crop type, height, coverage and heterogeneity [4,6,10,16–19]. The skylark does not prefer the fragmented landscapes by urbanized area, road network, hedgerows and heterogeneous land cultivation areas [7,20]. The agriculture is the dominant land use (matrix) of the European NATURA 2000 network, where the size and shape characteristics of different LULC patches, and the land cover heterogeneity can be essential for the protection of farmland bird species. Therefore, we hope that our results can be adding some new suggestions for the landscape planning and habitat design of national parks, NATURA 2000, and other protected areas. Our research also can provide important component for achieving the goals of the EU Birds directive [21].

The skylark is one of the most common farmland bird of rural landscapes in Eurasia, including Hungary. In the European Union, the Eurasian skylark has a declining trend in population between 2000 and 2018: Norway −47%, Lithuania −41%, France −38%, Czech Republic −29%, Hungary −24% and Germany −17%. Most individuals that breed in Central Europe spend the winter in the Mediterranean region, but small groups can stay in Hungary for winter [22]. This bird species have been introduced into the Nearctic, Australia and New Zealand [23,24]. From large-scale studies, habitat preferences, including for crop structure and heterogeneity are well-known. On the basis of small-scale regional-level studies, the regional-scale habitats and land-cover heterogeneity preference of a given species can be understood [10]. However, the connection between the spatial pattern of LULC patches (described with landscape indices), and skylark abundance is not clear.

In this study, we describe the landscape structure of rural landscapes with a very detailed (20 × 20 m minimal mapping unit) LULC map, the Hungarian Ecosystem Basemap (HEB). Comparing skylark abundance data with the HEB, we could identify preferred and nonpreferred skylark habitats, and calculate their landscape indices. The preferred habitat was separated into arable lands and grasslands because we wanted to analyze the effect of arable land and grassland landscape metrics on the skylark population. According to the pattern and process paradigm, which analyze the relationship between the landscape patterns spatial distribution and landscape processes, landscape indices are widely used as indicators of biodiversity and habitat changes [13,25–28]. After we identified preferred and non-preferred habitats for skylarks, we could calculate shape- and size-related class-level landscape metrics, and land-cover heterogeneity, and estimate the collective impact of these variables on skylark abundance [9–11,13,29,30].

The main goals of this study were to:

- identify skylark land-cover preferences on the basis of the local-scale LULC map;
- analyze the impact of landscape patterns of preferred and nonpreferred land-cover classes (habitats), and estimate the impact of all LULC-related variables (proportions, shape, and size characteristics of patches, heterogeneity) on skylark abundance; and
- estimate, based on our findings, the skylark population density inside the Natura 2000 Special Protection Area (SPA) of Hungary based on the HEB land cover categories.

According to our hypotheses the population density of skylark is predictable based on the preferred LULC categories of skylark and landscape indices (proportion of LULC categories and shape and size related landscape metrics). The methodology is adaptable for analyzing the impact of landscape composition on other farmland-bird populations, and for predicting the population density of the skylark, in protected areas, where field observation-based datasets are not available.

2. Materials and Methods

2.1. Study Area

Hungary is located in the Carpathian basin (45°43′ to 48°35′N and 16°06′ to 22°53′E) in central Europe, and is part of the Pannonian biogeographical region (Figure 1). The total area is 93,033 km*, and its elevation ranges from 77 to 1014 m a.s.l. The most important land-cover type (61%) is agricultural land [31]. A further 20.7% is natural and seminatural grasslands and forest, and 5.5% is built-up area. In the 1990s, a dramatic landscape change was mainly caused by land privatization. Agricultural lands with low quality and poor agroecological conditions were abandoned [32]. The common agricultural policy of the EU (strong decline of grazing livestock) and land abandonment caused the transformation of arable lands into non-cultivated lands, and the fast and spontaneous reforestation of open grasslands [33].

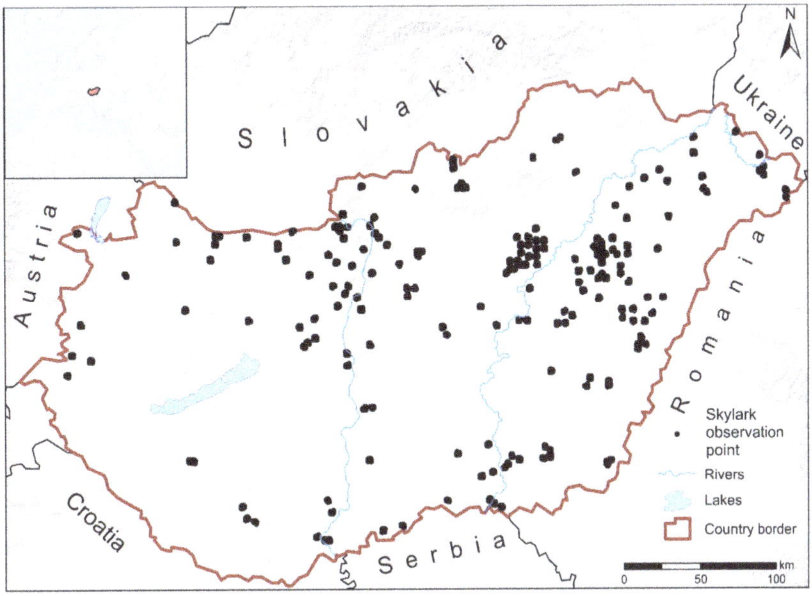

Figure 1. The spatial distribution of the MMM survey observation points in Hungary, where the Skylark occurred in 2015 (3049 observation points).

2.2. *Databases*

2.2.1. Skylark-Abundance Data

In Hungary, a countrywide bird-monitoring survey has been conducted every year (like in 2015) by approximately 800 field surveyors who add their field-observation datasets into the Hungarian Common Bird Monitoring Database (MMM) [34–36]. The volunteers were not randomly distributed across Hungary. The survey allowed that the observers choose their area of observation. Each observation point received two spring visits, and the abundance of birds was observed (by hearing and visually) within a 100-m radius of each point. There is a minimum 500 m distance between the observation points. The surveyors left a minimum of two weeks between visits in mid-April and mid-June. The count was accomplished between 5:00 and 10:00, when wind speed was less than 5 m/s and there was no rain. Each observation point contains the average number of observed birds which were counted at the point in the two spring visits [34,35]. In 2015, surveyors counted 6763 skylark individuals across 3049 field observation points (mean value: 2.22, maximum: 34, standard deviation: 4.38.). We used MMM survey points from 2015 in the study area because the HUB land-cover map was also available from that time scale. We analyzed the proportion and spatial configuration of the landscape in the 600 m radius surrounding of the MMM observation points. 600 m buffer zone was chosen, because many author found that landscape composition and land cover types have the highest impact on the abundance of this species within this radius [10,37]. Land use types also have an effect on abundance of skylark population within 600 m buffer radius [38]. We used a very detailed (20 × 20 m mapping units) country scale LULC HEB maps [39] for analyses of the LULC characteristics inside these buffer zones. Unfortunately, the more detailed country scale statistical datasets about the crop structure surroundings of the observation points were not available. Most of the MMM observation points (43%) is situated inside the NATURA 2000 SPA Protected areas, where the grasslands are mowed one-time every year after 15th of June.

2.2.2. Land-Cover Database—Hungarian Ecosystem Basemap

The digital LULC HEB was created by the Hungarian Ministry of Agriculture. The basis year of this database is 2015. This very high resolution LULC dataset was based on other LULC maps of the European Copernicus Program, such as Urban Atlas, Corine Land Cover and High-Resolution Layers, and Sentinel-2 images. The dataset has a 20 × 20 m resolution (minimal mapping unit) and three category levels. Six classes in Level 1, 22 classes in Level 2, and 56 classes in Level 3 (see Appendix A Table A1). The database also contained three additional LULC categories in Level 4. We used the second level for analysis, and regrouped the LULC classes to reduce the number and the likelihood of autocorrelation between them. Our dataset for statistical analyses contained the following main LULC categories inside the buffer zones (Figure 2.):

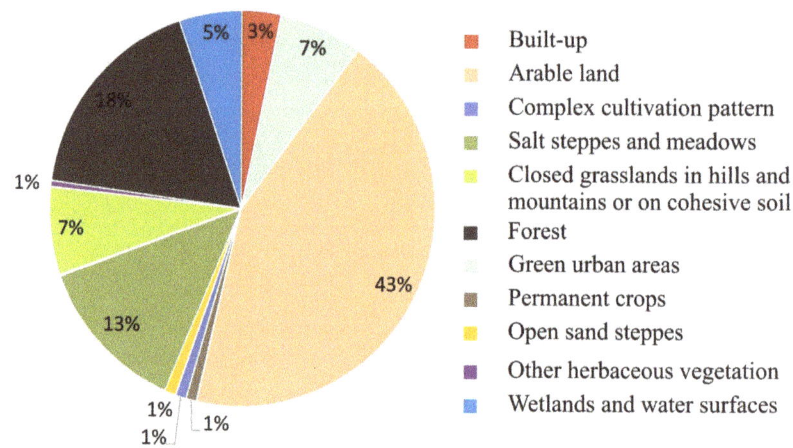

Figure 2. Proportion of the main land cover categories in the 600-m buffer zones, where the Skylark abundance were detected (3049 observation points) based on Hungarian Ecosystem Basemap.

In our investigations, we aggregated the LULC categories of the HEB database, such as "forest", "wetlands and water surfaces" LULC categories (Table A1). The HEB web map and its documentation are freely available (downloadable) on this website: http://alapterkep.termeszetem.hu/ (accessed on 15 February 2021). [39]. 46% of the country is arable land and cereals take the 62% of the arable lands. According to the country scale statistical datasets, the proportion of the crop structure in Hungary is 23% wheat, 26% grain maize, 14% sunflower, 7% barley, 5% rape and 7% fodder crops inside the arable lands (Hungarian Central Statistical Office [40]).

2.3. Landscape Metrics

The HEB database was applied to calculate size- and shape-related landscape metric parameters. Patch-level landscape indices were calculated for each LULC patch of the HEB database with the V-LATE 2 extension of Arc GIS 10.3 software [41]. Patch level metrics, created for individual land cover patches, characterize the spatial character and context of patches. These patch metrics serve primarily as the computational basis for developing a landscape metric. During our landscape metrics analyses, we calculated the following patch-level landscape metrics, which represent size and shape characteristics of land-cover patches (Table 1). The mean patch size (MPS) has been widely applied in landscape ecology, since it is commonly agreed that the occurrence and abundance of different species and species richness strongly correlates with the mean patch size. The shape complexity of individual LULC types was quantified by using landscape metrics (MFRACT). We applied the Shannon Diversity Index (SDI) to determine the landscape heterogenety [25]. We calculated these landscape indices (MPS, MFRACT, SDI) inside the 600 m radius buffer zones.

Table 1. Descriptions and calculations of the applied landscape indices [26,42,43].

Structural Feature	Index	Name and Description	Calculation
Size and shape related metrics	MPS	Mean patch size is computed by dividing the area of the patches of the total landscape (or class) by the number of patches.	$MPS = \frac{\sum_{j=1}^{n} a_{ij}}{n_i}$ where a_{ij} represents the area of the j^{**} patch in the i^{**} class, n_i represents the number of patches in the i^{**} class, n represents the number of patches (>0).
	MFRACT	Mean fractal dimension index equals 2 times the logarithm of the patch perimeter (m) divided by the logarithm of patch area (m*).	$MFRACT = \frac{\sum_{j=1}^{n}\left(\frac{2\ln p_{ij}}{\ln a_{ij}}\right)}{n_i}$ where p_{ij} represents the perimeter of the j^{***}patch in class i^{**}, a_{ij} represents the area of the j^{***}patch in class i^{**}, n_i represents the number of patches in the i^{**} class, n represents the number of patches (1–2).
Landscape Heterogeneity	SDI	The Shannon diversity index (SDI) provides more information about area composition than simply area richness (i.e., the number of land-cover types present).	$SDI = -\sum_{i}^{m}(P_i * \ln(P_i))$ where (m) represents the number of different land-cover types, P_i = the relative abundance of different land-cover types in each BMMU quadrant or LUCAS transect.

2.4. Statistical Analyses

To understand the relationship between LULC types and skylark abundance, first we had to identify those LULC categories which are selected (used as habitat) by skylark or are avoided. We applied a preliminary test to identify the group of correlated land-cover and landscape index variables using variance inflation factors (VIFs), and the explanatory variables were not linearly related. VIF values were between 0 and 1.9, which shows that the multicollinearity is low between the variables (LULC types and indices). The arable-land category was ignored from statistical analyses (model) because in Hungary and other European countries, the agricultural land is the matrix (dominant LULC type) in the landscape, so the proportion of this category shows strong autocorrelations with other LULC types. We used generalized linear models (GLM) to determine the impact of land cover and landscape structure (composition) on skylark abundance. We applied negative binomial models (link = log) to account the overdispersion of skylark-abundance data (tested by overdispersiontest function of AER package in R). Models with all possible combinations of explanatory variables were generated, and we established Akaike's information criterion to rank them with the "dredge" function from the MuMin package in R [44]. We used model averaging for competitive models (delta AICc < 2) to include uncertainty arising from the high number of candidate models (Table A3) [45]. The significance of the variables was estimated by the LmerTest package [46]. We constructed two groups from the LULC categories of the HEB database based on GLM results, namely, preferred (significant positive relation) and nonpreferred (significant negative relation) land-cover types. We analyzed the relationship between the landscape metrics of the preferred (as habitats) and nonpreferred land-cover types, and the skylark abundance data with negative binominal GLM and model averaging. In the next step in our investigation, we analyzed the shape and size characteristics of those LULC types which showed significant positive relation with skylark abundance. These land-cover types were separated into arable lands and grasslands because we wanted to analyze the effect of arable land and grassland landscape metrics on the skylark population. In this model, the arable land category has been used. The distribution of landscape metric variables was not normal, so logarithmic transformation was used to normalize the data. These variables were in different dimensions, so we created a range function in R that transformed the variable values into a number between 0 and 1:

$$Range\ function = \frac{x - \min(x, na.rm = T)}{\max(x, na.rm = T) - \min(x, na.rm = T)},$$

where *Range function* is a number that describes the given number between 0 and 1, *na.rm* = *T* means that NA values were removed, *min* is the minimal value of the list, and *max* is the maximal value of the list. On the basis of the output of the statistical model, we could describe the optimal landscape configurations for this species.

2.5. Model Validation

We calculated the predicted marginal effects (ggeffects package in R) of the preferred land-cover types and their landscape metrics on the skylark population [47]. To validate our model, we set up a training and a testing group (66.6% and 33.3% proportion, respectively) with random sampling (sample.split function from caTools 1.17 package) on the basis of our dataset in R statistics software. We used the predict function from the car package to calculate the estimated skylark-abundance data. Model accuracy was measured by three indices: Spearman's rank correlation to show the relationship between observed and predicted values, mean absolute error to show the distance of the predicted values from the observed values [48], and mean absolute percentage error to show the percentage of error between observed and predicted values [49].

2.6. Prediction of Skylark Population in Natura 2000 SPAs

We could estimate skylark population density using the 600 m buffer areas and the HEB dataset. The centers of the buffer zones were in a regular grid (1200 × 1200 m) inside the Natura 2000 SPA dataset. We used the Natura 2000 SPA areas as the basis of our prediction site, because of the Eurasian skylark is a very common indicator species of agrarian landscapes (Natura 2000 Annex I. list). In Hungary the Natura 2000 SPA areas are typical agrarian landscapes which contain Urban areas (1.5%), Croplands (31.7%), Grasslands and other herbaceous vegetation (21.7%), Forest and woodlands (27.8%) and wetland and water surfaces (17.2%). Mowing of the grasslands inside the Natura 2000 sites is regulated by the law. The mowing machine should cut the grass 10 cm above the soil surface. Mowing should not begin before 1 of July, to protect the ground nesting birds. The number of the animals and the method (it is different based on the grassland type) are also regulated by the law. Prediction was performed based on the model results that analyzed the connection between the preferred area and the landscape metrics. Landscape indices were calculated inside the Natura 2000 SPAs. The estimated skylark population was calculated by the predict function in R software. Figure 3 shows the spatial distribution of the 600 m buffer zones.

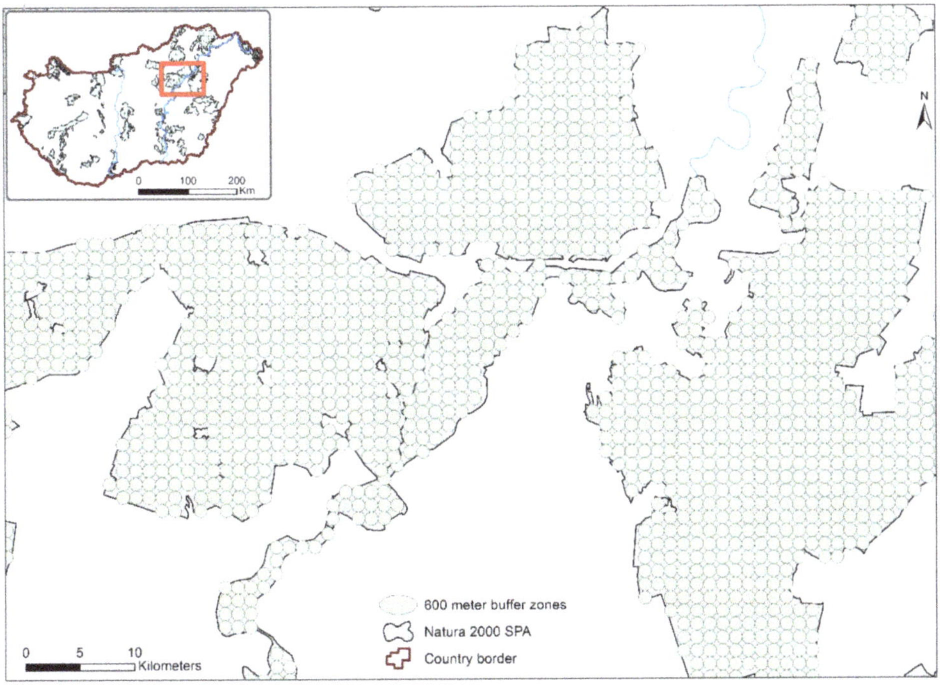

Figure 3. Example the spatial distribution of the 600-m buffer zones inside a Natura 200 Spa protected area of Hungary.

3. Results

3.1. Relationship between Land-Cover Proportions and Skylark Abundance

Based on GLM results, we identified two main groups (classes) of the LULC categories of the HEB database. Preferred LULC categories that were considered the habitats of the Eurasian skylark because they showed significant positive relation with skylark abundance were those such as salt steppes and meadows, and closed grasslands in hills and mountains. The closed-grasslands LULC category showed the highest significant relation, thereby having the most important effect on skylark abundance. The arable-land LULC category is also a preferred category according to the literature [11,18,29,50]. The nonpreferred group (class) of LULC categories contains land-cover types with significant negative relations with skylark abundance: built-up land, green urban areas, complex cultivation patterns, forests, and wetlands and water surfaces. The complex-cultivation-pattern LULC category had the strongest negative association with the skylark population, followed by wetland and water surfaces, and green urban areas. The relative importance of the significant variables was 100% in all cases (Table 2).

Table 2. Summary table for LULC categories, which shows the GLM results after multimodel averaging of best candidate models showing relative importance of each explanatory variable on Skylark abundance, estimated parameter values ± Standard deviation. (For detailed descriptions of the LULC categories see Table A1).

Variable	Estimates	Standard Deviation	Conf. Int (95%)	*p*-Value	Relative Importance (%)	*VIF*
Built-up	−0.019 *	0.008	−0.035–0.003	0.022	100	1.88
Green urban areas	−0.024 ***	0.005	−0.034–0.014	<0.001	100	1.93
Permanent crops	−0.014	0.013	−0.040–0.013	0.308	24	1.03
Complex cultivation pattern	−0.034 *	0.015	−0.064–0.005	0.021	100	1.05
Open sand steppes	−0.014	0.012	−0.037–0.009	0.228	19	1.02
Salt steppes and meadows	0.059 ***	0.002	0.054–0.063	<0.001	100	1.17
Open rocky grasslands	−0.045	0.114	−0.269–0.180	0.697	100	1.06
Closed grasslands in hills and mountains or on cohesive soil	0.067 ***	0.004	0.058–0.076	<0.001	100	1.03
Other herbaceous vegetation	−0.019	0.075	−0.165–0.128	0.805	80	1.07
Forests	−0.021 ***	0.002	−0.025–0.016	<0.001	100	1.11
Wetlands and water surfaces	−0.030 ***	0.006	−0.041–0.018	<0.001	100	1.02

Number of MMM observations (data pairs): 3049, * $p < 0.05$, ** $p < 0.01$, *** $p < 0.001$, Positive significant relation with skylark abundance, Negative significant relation with skylark abundance, No significant relation with skylark abundance.

3.2. Relationship between Landscape Structure (Compositon) and Skylark Abundance

The landscape metrics that describe the shape and size characteristics of the preferred and nonpreferred LULC classes showed different directions of significant relation with skylark abundance (Table 3). The metrics that describe the shape complexity and size of the LULC patches of preferred LULC categories of the HEB database showed significant positive relations with skylark abundance. The shape complexity (MFRACT index) of the preferred LULC patches has stronger influence on the skylark abundance than the mean patch size (MPS). The shape complexity and size of the nonpreferred LULC categories had significant negative relation with skylark abundance in this case, MPS had higher association with skylark abundance. (Table 3). Land-cover heterogeneity, described with SDI, had a significant negative effect on skylark abundance, which showed that this species prefers a homogeneous landscape.

Table 3. Summary table for landscape metrics, which shows the GLM results after multimodel averaging of best candidate models showing relative importance of each explanatory variable on Skylark abundance, estimated parameter values ± Standard deviation.

	Variable	Estimates	Standard Deviation	Conf. Int (95%)	*p*-Value
Shape and size related landscape metrics	MPS of preferred LC types	0.4345 ***	0.0001	0.2324–0.6156	<0.001
	MFRACT of preferred LC types	1.1635 ***	0.3349	0.5072–1.8199	0.001
	MPS of non-preferred LC types	−1.9126 ***	0.0004	−2.7145–1.1237	<0.001
	MFRACT of non-preferred LC types	−1.1993 **	0.4205	−2.0236–0.3751	0.004
Landscape heterogeneity	Shannon Diversity Index of landscape	−1.3711 ***	0.1639	−1.6923–1.0500	<0.001

Number of MMM observations (data pairs): 3049, * $p < 0.05$, ** $p < 0.01$, *** $p < 0.001$.

3.3. Impact of Preferred Land-Cover Categories and Their Landscape Metrics

Total grassland proportion had the highest association with skylark abundance, as shown in Table 2; the average size of arable-land patches (MPS) was more important from an abundance point view of this species than the mean patch size (MPS) of grassland patches. The complexity of grassland patches (MFRACT) had a significant positive association with skylark abundance, while the shape characteristics of arable land had no significant relationship with skylark abundance. The predicted marginal-effect graphs visualize the above-described connections between proportions of LULC categories, size- and shape-related landscape indices, and the estimated population density changes of the skylarks (Figure 4). According to the modeled population density changes, in the case of 100% grassland coverage of a hypothetical landscape, we could find about 4–6 skylark individuals/km*. While the connection between the change in proportions of different land-cover types showed a near exponential curve, landscape metrics showed almost flat linear connections with estimated skylark abundance.

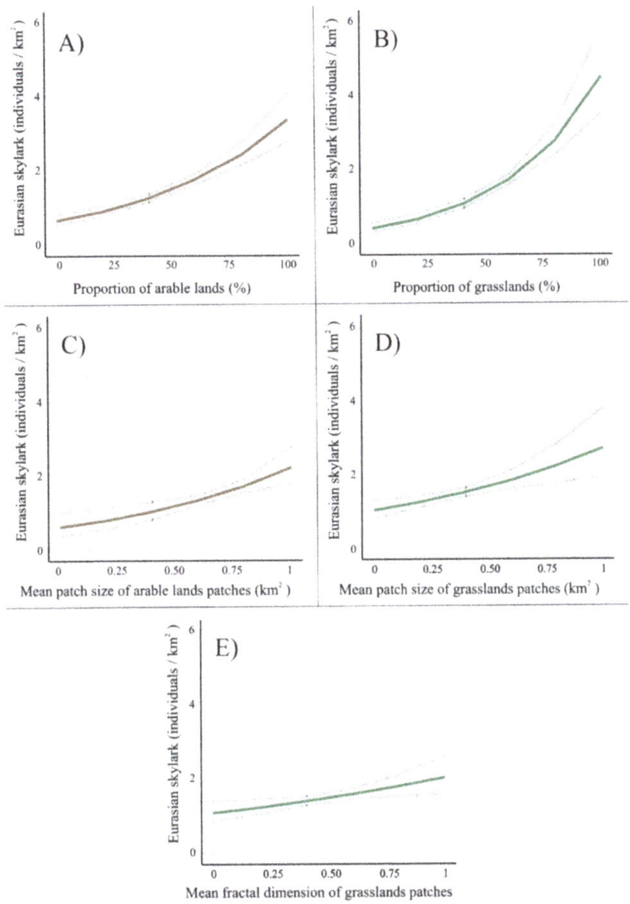

Figure 4. Predicted marginal effects between the skylark individuals / km2 proportions and landscape metrics of arable and grasslands. The confidence intervals (95%) of the prediction are shown between the dotted lines. ((**A**), Connection between the proportion of arable land and estimated population density of skylark, (**B**), Connection between the proportion of grassland and estimated population density of skylark, (**C**), Connection between the MPS of arable land and estimated population density of skylark, (**D**), Connection between the MPS of grassland and estimated population density of skylark, (**E**) Connection between the MFRACT of grassland and estimated population density of skylark).

On the basis of our results (Table A2), we could create an equation that describes and estimates the skylark population in a given landscape:

$$Skylark_{population} = -3.24 + 1.29 * MPS_{arable\ land} + 0.97 * MPS_{grassland} + 0.63 * MFRACT_{grassland} + 1.65 * Area_{arable\ land} + 2.4 * Area_{grassland}$$

where $Skylark_{population}$ is the skylark number density (individual/km*), $MPS_{arable\ land}$ is the mean patch size of arable land, $MPS_{grassland}$ is the mean patch size of grasslands, $MFRACT_{grassland}$ is the mean fractal dimension of grasslands, $Area_{arable\ land}$ is the proportion of arable land, and $Area_{grassland}$ is the proportion of grasslands.

3.4. Model Validation

According the validation of our results, there was a significant Spearman's correlation between the observed and predicted skylark abundance values. Mean absolute error shows the distance between the predicted and observed abundance values of this species, which is +− 2.12. Mean absolute percentage error (MAPE) shows the prediction accuracy of the model in percentage; in this case, it was 37.77%. The accuracy of this model based on the MAPE was 62.23% (Table 4). If the model contains just the land cover types, the MEA is 2.95; MAPE is 46.56% and the Spearman correlation coefficient is 0.493.

Table 4. Summary table of the correlation and error indices, which show the accuracy of the predicted values, based on land cover types and land cover types + landscape indices.

	Spearman's Rho	Mean Absolute Error	Mean Absolute Percentage Error	Number of Data Pairs
Land cover types + landscape metrics	0.504 **	2.12	37.77%	949
Land cover types	0.493 **	2.95	46.56%	

** $p < 0.01$.

3.5. Prediction of Skylark Population of Natura 2000 Special Protection Areas of Hungary

The spatial distribution of the predicted skylark population in each 600-m zone of the Natura 2000 SPAs of Hungary was very diverse (Figure 5). The total investigated Natura 2000 SPA was 13,514 km** which cover the most valuable agroecosystems and rural landscapes of Hungary. Based on model prediction (predict function in R) inside these protected areas, approximately 23,746 skylark individuals were predicted. The density of this species is the highest in the agricultural-landscape-dominated areas of the great Hungarian plain.

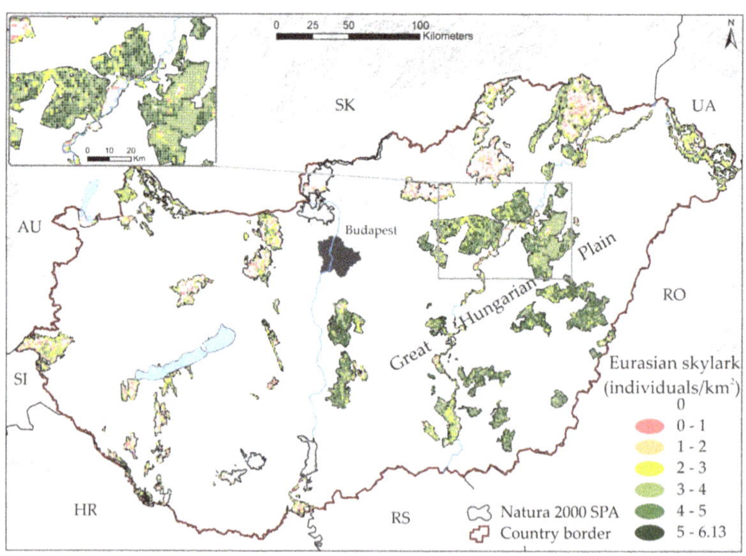

Figure 5. Predicted Eurasian skylark population (individuals/km*) in the 600 m buffer zones inside the Natura 2000 SPA area.

4. Discussion

There are several publications analyzing the relationship between skylark and LULC [4,51–54] in local small study areas, but our very detailed LULC dataset (HEB) offers a unique opportunity to obtain regional (country)-scale information about this relationship In our study, we considered both datasets describing proportions of LULC categories and landscape indices that describe the shape and size characteristics of preferred (habitat) and nonpreferred LULC categories. Based on our research findings, population density (individuals/km*) could be estimated because there was a significant statistical relationship between proportions, the shape and size characteristics of different LULC types, and the abundance of this farmland bird. One new finding from our research is that, for the estimation of skylark population density, it is necessary to consider landscape indices together with the proportions of different LULC categories because shape (mean fractal dimension) and size (mean shape size) characteristics of these LULC categories also have significant association with skylark abundance. Based on our finding we have predicted the number skylarks inside the Natura 2000 SPA areas in Hungary.

4.1. Impact of Proportions of LULC Categories on Skylark Abundance

We could select two LULC groups (classes) from the land-cover types of a very detailed (20 × 20 m resolution) LULC map. Nonpreferred types had negative significant relation with skylark abundance. These were built-up and green urban areas, which negatively affected the population because of the lack of openness and the high proportion of constructed surfaces. Our findings are confirmed by other international publications [4,6,10]. The complex cultivation pattern land-cover type has negative significant relation with skylark data. Other authors underlines that the skylarks do not prefer heterogeneous agricultural lands because this rural landscape contains many different LULC patches, including also those that are not preferable to the skylark, like vineyards, fruit and berry plantations (because of its height, they obscure the view) [10,11,16,17]. Small parcels of, annual crops, city gardens pastures, fallow lands and/or permanent crops somewhere with scattered houses. Forest and wetland LULC categories are well-known nonpreferred land-cover types of the skylark. The skylark is a typical farmland bird; therefore, it is not a surprise that wetland areas, water bodies, and water courses are not suitable habitat types for this species. The main reason of the negative significant relation of the forest is the lack of openness, which is very important for the skylark [10,11,16,55]. In our research we were not take difference between the type of forests, because according to previous studies every types of forest areas are not habitats of this species.

In the estimation of skylark population density, the preferred land-cover types had higher weights (were more important) than those of the nonpreferred LULC categories. Arable land is a well-known habitat type of this farmland bird species according to the international literature [11,18,19,56,57]. Unfortunately, in Hungary is no available detailed country scale spatial statistical data about the cultivated crop types inside the arable lands (cropland) areas. According to the available most detailed Hungarian LULC dataset, the HEB dataset the 57% of Hungary is covered by agricultural fields and its 81% is arable land (Cropland). Grassland and pasture areas are also preferred LULC categories for skylark, [7,8,10,11,58–61]. The HEB dataset allow us to analyze the impact of different types of grassland on skylark abundance. We did not find significant statistical relations with open sand steppes and open rocky grasslands because the number of 600 m circle radius observation points of LULC categories have been low, and these landscape conditions (too-fragmented grassland areas with very short and very sparse vegetation) are not suitable for breeding skylarks [57,62]. There was a significant positive relation between skylark abundance, and the LULC categories of salt steppes and meadows, and closed grasslands. Each LULC category is suitable for nesting breeding skylarks because of the medium vegetation height and optimal proportion inside the 600 m radius circles. Our results are similar with those of others, who described strong relationship between closed grasslands and meadows and skylark abundance, the reason of this relation could be the larger amount of food [63–66]. According to our findings for the prevention of the farmland bird habitats, the EU agri-environmental policy should pay more attention to the management of salt steppes and meadows, and closed grasslands. To increase the population density of skylark, the mean patch size and the proportion of these land cover types (compare to all) in the landscape should increase. In case of the protected grassland areas, one of the biggest ecological problem is the spontaneous spreading of the bush vegetation, which can reduce the skylark habitats. If we want to stop this process, and keep the openness of the landscapes, we should reduce the size and the shape of the bush and forest patches inside these grassland areas. Therefore, we must eradicate the spontaneously spread bush vegetation (which often full of invasive species) by the proper way of grazing or haymaking, the grasslands can keep its size, shape, and openness characteristics in the protected landscapes. This kind of management of protected areas can preserve not only the vegetation diversity of grasslands but it has also important key factor in the skylark habitat protection.

4.2. *Impact of Land-Cover Categories and Their Landscape Metrics*

The landscape metrics of the preferred LULC classes showed positive significant relation with skylark abundance, meaning that, if arable-land and grassland proportion and shape complexity was higher, then the skylark population would also be higher. The landscape metrics of the nonpreferred LULC classes showed negative significant relation with the skylark population, meaning that, in landscapes with small size and in compact-shape nonpreferred LULC categories, skylark population density (abundance) would be higher.

LULC landscape heterogeneity has a negative effect on the skylark in this scale, where one land cover patch can contain more parcels. If landscape heterogeneity increases, the skylark population declines. This species prefer the homogenous LULC structures, which is in accordance with the results of other authors [10,11,16,17,62].

The grassland proportion had the highest association with the skylark population. This species usually nests and feeds in grasslands. The proportion of arable land has a high association with skylark abundance, but the level of its significance is lower. In the case of the MPS, the opposite phenomenon was observed: the MPS of arable lands (arable land patches of HEB) had a higher effect on skylark abundance than that of grassland. The skylark does not prefer small size arable lands (parcels) and grassland fields in that scale, where one arable land patch can contain more parcels [7,51,60,64]. According to Uuemaa et al. 2009 most bird species react more strongly to the composition land cover than to the configuration of landscapes [25]. Our results also show that the LULC proportions and mean patch sizes have stronger impacts on the abundance of this species, than the shape (fractal dimension index) characteristics of the habitat patches. The mean-absolute-percentage-error value (37.77%) was acceptable since, for a more precise prediction, we would have to use more variables (e.g., species and quantity of insects, used pesticides, parcel management) that are not accessible in country-scale analysis. We can determine that the landscape indices improved the model accuracy, based on the Table 4.

4.3. *Predicted Population Inside the Natura 2000 SPAs*

In Hungary, the latest estimated country-wide Eurasian skylark population is from 1999–2002. There is no spatially detailed population estimate. This study is the first estimate for Natura 2000 protected areas in Hungary. There some early 2000s studies about the skylark densities in Europe (Table 5).

The studies listed above do not use the shape and size related landscape indices for estimation of the skylark abundance (density). With the combination of the detailed point-based bird census data, detailed country-wide LULC dataset and landscape indices we can get a more precise prediction of skylark population. Our results are comparable with these previous estimations and the density values are similar [63,67–69].

Table 5. Summary table of studies, which predicted the Eurasian skylark density inside European study areas.

Study Area	Estimated Skylark Density (Individuals/km *)	Reference
Natura 2000 SPA in Hungary	0–6.13	This study
Great Britain	1.97–7.45	Browne et al. 2000 [67]
Small study area in France	3.28–3.69	Eraud and Boutin 2002 [68]
Spain	~5.21	Suárez et al. 2003 [63]
Ireland	1.72	Copland et al. 2012 [69]
Northwest Ireland	4.87	Copland et al. 2012 [69]

5. Conclusions

Landscape composition (proportions, and shape and size characteristics of LULC categories) has significant association with the skylark population. The salt steppes and meadows, and closed grassland serve as habitat for the Eurasian skylark. This study provides new information about the relationship between landscape metrics of the habitat types (shape and size characteristics of patches) and skylark abundance. Fractal dimension index, which describes the shape complexity of grassland patches has a positive impact on the skylark abundance, while the shape complexity of non-habitat types shows opposite relationships with the skylark density. We analyzed them together and could estimate the association of these landscape composition variables (proportions, shape and size characteristics of LULC classes) with skylark abundance. We could estimate skylark population density inside Natura 2000 SPAs in Hungary.

The outcomes of this study can be used for further land use planning, and the habitat design of Natura 2000 SPAs and other protected areas of the rural landscapes. According to our findings, inside the protected areas should increase the proportion, the average size and shape complexity of those LULC types (arable land, salt steppes and meadows, and closed grassland), which shows positive relations with the abundance data of skylark. It is feasible by stopping the spontaneous reforestation and eradicating the spontaneously spread vegetation (especially invasive bush species). The grazing or mowing, the protected grasslands can preserve the size, shape and openness characteristics of these skylark habitats. This kind of environmental management forms help to conserve the habitat types of skylark. The skylark is an area sensitive species and it is an indicator species of farmlands, so the shown methodology is adaptable for analyzing the impact of landscape composition on other farmland bird populations [56,70–72]. The skylark is considered as indicator for monitoring of agricultural landscapes, because its abundance shows strong relationships with other farmland bird species [73].

Author Contributions: Conceptualization, N.C. and P.S.; methodology, N.C.; validation, N.C.; writing—original draft preparation, N.C. and P.S.; writing—review and editing, N.C. and P.S.; visualization, N.C.; supervision, P.S.; All authors have read and agreed to the published version of the manuscript.

Funding: This research was funded by the "UNKP-20-3-SZTE-515 New national excellence program of the Ministry for Innovation and Technology".

Institutional Review Board Statement: Not applicable.

Informed Consent Statement: Not applicable.

Acknowledgments: We thank the anonymous referees for their valuable recommendations and suggestions. We also thank Károly Nagy and MME (Bird Monitoring Centre, MME/BirdLife Hungary) for data provision.

Conflicts of Interest: The authors declare no conflict of interest.

Appendix A

Table A1. The LULC categories of the Hungarian Ecosystem Basemap, and the investigated LULC categories.

HEB LULC Categories			**The Investigated LULC Categories**	
Level 1	Level 2 Code	Level 2 (~EUNIS 2)	Level 2 Code	Level 2
Urban	11 12 13	Buildings Roads and railways Other paved or non-paved artificial areas	10	Built-up
	14	Green urban areas	14	Green urban areas
Croplands	21	Arable land	21	Arable land
	22	Permanent crops	22	Permanent crops
	23	Complex cultivation pattern	23	Complex cultivation pattern
Grasslands and other herbaceous vegetations	31	Open sand steppes	31	Open sand steppes
	32	Salt steppes and meadows	32	Salt steppes and meadows
	33	Open rocky grasslands	33	Open rocky grasslands
	34	Closed grasslands in hills and mountains or on cohesive soil	34	Closed grasslands in hills and mountains or on cohesive soil
	35	Other herbaceous vegetation	35	Other herbaceous vegetation
Forests and woodlands	41 42 43 44 45 46	Forests without excess water Natural riverine (gallery) forests Other forests with excess water Plantations Non-wooded areas registered as forest, or areas under reforestation Other ligneous vegetation, woodlands	40	Forest
Wetlands	51 52	Herbaceous-dominated wetlands Woodland-dominated wetlands (uncertain translation)	50	Wetlands and water surfaces
Rivers and lakes	61 62	Water bodies Water courses		

Table A2. Summary table for landscape metrics and LULC categories, which shows the GLM results after multimodel averaging of best candidate models showing relative importance of each explanatory variable on Skylark abundance, estimated parameter values ± Standard deviation. MPS is Mean Patch Size and MFRACT is Mean Fractal dimension.

Predictors	Estimate	Standard deviation	Conf. Int (95%)	p-Value	Relative importance (%)
(Intercept)	−3.2352 ***	0.3579	−3.9005–−2.5772	<0.001	
MPS of arable lands	1.2850 ***	0.3588	0.6528–1.9195	<0.001	100
MPS of Grasslands	0.9689 ***	0.2755	0.4145–1.5358	<0.001	100
MFRACT of arable lands	−0.1719	0.2928	−0.7136–0.3745	0.557	31
MFRACT of grasslands	0.6255 **	0.2409	0.1657–1.0845	0.009	100
Total area of arable lands	1.6482 ***	0.1916	1.2788–2.0202	<0.001	100
Total area of grasslands	2.4023 ***	0.2731	1.8781–2.9262	<0.001	100

Number of MMM observations (data pairs): 1897, * $p < 0.05$, ** $p < 0.01$, *** $p < 0.001$.

Table A3. Summary table of component models from model averaging.

Variables	df	logLik	AICc	Delta	Weight
1/2/4/6/7/8/9/10/11	11	−4658.47	9339.02	0	0.38
1/2/3/4/6/7/8/9/10/11	12	−4657.9	9339.91	0.89	0.24
1/2/4/5/6/7/8/9/10/11	12	−4658.14	9340.39	1.37	0.19
1/2/4/6/7/8/10/11	10	−4660.2	9340.48	1.46	0.18

1 Built-up, 2 Green urban areas, 3 Permanent crops, 4 Complex cultivation pattern, 5 Open sand steppes, 6 Salt steppes and meadows, 7 Open rocky grasslands, 8 Closed grasslands in hills and mountains or on cohesive soil, 9 Other herbaceous vegetation, 10 Forest, 11 Wetlands and water surfaces.

References

1. FAO. *World Reference Base for Soil Resources 2014 International Soil Classification system for Naming Soils and Creating Legends for Soil Maps*; Update 2015; World Soil Resources Reports; FAO: Rome, Italy, 2014.
2. Morelli, F.; Benedetti, Y.; Callaghan, C.T. Ecological specialization and population trends in European breeding birds. *Glob. Ecol. Conserv.* **2020**, *22*, e00996. [CrossRef]
3. Toth, G.; Huzui-Stoiculescu, A.; Toth, A.-I.; Stoiculescu, R. How Do Natura 2000 Areas Intersect with Peoples' Livelihood Strategies in High Nature Value Farmlands in Southern Transylvania? *Land* **2020**, *9*, 484. [CrossRef]
4. Guerrero, I.; Morales, M.B.; Oñate, J.J.; Geiger, F.; Berendse, F.; de Snoo, G.; Eggers, S.; Pärt, T.; Bengtsson, J.; Clement, L.W.; et al. Response of ground-nesting farmland birds to agricultural intensification across Europe: Landscape and field level management factors. *Biol. Conserv.* **2012**, *152*, 74–80. [CrossRef]
5. Verhulst, J.; Báldi, A.; Kleijn, D. Relationship between land-use intensity and species richness and abundance of birds in Hungary. *Agric. Ecosyst. Environ.* **2004**, *104*, 465–473. [CrossRef]
6. Gottschalk, T.K.; Dittrich, R.; Diekötter, T.; Sheridan, P.; Wolters, V.; Ekschmitt, K. Modelling land-use sustainability using farmland birds as indicators. *Ecol. Indic.* **2010**, *10*, 15–23. [CrossRef]
7. Moreira, F.; Beja, P.; Morgado, R.; Reino, L.; Gordinho, L.; Delgado, A.; Borralho, R. Effects of field management and landscape context on grassland wintering birds in Southern Portugal. *Agric. Ecosyst. Environ.* **2005**, *109*, 59–74. [CrossRef]
8. Piha, M.; Pakkala, T.; Tiainen, J. Habitat preferences of the Skylark Alauda Arvensis in southern Finland. *Ornis Fenn.* **2003**, *80*, 97–110.
9. Gil-Tena, A.; De Cáceres, M.; Ernoult, A.; Butet, A.; Brotons, L.; Burel, F. Agricultural landscape composition as a driver of farmland bird diversity in Brittany (NW France). *Agric. Ecosyst. Environ.* **2015**, *205*, 79–89. [CrossRef]
10. Szilassi, P.; Csikos, N.; Galle, R.; Szep, T. Recent and Predicted Changes in Habitat of the Eurasian Skylark Alauda arvensis Based on the Link between the Land Cover and the Field Survey Based Abundance Data. *Acta Ornithol.* **2019**, *54*, 59. [CrossRef]
11. Csikós, N.; Szilassi, P. Impact of Energy Landscapes on the Abundance of Eurasian Skylark (Alauda arvensis), an Example from North Germany. *Sustainability* **2020**, *12*, 664. [CrossRef]
12. Link, P.M.; Schleupner, C. Agricultural land use changes in Eiderstedt: Historic developments and future plans. *Coastline Rep.* **2007**, *9*, 928–2734.
13. Borges, F.; Glemnitz, M.; Schultz, A.; Stachow, U. Assessing the habitat suitability of agricultural landscapes for characteristic breeding bird guilds using landscape metrics. *Environ. Monit. Assess.* **2017**, *189*, 166. [CrossRef]
14. Moser, D.; Zechmeister, H.G.; Plutzar, C.; Sauberer, N.; Wrbka, T.; Grabherr, G. Landscape patch shape complexity as an effective measure for plant species richness in rural landscapes. *Landsc. Ecol.* **2002**, *17*, 657–669. [CrossRef]
15. Schindler, S.; von Wehrden, H.; Poirazidis, K.; Wrbka, T.; Kati, V. Multiscale performance of landscape metrics as indicators of species richness of plants, insects and vertebrates. *Ecol. Indic.* **2013**, *31*, 41–48. [CrossRef]
16. Berg, Å.; Wretenberg, J.; Zmihorski, M.; Hiron, M.; Pärt, T. Linking occurrence and changes in local abundance of farmland bird species to landscape composition and land-use changes. *Agric. Ecosyst. Environ.* **2015**, *204*, 1–7. [CrossRef]
17. Redlich, S.; Martin, E.A.; Wende, B.; Steffan-Dewenter, I. Landscape heterogeneity rather than crop diversity mediates bird diversity in agricultural landscapes. *PLoS ONE* **2018**, *13*, e0200438. [CrossRef]
18. Praus, L.; Weidinger, K. Breeding Biology of Skylarks Alauda arvensis in Maize and Other Crop Fields. *Acta Ornithol.* **2015**, *50*, 59–68. [CrossRef]
19. Hoffmann, J.; Wittchen, U.; Berger, G.; Stachow, U. Moving window growth—A method to characterize the dynamic growth of crops in the context of bird abundance dynamics with the example of Skylark (Alauda arvensis). *Ecol. Evol.* **2018**, *8*, 8880–8893. [CrossRef]
20. Loretto, M.C.; Schöll, E.M.; Hille, S. Occurrence of Eurasian Skylark Alauda arvensis territories in relation to urban area and heterogeneous farmland. *Bird Study* **2019**, *66*, 273–278. [CrossRef]
21. The Birds Directive—Environment—European Commission. Available online: https://ec.europa.eu/environment/nature/legislation/birdsdirective/index_en.htm (accessed on 15 February 2021).
22. Csörgo, T.; Karcza, Z.; Halmos, G.; Magyar, G.; Gyurácz, J.; Szép, T.; Bankovics, A.; Schmidt, A.; Schmidt, E. *Hungarian Bird Migration Atlas*; Kossuth Kiadó: Budapest, Hungary, 2009; ISBN 978-963-09-5865-3.
23. Cramp, S. *The Birds of the Western Palearctic*; Oxford University Press: Oxford, UK, 1988.
24. Campbell, R.W.; Van Damme, L.M.; Johnson, S.R.; Donald, P.; Garcia, E. Eurasian Skylark (Alauda arvensis). In *Birds of the World*; Cornell Lab of Ornithology: Ithaca, NY, USA, 2020.
25. Uuemaa, E.; Antrop, M.; Roosaare, J.; Marja, R.; Mander, Ü. Landscape Metrics and Indices: An Overview of Their Use in Landscape Research. *Living Rev. Landsc. Res.* **2009**, *3*, 28. [CrossRef]
26. Uuemaa, E.; Mander, Ü.; Marja, R. Trends in the use of landscape spatial metrics as landscape indicators: A review. *Ecol. Indic.* **2013**, *28*, 100–106. [CrossRef]
27. Szilassi, P.; Bata, T.; Szabó, S.; Czúcz, B.; Molnár, Z.; Mezősi, G. The link between landscape pattern and vegetation naturalness on a regional scale. *Ecol. Indic.* **2017**, *81*, 252–259. [CrossRef]
28. Walz, U. Landscape structure, landscape metrics and biodiversity. *Living Rev. Landsc. Res.* **2011**, *5*, 3. [CrossRef]
29. Schlager, P.; Ruppert-Winkel, C.; Schmieder, K. Assessing the potential impacts of bioenergy cropping on a population of the ground-breeding bird Alauda arvensis: A case study from southern Germany. *Landsc. Res.* **2020**, *45*, 1000–1017. [CrossRef]

30. Vögeli, M.; Serrano, D.; Pacios, F.; Tella, J.L. The relative importance of patch habitat quality and landscape attributes on a declining steppe-bird metapopulation. *Biol. Conserv.* **2010**, *143*, 1057–1067. [CrossRef]
31. Farkas, J.Z.; Lennert, J. Modelling and predicting of the land use change in Hungary. In *Climate Change—Society—Economy: Long-Term Processes and Trends in Hungary*; Czirfusz, M., Hoyk, E., Suvák, A., Eds.; Publikon: Pécs, Hungary, 2015; pp. 193–221. ISBN 978-615-5457-62-3.
32. Tryjanowski, P.; Hartel, T.; Bldi, A.; Szymański, P.; Tobolka, M.; Herzon, I.; Goławski, A.; Konvička, M.; Hromada, M.; Jerzak, L.; et al. Conservation of farmland birds faces different challenges in Western and Central-Eastern Europe. *Acta Ornithol.* **2011**, *46*, 1–12. [CrossRef]
33. Báldi, A.; Faragó, S. Long-term changes of farmland game populations in a post-socialist country (Hungary). *Agric. Ecosyst. Environ.* **2007**, *118*, 307–311. [CrossRef]
34. Szép, T.; Gibbson, D. Monitoring of common breeding birds in Hungary using a randomised sampling design. *RING* **2000**, *22*, 45–55.
35. Szép, T.; Nagy, K. The GIS UTM Quadrants based Hungarian Bird Monitoring database of the BirdLife Hungary. *Természetvédelmi Közlemények* **2001**, *9*, 31–37.
36. Szép, T.; Nagy, K.; Nagy, Z.; Halmos, G. Population trends of common breeding and wintering birds in hungary, decline of long-distance migrant and farmland birds during 1999–2012. *Ornis Hungarica* **2012**, *20*, 13–63. [CrossRef]
37. Engel, J.; Huth, A.; Frank, K. Bioenergy production and Skylark (*Alauda arvensis*) population abundance—a modelling approach for the analysis of land-use change impacts and conservation options. *GCB Bioenergy* **2012**, *4*, 713–727. [CrossRef]
38. Miguet, P.; Gaucherel, C.; Bretagnolle, V. Breeding habitat selection of Skylarks varies with crop heterogeneity, time and spatial scale, and reveals spatial and temporal crop complementation. *Ecol. Modell.* **2013**, *266*, 10–18. [CrossRef]
39. NÖSZTÉP. Available online: http://alapterkep.termeszetem.hu/ (accessed on 15 February 2021).
40. STADAT–4.1.21. Harvested Area, Total Production and Average Yield of Main Field Crops (1990–). Available online: https://www.ksh.hu/docs/eng/xstadat/xstadat_annual/i_omn007a.html (accessed on 15 February 2021).
41. Lang, S.; Tiede, D. vLATE Extension for ArcGIS 2003. Available online: https://sites.google.com/site/largvlate/gis-tools/v-late (accessed on 15 February 2021).
42. Blaschke, T. The role of the spatial dimension within the framework of sustainable landscapes and natural capital. *Landsc. Urban Plan.* **2006**, *75*, 198–226. [CrossRef]
43. Forman, R.T.T. *Land Mosaics: The Ecology of Landscapes and Regions*; Cambridge University Press: Cambridge, UK, 1995; ISBN 0521479800.
44. Barton, K. Package MuMIn, Version 1.43.15; 2015. Available online: https://cran.r-project.org/web/packages/MuMIn/MuMIn.pdf (accessed on 15 March 2021).
45. Burnham, K.P.; Anderson, D.R. *Model Selection and Multi-Model Inference: A Practical Information—Theoretical Approach*; Springer: Berlin/Heidelberg, Germany, 2002; Volume 3, pp. 2–3.
46. Kuznetsova, A.; Brockhoff, P.B.; Christensen, R.H.B. Package 'lmerTest.'. *R Found. Stat. Comput.* **2020**, *3*, 1.
47. Lüdecke, D. ggeffects: Tidy Data Frames of Marginal Effects from Regression Models. *J. Open Source Softw.* **2018**, *3*, 772. [CrossRef]
48. Willmott, C.; Matsuura, K. Advantages of the mean absolute error (MAE) over the root mean square error (RMSE) in assessing average model performance. *Clim. Res.* **2005**, *30*, 79–82. [CrossRef]
49. De Myttenaere, A.; Golden, B.; Le Grand, B.; Rossi, F. Mean Absolute Percentage Error for regression models. *Neurocomputing* **2016**, *192*, 38–48. [CrossRef]
50. Chamberlain, D.E.; Wilson, A.M.; Browne, S.J.; Vickery, J.A. Effects of habitat type and management on the abundance of skylarks in the breeding season. *J. Appl. Ecol.* **1999**, *36*, 856–870. [CrossRef]
51. Perkins, A.J.; Whittingham, M.J.; Bradbury, R.B.; Wilson, J.D.; Morris, A.J.; Barnett, P.R. Habitat characteristics affecting use of lowland agricultural grassland by birds in winter. *Biol. Conserv.* **2000**, *95*, 279–294. [CrossRef]
52. Nagy, S.; Nagy, K.; Szép, T. Potential Impact of EU Accession on Common Farmland Bird Populations in Hungary. *Acta Ornithol.* **2009**, *44*, 37–44. [CrossRef]
53. Gevers, J.; Høye, T.T.; Topping, C.J.; Glemnitz, M.; Schröder, B. Biodiversity and the mitigation of climate change through bioenergy: Impacts of increased maize cultivation on farmland wildlife. *GCB Bioenergy* **2011**, *3*, 472–482. [CrossRef]
54. Wretenberg, J.; Lindström, Å.; Svensson, S.; Pärt, T. Linking agricultural policies to population trends of Swedish farmland birds in different agricultural regions. *J. Appl. Ecol.* **2007**, *44*, 933–941. [CrossRef]
55. Sauerbrei, R.; Ekschmitt, K.; Wolters, V.; Gottschalk, T.K. Increased energy maize production reduces farmland bird diversity. *GCB Bioenergy* **2014**, *6*, 265–274. [CrossRef]
56. Hoffmann, J.; Wittchen, U.; Stachow, U.; Berger, G. Moving Window Abundance - A method to characterise the abundance dynamics of farmland birds: The example of Skylark (Alauda arvensis). *Ecol. Indic.* **2016**, *60*, 317–328. [CrossRef]
57. Dietzen, C.; Edwards, P.J.; Wolf, C.; Ludwigs, J.-D.; Luttik, R. Focal species of birds in European crops for higher tier pesticide risk assessment. *Integr. Environ. Assess. Manag.* **2014**, *10*, 247–259. [CrossRef]
58. Csikós, N. Impact of land cover change on the abundance of farmland birds. In *Proceedings of the 26th International Symposium on Analytical and Environmental Problems*; Alapi, T., Berkecz, R., Ilisz, I., Eds.; University of Szeged: Szeged, Hungary, 2020; pp. 113–117.

59. Koleček, J.; Reif, J.; Weidinger, K. The abundance of a farmland specialist bird, the skylark, in three European regions with contrasting agricultural management. *Agric. Ecosyst. Environ.* **2015**, *212*, 30–37. [CrossRef]
60. Hamer, T.L.; Flather, C.H.; Noon, B.R. Factors associated with grassland bird species richness: The relative roles of grassland area, landscape structure, and prey. *Landsc. Ecol.* **2006**, *21*, 569–583. [CrossRef]
61. Reif, J.; Hanzelka, J. Grassland winners and arable land losers: The effects of post-totalitarian land use changes on long-term population trends of farmland birds. *Agric. Ecosyst. Environ.* **2016**, *232*, 208–217. [CrossRef]
62. Báldi, A.; Batáry, P.; Erdos, S. Effects of grazing intensity on bird assemblages and populations of Hungarian grasslands. *Agric. Ecosyst. Environ.* **2005**, *108*, 251–263. [CrossRef]
63. Suárez, F.; Garza, V.; Morales, M.B. The role of extensive cereal crops, dry pasture and shrub-steppe in determining skylark Alauda arvensis densities in the Iberian Peninsula. *Agric. Ecosyst. Environ.* **2003**, *95*, 551–557. [CrossRef]
64. Donald, P.F.; Green, R.E.; Heath, M.F. Agricultural intensification and the collapse of Europe's farmland bird populations. *Proc. R. Soc. B Biol. Sci.* **2001**, *268*, 25–29. [CrossRef] [PubMed]
65. Wolff, A.; Paul, J.-P.; Martin, J.-L.; Bretagnolle, V. The benefits of extensive agriculture to birds: The case of the little bustard. *J. Appl. Ecol.* **2001**, *38*, 963–975. [CrossRef]
66. Brotons, L.; Wolff, A.; Paulus, G.; Martin, J.L. Effect of adjacent agricultural habitat on the distribution of passerines in natural grasslands. *Biol. Conserv.* **2005**, *124*, 407–414. [CrossRef]
67. Browne, S.; Vickery, J.; Chamberlain, D. Densities and population estimates of breeding skylarks alauda arvensis in britain in 1997. *Bird Study* **2000**, *47*, 52–65. [CrossRef]
68. Eraud, C.; Marie Boutin, J. Density and productivity of breeding Skylarks Alauda arvensis in relation to crop type on agricultural lands in western France. *Bird Study* **2002**, *49*, 287–296. [CrossRef]
69. Copland, A.S.; Crowe, O.; Wilson, M.W.; O'Halloran, J. Habitat associations of Eurasian Skylarks *Alauda arvensis* breeding on Irish farmland and implications for agri-environment planning. *Bird Study* **2012**, *59*, 155–165. [CrossRef]
70. Butler, S.J.; Freckleton, R.P.; Renwick, A.R.; Norris, K. An objective, niche-based approach to indicator species selection. *Methods Ecol. Evol.* **2012**, *3*, 317–326. [CrossRef]
71. Wakeham-Dawson, A. Hares and Skylarks as Indicators of Environmentally Sensitive Farming on the South Downs, Open Research Online. Ph.D. Thesis, The Open University, Kents Hill, UK, 1995.
72. Achtziger, R.; Stickroth, H.; Zieschank, R. Nachhaltigkeitsindikator für die Artenvielfalt–ein Indikator für den Zustand der Natur und Landschaft in Deutschland. *Angew. Landschaftsökologie* **2004**, *63*, 137.
73. Gippoliti, S.; Battisti, C. More cool than tool: Equivoques, conceptual traps and weaknesses of ecological networks in environmental planning and conservation. *Land Use Policy* **2017**, *68*, 686–691. [CrossRef]

land

MDPI

Article

Structural Variations in the Composition of Land Funds at Regional Scales across Russia

Vasilii Erokhin [1], Tianming Gao [1,*] and Anna Ivolga [2]

1 School of Economics and Management, Harbin Engineering University, Harbin 150001, China; basilic@list.ru
2 Faculty of Social and Cultural Service and Tourism, Stavropol State Agrarian University, Stavropol 355017, Russia; annya_iv@mail.ru
* Correspondence: gaotianming@hrbeu.edu.cn; Tel.: +86-136-6461-9594

Received: 12 May 2020; Accepted: 17 June 2020; Published: 17 June 2020

Abstract: In recent decades, Russia has experienced substantial transformations in agricultural land tenure. Post-Soviet reforms have shaped land distribution patterns but the impacts of these on agricultural use of land remain under-investigated. On a regional scale, there is still a knowledge gap in terms of knowing to what extent the variations in the compositions of agricultural land funds may be explained by changes in the acreage of other land categories. Using a case analysis of 82 of Russia's territories from 2010 to 2018, the authors attempted to study the structural variations by picturing the compositions of regional land funds and mapping agricultural land distributions based on ranking "land activity". Correlation analysis of centered log-ratio transformed compositional data revealed that in agriculture-oriented regions, the proportion of cropland was depressed by agriculture-to-urban and agriculture-to-industry land loss. In urbanized territories, the compositions of agricultural land funds were predominantly affected by changes in the acreage of industrial, transportation, and communication lands. In underpopulated territories in the north and far east of Russia, the acreages of cropland and perennial planting were strongly correlated with those of disturbed and barren lands. As the first attempt at such analysis in Russia, the conversion of cadastral classification data into land-rating values enabled the identification of region-to-region mismatches between the cadaster-based mapping and ranking-based distribution of agricultural lands.

Keywords: agricultural land; cropland; land category; land fund; territory; Russia

---

1. Introduction

Structural alterations in land use have been intrinsically associated with a growing demand for food [1,2]. Increasingly, contemporary processes of progressing urbanization and industrialization have been aggravating the conflicts between different functional land types [3]. As land systems represent a critical intersection between economic and ecological systems [4], land distribution patterns are becoming more vulnerable to a variety of environmental and social issues. Out of the one-fifth of the world's total land surface, which is potentially suitable for crop production, more than half is already actively cultivated [5]. Further agricultural expansion is hampered by natural and geographical factors [6], pervasive land-use change impacts [7], high economic costs [8], and infrastructure constraints [9]. At the same time, according to DeFries et al. [10], Ajani [11], and Lambin [12], agricultural production tends to face increasing competition for land with other types of land use. Over recent decades, many scholars and practitioners, including Platt [13], Briggs and Yurman [14], Vining et al. [15], and Sioen et al. [16], among others, have been reporting the irreversible removal of substantial areas of land previously used for agriculture to urban, industrial, infrastructure, and other types of use instead. Urbanization and industrialization intensify competition between agricultural and non-agricultural land-use practices [17]. Along with industrial development and urban sprawl, there are significant alterations of land use far beyond city limits that result in arable land loss [18].

Generally, at a regional scale, agricultural lands do not strictly compete with other categories for the same land areas due to the specific climate, soil, and topographical requirements for farming. However, in land-abundant and climate-diverse countries, the geographical distribution of agricultural land use tends to adjust to better match land quality [19]. Russia is aa good example of aa country that can be used to demonstrate this fact. Agriculture abandonment in vast northern and eastern areas has occurred in parallel with a concentration of intensive agriculture in fertile lands in the southern, western, and central regions of the country. In Russia, agricultural lands only represent 12.96% of the total national land fund (cropland at 7.16%, rangeland at 3.99%, hayfields at 1.40%, fallow at 0.28%, and perennial plantings at 0.11%). Per-territory concentrations of agricultural land vary from 75.32% in the Southern Federal District and 70.96% in the North Caucasian Federal District to only 4.05% in the Northwestern Federal District and 1.30% in the Far Eastern Federal District.

We clarified the definitions of the main terms used in this study as follows:

- District—A type of supraregional administrative division of Russia, which includes several territories based on a geographical principle (currently, eight federal districts exist).
- Land distribution—how lands of particular categories are spread out in a country, district, or territory.
- Land fund—the total of available land resources in a country, district, or territory.
- Land fund composition—a division of a land fund into land categories.
- Land use—the total of arrangements, activities, and inputs that people undertake in a certain land cover type.
- Territory—an umbrella term to designate various types of administrative divisions of the Russian Federation (oblasts, krais, republics, autonomous districts, and autonomous republics).

The disproportions of agricultural land distribution are, to some extent, caused by economic factors, not only geographic and natural conditions. Similar to most post-socialist countries, Russia has experienced dramatic changes in land ownership and land tenure since the early 1990s. Among the principal transformations, Lerman and Shagaida [20] have outlined the privatization of agricultural land, rights to agricultural land for individual landowners, and the removal of prohibitions on buying and selling land. The land market has responded positively to the liberalization with an increase in transactions between individual landowners [20]. However, the domination of shared and joint land ownership has weakened the role of the state in controlling land use [21] and has increased the fragmentation of public land property into many scattered units [22]. Almost twelve million land shares (certificates) were distributed between rural individuals and former employees of collective and state farms [23]. According to Trukhachev et al. [23], Lerman and Shagaida [20], Rozhkov [24], and Visser et al. [25], land reform in Russia has significantly contributed to structural variations in the composition of land funds. The proportion of agricultural land in the total land fund has been declining due to a loss of arable land, particularly in the vast areas of the Far Eastern Federal District and the Siberian Federal District [26]. From 1990 to 2000, the rate of land abandonment in Russia was above 30%, one of the highest among the economies in transition [27]. Milanova [28] reported a decrease in the cropped area for all crops during the 1990s due to the changes in land tenure and stagnation of the agricultural sector. A drastic decline in livestock production resulted in a reduction of hayfields and rangelands. Vast areas of arable land were abandoned due to land degradation. In some territories in the central, northern, and eastern parts of the country, humus content dropped by 50%. Prishchepov et al. [29] revealed the correlation between the spatial distribution of abandoned croplands and natural factors, such as inadequate precipitation and shorter growing periods, in both Siberia and eastern parts of the country. As many farms were situated in the boreal zone, some of the abandoned lands have experienced shrub and tree encroachment [30].

Many experts report an aggravated environmental degradation of agricultural lands due to over-exploitation [31,32]. The changes in land cover and land use in forest-steppe and steppe vegetation zones (agriculture-oriented territories of southern Russia, the European center, and southern parts of Ural and Siberia) have been driven by extensive farming. Milanova [28] and Milanova et al. [33] reported that up to 90% of lands in some territories were converted to crop production. However, where environmental concerns of land use are mentioned in either federal or regional legislation, they predominantly relate to reducing industrial emissions or waste disposal in urban and suburban areas, not to agricultural land use [34]. Over 40 million hectares of cropland is now abandoned in Russia, and another 58 million is eroded. Land degradation, along with desertification due to irrational land use, poses serious environmental, economic, and social threats in the long-term. Griewald et al. [34] argued that the land use context in Russia did not support a transition towards sustainable land management, i.e., a "use of land resources, including soils, water, animals, and plants, for the production of goods to meet changing human needs, while simultaneously ensuring the long-term productive potential of these resources and the maintenance of their environmental functions" [35]. The urban expansion causes shrinkage of arable and other categories of agricultural land [36], which are transferred to various non-agricultural types of land use. A considerable amount of agricultural land loss due to urbanization and industrialization takes place on fertile soil [37] and irrigated lands [38]. In return, the increase in agricultural land acreage occurs on soils that are lower in terms of their fertility. Prishchepov et al. [39], Brueckner [40], and Brown et al. [41] raised concern over the growing concentration of arable land in smaller and more fragmented locations in proximity to urban and industrialized areas. Erma et al. [42] reported many cases where residential settlements occupied agricultural land in southern and central parts of the country, which are known as the breadbasket regions of Russia.

With increased variability in the composition of land funds, a reliance on research in this area has become more critical. In a series of empirical studies, many authors, including Verburg et al. [43], Van Doorn and Bakker [44], Nainggolan et al. [45], and Diogo and Koomen [46], among others, have attempted to construct hypotheses about the relationship between proximate driving forces and agricultural land-use patterns. The problem is that the established hypotheses do not adequately explain the causality between land-use processes and the compositions of land funds at different regional scales. In transition economies, including Russia, where land reforms have dramatically changed the distribution of the land inventory in recent decades [42], variations in agricultural lands due to the pressure of non-agricultural land use have remained under-investigated. The composition of agricultural land funds has commonly been considered out of a non-agricultural context [47,48], instead of exploring the interactions between the proportions of agricultural, urban, infrastructure, and industrial lands. Most of the studies have applied a proportion of agricultural land in a land fund as a core territorial specification without further testing for alternative non-agricultural land use variables [4]. Therefore, in regional studies, a knowledge gap has emerged in terms of how the variations in the compositions of land funds may be tracked with an aim to optimize agricultural land use. A more explicit focus on the relationships between land categories is required to be able to explain and predict land system dynamics in diverse locations [49]. With this background, in the case of Russia, this study aimed to contribute to the body of knowledge on regional scale land uses by identifying structural variations in the compositions of territory land funds and revealing the interdependencies between the proportions of agricultural, on the one side, and urban, industrial, and other types of land on the other.

2. Materials and Methods

This study was a quantitative study that was performed based on the data obtained from land registers from 82 out of 85 of the administrative entities of Russia (further detailed in Section 2.6). Russian public statistics report thirteen land categories within land funds, including five agricultural and eight non-agricultural ones. As we aimed to study structural variations in the compositions of land funds by identifying the changes in the proportions of different lands, all thirteen land categories were considered here (the definitions are given in Section 2.1). The overarching methods adopted in this study included a ranking of the territories on the degree of agricultural land activity (see Sections 2.2 and 2.3), centered log-ratio transformation of compositional land share data to an unconstrained space

(Section 2.4), and correlation analysis to reveal the variations in the proportion of land categories within the groups of territories (Section 2.5). In total, the study algorithm followed five stages (Table 1), which are further addressed in Sections 2.1–2.5 of the paper.

Table 1. Study flow algorithm.

Stage	Method	Section in the Paper for Methods	Results	Section in the Paper for Results
1	Merging of agricultural census data with operative land cadaster information.	Section 2.1	Establishment of an array of thirteen categories of agricultural (five variables) and non-agricultural (eight variables) land.	-
2	Computation of the shares of land categories in the land funds across Russia's territories.	Section 2.2	Map of the spatial distribution of agricultural lands in Russia per territories.	Section 3.1
3	Ranking of the shares of land categories in territory land funds.	Section 2.3	Rating scores and scales to measure the degree of agricultural land activity.	Section 3.2
4	Centered log-ratio transformation of compositional land shares data to an unconstrained space and correlation analysis of the obtained standard multivariate data.	Section 2.4	Four centered log-ratio-transformed correlation matrices based on the level of agricultural land activity.	Section 3.3
5	Computation of the coefficient of correlation variance.	Section 2.5	Identification of strong synergies between the variations of the proportions of agricultural and non-agricultural land categories in the land funds.	Section 3.3

Source: Authors' development.

2.1. *Stage 1: Land Categories*

As the structural features of land classification frameworks largely depend on the purpose of classification [50], various country specific approaches exist to categorize agriculture and other types of land. In Russia, Shagaida [51], Nosov [52], and Macht et al. [53] have contributed to the identification of various categories of agricultural lands. The majority of the studies, however, have paid inadequate attention to revealing variations in land fund compositions due to the specific needs for farming, residential construction, or industrial and infrastructure development in particular locations. For instance, Zhang et al. [3] applied an ecological-living-production classification system, to demonstrate the distribution of agricultural land across arable land, pastures, timberland, aquaculture land, and orchards, but they did not reveal the variations in the spatial concentration of particular land categories. Loshakov [54] developed an approach for the categorization of agricultural lands based on the productive qualities of soils but did not consider mismatches between agro-climatic zoning and land registers.

While the adjustments to land classification systems may be useful in achieving some specific technical, geographical, environmental, or economic goals, there are situations in which various existing approaches should be merged [55]. Many of the systems have a limitation in their ability to demonstrate the interrelationships between the categories of land cadasters for agricultural production. In general, classification concepts do not correctly emphasize per-category changes in the composition of a land fund. This is also one of the inherent vices of state statistics reporting on land fund structures in many countries. Notably, the Federal State Statistics Service of the Russian Federation (Rosstat) generalizes land into three broad categories (namely, agricultural, woodlands, and water reserve lands) [56]. Separate forms also report urban lands and lands for industrial, transportation, and communication infrastructure purposes; however, these forms exist at a national scale, not a regional scale. More detailed classification for five categories of agricultural land (croplands, hayfields, rangelands, perennial plantings, and fallows) is available in agricultural census report forms [57]. However, since the agricultural census is conducted decennially, intercategory variations cannot be effectively tracked on an annual basis.

One of the possible solutions to this discontinuity problem is to supplement census data with operative land cadaster information [58]. In Russia, the Federal Service for State Registration, Cadastre, and Cartography (Rosreestr) continually monitors land fund compositions per territories across seven categories of land, including agricultural land, residential land, industrial land, specially protected territories, woodlands, water fund lands, and reserve lands [59]. Among several classification schemes used by Rosreestr, one breaks agricultural lands into five categories, similar to Rosstat's decennial census, but instead on an annual basis. The usage of this data may allow the creation of a better time-sensitive model to represent changes in the proportion of land categories within different regions.

In this study, simple classifications determining the allocation of land between agriculture, urban, and nature were merged with more comprehensive ones, in which cadaster synergies could be detailed for a wider range of agricultural, industrial, urban and built-up, forest, and water reserve lands. The array included the categories of urban and infrastructure lands (obtained from separate sections of Rosstat's reports), as well as wetlands, disturbed lands, and barren lands (all reported by Rosreestr's alternative classification of utilized lands). In total, the authors' model merged thirteen land categories, including five agricultural ($L_{(1–5)}$) and eight non-agricultural ($L_{(6–13)}$) categories (Table 2). As reported by Rosstat [56,57] and Rosreestr [59], the categories were mutually exclusive and exhaustive. That is, each location within the T_j territory could be classified into one and only one L_i category.

Table 2. Land categories in the study.

Codes	Land Categories	Definitions
L_1	Croplands	Land systematically cultivated for crop production, including perennial grasses, clean fallow, and land under greenhouses.
L_2	Fallows	Land previously used as cropland but left unseeded for more than one year and not included in clean fallow.
L_3	Perennial plantings	Land under homogeneous stands of arboreal plants, bushes, and herbaceous plants used for the production of horticultural, technical, and medical products.
L_4	Hayfields	Fields where herbaceous plants are systematically grown for hay.
L_5	Rangelands	Land systematically and predominantly used for livestock grazing, including lands appropriate for livestock grazing but not used as hayfields or fallow.
L_6	Woodlands	Land that is mostly covered with woods or dense growths of trees and shrubs.
L_7	Forest ranges	Forest plantings on military lands, urban lands, and lands of rural settlements.
L_8	Water reserve lands	Land covered by surface water in water bodies (seas, lakes, ponds, water storage reservoirs) and land under waterworks and other facilities located within water bodies.
L_9	Residential and industrial lands	Areas of intensive use in cities, towns, and villages with much of the land covered by residential and industrial structures (those occupied by residential real estates, administrative buildings, shopping centers, industrial and commercial complexes), including in the locations isolated from urban areas.
L_{10}	Lands under transportation and communication infrastructure	Land under railways and highways, right-of-ways, cuttings in forests, livestock alleyways, and other routes of communication, as well as areas involved in processing, treatment, and transportation of water, gas, oil, and electricity.
L_{11}	Wetlands	Swampy or marshy areas saturated with moisture where the water table is at, near, or above the surface of the soil all year or for varying periods during the year, including during the growing season.
L_{12}	Disturbed lands	Land from which vegetation, topsoil, or overburden is removed or other damage is made as a result of economic and other human activities or natural processes and which is not reclaimed under the reclamation plan.
L_{13}	Barren lands	Land of limited ability to support life and incapable of producing crops or any useful vegetation.

Source: Authors' development based on Rosstat [56,57] and Rosreestr [59].

2.2. Stage 2: Composition of Land Funds

As the keynote idea is to reveal the variations in the compositions of the land funds across diverse territories, a kind of assessment scale should be applied. There have been many attempts to find a reliable approach for the conversion of cadastral classification data into land-rating values. Land classification systems based on rankings have been in use since the 1980s when Wright et al. [60] and Cocks et al. [61] first applied simple additive linear models of factor weights to the evaluation of land utility for crop production. In the realm of building a relevant ranking framework, one of the major challenges is determining how to align categorization (public statistics) with functional scales. In agriculture, variations between the proportions of lands are hard to identify [62] and thus cannot be effectively linked with territory fragmentations of agricultural production [63]. The immediacy of the problem was convincingly demonstrated by Grčman et al. [64], who found the difference between land-rating values based on precise calculations and those based on official information (specifically, for agricultural land with lower production potential).

Another challenge is that the ranking systems are not comparable and, therefore, inapplicable across a variety of agricultural and non-agricultural lands [65,66], and even across croplands, fallows, and pastures [67,68]. There have been attempts to overcome this problem by finding an integral parameter that would allow the adjustment of agricultural- and non-agricultural-oriented ranking systems to be comparable. In terms of land fund compositions, one of the most promising foundations of ranking is the contribution of a land category to the total land acreage per territory [69] (Equation (1)). The applicability of this parameter for building category-based land assessment frameworks was successfully tested by Mazurkin and Mihailova [70], Buckett [71], Artamonova et al. [72], Stupen et al. [73], Shishkina et al. [74], and Yerseitova et al. [75].

$$A_{jLi} = \frac{S_{jLi}}{S_j} \tag{1}$$

where A_{jLi} = share of land category L_i in the land fund in territory T_j; S_{jLi} = area of L_i in territory T_j; S_j = total land acreage of territory T_j.

The shares of the $L_{(1-13)}$ land categories in the land funds were computed across $T_{(1-82)}$ territories (Appendix B, Tables A9–A16).

2.3. Stage 3: Agricultural Land Activity

Further, the A_{jLi} values are ranked across the arrays of L_i land categories and T_j territories to calculate a parameter of land activity. Agricultural land activity is a degree of orientation of a land fund composition toward an agricultural type of land use. It is an indicator of how a proportion of $L_{(1-5)}$ to $L_{(6-13)}$ serves the purpose of agricultural production in particular geographic and economic conditions at a regional scale. Land activity is a score of a T_j territory, obtained based on the proportions of various land categories within a land fund. Higher contributions of $L_{(1-5)}$ to total acreage result in higher agricultural land activity scores. The activity-rank correspondence is straightforward, where the higher is A_{jLi} value, the higher is R_{ji} score. A high rank demonstrates an orientation of land fund composition towards agricultural specialization. Since the prevalence of non-agricultural lands is considered as a spatial constraint for the allocation of agricultural land uses, higher proportions of $L_{(6-13)}$ within a land fund result in lower agricultural land activity scores. For these land categories, the activity-rank relationship is inverse, where the higher is A_{jLi} value, the lower is the R_{ji} score. For j territories included in the study, the R interval was $[0; j-1]$. In our model, as $A_{jL(1-5)}$ tended to 1, R tended to $(j-1)$, while as $A_{jL(6-13)}$ tended to 1, R tended to 0.

Then, we assessed the significance of derived estimates. R_{ji} scores were used to identify the quartiles of A_{jLi} (Figure 1). The $[\sum R_{jmin}; \sum R_{jmax}]$ interval was divided into the quartiles by finding the n multiplier, where $n = \frac{\sum R_{jmax} - \sum R_{jmin}}{4}$.

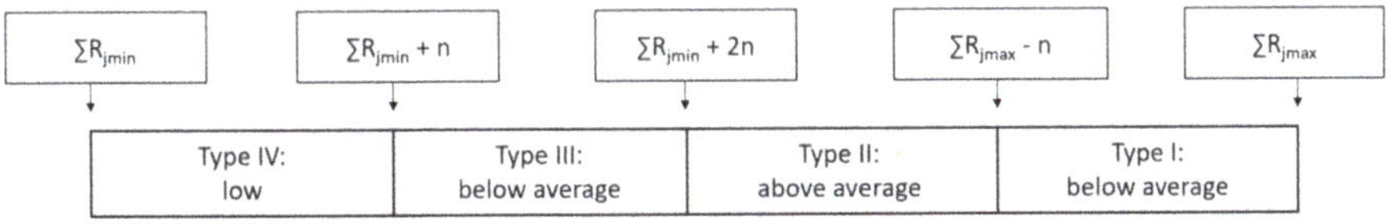

Figure 1. Scale to classify T_j territories on the degree of agricultural land activity. Source: Authors' development.

The quartile-based approach was used by Mazurkin [69] for the ranking of territories based on absolute values of land activity parameters. It also agrees with Kotykova et al. [76] and Zhildikbaeva et al. [77], who compared the deviations of land category estimates from their highest level on a territory-by-territory basis. In this study, such a method for the classification of rankings allowed consideration of the information in the percentage areas measured for each L_i in each T_j.

2.4. Stage 4: Revealing Structural Variations of Agricultural and Non-Agricultural Land Categories

Since the early years of Russia's land reform, structural variations in the compositions of land funds have progressed in response to socioeconomic and anthropogenic processes. To identify these variations between various land categories across four types of territories, this study employed factor analysis. It enables the transformation of land fund data into meaningful information [43,78] and revelation of variations in the structure of the use of territory land funds. According to Alcamo et al. [79] and Lavalle et al. [80], the integration of proximate and underlying factors may capture both the spatial distribution and the variety of land categories claimed for different land-based activities. The employment of factor analysis tools at a regional scale by Bakker et al. [81], Van Doorn and Bakker [44], and Hatna and Bakker [82] demonstrates the appropriateness of the method for cross-territory comparisons.

Among numerous factor analysis approaches, correlation analysis is one of the most suitable approaches to reveal variations in land fund compositions [83,84]. Since the A_{jLi} data are compositional, i.e., they add up to a constant value of 1 or 100% of a land fund, they need a special treatment prior to correlation analysis [85]. Aitchison [86] named land fund compositions among the typical datasets associated with challenging problems in compositional data analysis. In a compositional vector that consists of several parts summing up to a constant, the relevant information is contained only in the ratios between these parts [87] (Equation (2)).

$$x = (x_1, \ldots, x_D)^t,\ x_i > 0,\ i = 1, \ldots, D,\ \sum_{i=1}^{D} x_i = k \quad (2)$$

where D = number of compositions, and k = a positive constant value, i.e., the sum of D compositions.

If correlation analysis is applied directly to the A_{jLi} data, this can give misleading results [88] and form undesirable properties, like scale dependence [89]. The best way to analyze data with constant sum constraints is by first transforming them into an unconstrained space [88], where standard data analysis tools can then be employed [90]. Several log-ratio transformations have been introduced by Aitchison [89,91], Pawlowsky-Glahn et al. [87,90], Filzmoser and Hron [85], Long and Wang [92], and Van den Boogaart and Tolosana-Delgado [93]. Commonly used methods include using the additive log-ratio (alr), isometric log-ratio (ilr), and centered log-ratio (clr). Additive log-ratio transformation is based on log-ratios to a single reference variable. It is the simplest way to transform compositional data. However, it does not preserve distances between variables; i.e., it is not isometric [85]. Isometric log-ratio transformation is built on the choice of an orthonormal basis and thus solves the isometry problem. However, according to Egozcue et al. [94] and Egozcue and Pawlowsky-Glahn [95], base compositional parts are only related to isometric log-ratio transformed variables through non-linear functions. In our case, this meant that the computed correlations between the proportions of land categories could not be interpreted in the sense of the A_{jLi} data.

For this study, we employed centered log-ratio transformation (Equation (3)). Distinct from the additive log-ratio method, the centered log-ratio method is based on the geometric mean of all variables. It allows for the selection of a ratio variable to be avoided [85]. In contrast with the isometric log-ratio method, the centered log-ratio method simplifies the interpretation of the transformed variables because one could think of them in terms of the original variables [85,96].

$$y = [y_1, \ldots, y_D] = \left[\ln \frac{x_1}{\sqrt[D]{\prod_{i=1}^{D} x_i}}, \ldots, \ln \frac{x_D}{\sqrt[D]{\prod_{i=1}^{D} x_i}} \right] \quad (3)$$

where $x = A_{jLi}$ share of land category L_i in the land fund in territory T_j; y = transformed A_{jLi} compositions ATR_{jLi}; D = number of compositions, i.e., L_i land categories.

The A_{jLi} compositions were transformed into ATR_{jLi} data across all T_j territories using CoDaPack. This open-access software is one of the easiest-to-use applications that is commonly employed for compositional data transformation (for instance, see Thió-Henestrosa and Martín-Fernández [97], Egozcue and Pawlowsky-Glahn [98], and Muriithi [99]). The centered log-ratio-transformed data that were obtained were standard multivariate data that enabled us to use correlation analysis. Correlation matrices were built separately for the four groups of territories earlier ranked by the type of agricultural land activity. Correlation analysis was carried out here using the Excel Data Analysis ToolPak.

2.5. Stage 5: Significance of Correlations

When conducting correlation analysis for land systems, most scholars have faced a challenge similar to what we outlined earlier concerning ranking scales, namely, determining the significance of synergies between variables. Among various methods, the coefficient of correlation variance seems to be the most appropriate for dealing with interdependent multitudes of land categories [69,70] (Equation (4)).

$$C_{cv} = \frac{\sum ATR_{jLi}}{ATR_{max} \times N_L \times N_T} \quad (4)$$

where C_{cv} = coefficient of correlation variance; $\sum ATR_{jLi}$ = sum of transformed A_{jLi} values of L_i land categories in T_j territories in the group; ATR_{max} = the highest value of ATR_{jLi} in the group; N_L = number of land categories in the array; N_T = number of territories in the array.

The C_{cv} value was applied across four correlation matrices (types of land activity) to remove weak interdependencies and reveal strong synergies between the proportions of the $L_{(1–5)}$ and $L_{(6–13)}$ land categories in a land fund.

2.6. Territories and Data

Russia is a federation comprised of 85 administrative entities, or territories, as defined in the Section 1. Our study included 82 of them (mapped in Figures 2 and 3). The three municipal areas of Moscow, Saint Petersburg, and Sevastopol were excluded from the array as they are areas in which the proportion of agricultural land in the territory land fund is of negligible importance. For each territory, land cadaster data were derived from the annual reports from Rosreestr [59] and Rosstat [56,57] during 2010–2018. In Russia, these data are reported across thirteen land categories in thousand hectares. Appendix A summarizes the data of the total acreages of the territories included in the study, along with the acreages of the thirteen land categories. The study was built on the mean acreages of $L_{(1–13)}$ land categories during 2010–2018 (Appendix A, Tables A1–A8). The proportions of the $L_{(1–13)}$ land categories in regional land funds across $T_{(1–82)}$ territories are provided as percentages in Appendix B, Tables A9–A16. The variations in the proportions are provided as differences between 2010 and 2018 in Appendix B, Tables A9–A16. The consideration of the Republic of Crimea as a part of the array was determined by the current position of the territory as being de-facto controlled by Russia. In no way, these results reflect the authors' attitude to the international status of the area. For the Republic of Crimea, we used the mean data of the land acreage and land categories' proportions from 2015–2018.

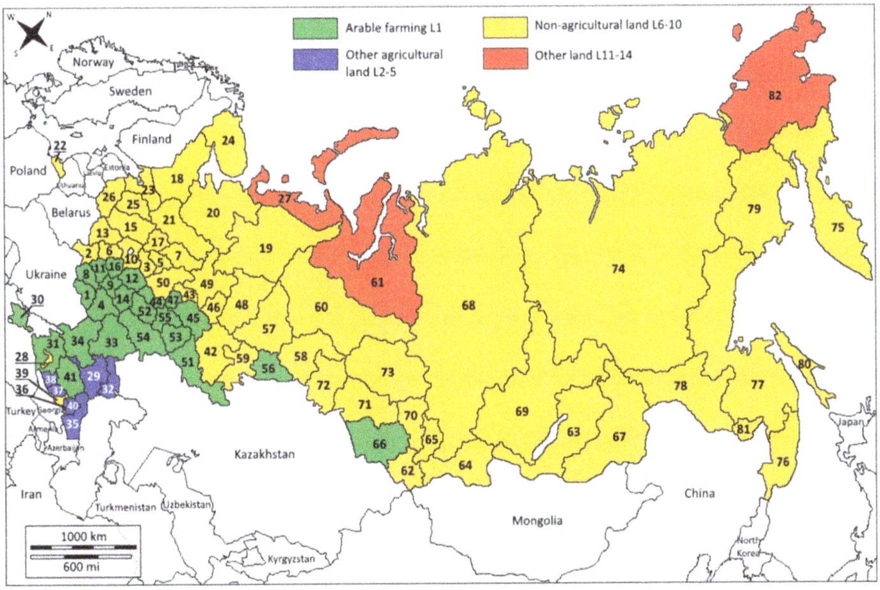

Figure 2. Spatial distribution of agricultural lands in Russia. Note: 1 = Belgorod; 2 = Bryansk; 3 = Vladimir; 4 = Voronezh; 5 = Ivanovo; 6 = Kaluga; 7 = Kostroma; 8 = Kursk; 9 = Lipetsk; 10 = Moscow Oblast; 11 = Orel; 12 = Ryazan; 13 = Smolensk; 14 = Tambov; 15 = Tver; 16 = Tula; 17 = Yaroslavl; 18 = Karelia; 19 = Komi; 20 = Arkhangelsk; 21 = Vologda; 22 = Kaliningrad; 23 = Leningrad; 24 = Murmansk; 25 = Novgorod; 26 = Pskov; 27 = Nenets; 28 = Adygeya; 29 = Kalmykia; 30 = Crimea; 31 = Krasnodar; 32 = Astrakhan; 33 = Volgograd; 34 = Rostov; 35 = Dagestan; 36 = Ingushetia; 37 = Kabardino-Balkaria; 38 = Karachaevo-Cherkessia; 39 = North Osetia-Alania; 40 = Chechnya; 41 = Stavropol; 42 = Bashkortostan; 43 = Mari El; 44 = Mordovia; 45 = Tatarstan; 46 = Udmurtia; 47 = Chuvashia; 48 = Perm; 49 = Kirov; 50 = Nizhny Novgorod; 51 = Orenburg; 52 = Penza; 53 = Samara; 54 = Saratov; 55 = Ulyanovsk; 56 = Kurgan; 57 = Sverdlovsk; 58 = Tyumen; 59 = Chelyabinsk; 60 = Khanty-Mansi; 61 = Yamal-Nenets; 62 = Altay Republic; 63 = Buryatia; 64 = Tyva; 65 = Khakasia; 66 = Altay; 67 = Zabaikalsk; 68 = Krasnoyarsk; 69 = Irkutsk; 70 = Kemerovo; 71 = Novosibirsk; 72 = Omsk; 73 = Tomsk; 74 = Sakha Yakutia; 75 = Kamchatka; 76 = Primorye; 77 = Khabarovsk; 78 = Amur; 79 = Magadan; 80 = Sakhalin; 81 = Jewish AO; 82 = Chukotka. The Republic of Crimea was included in the study due to its current position as a territory under the de-facto control of Russia. This in no way reflects the authors' attitude to the international status of the area. Source: Authors' development.

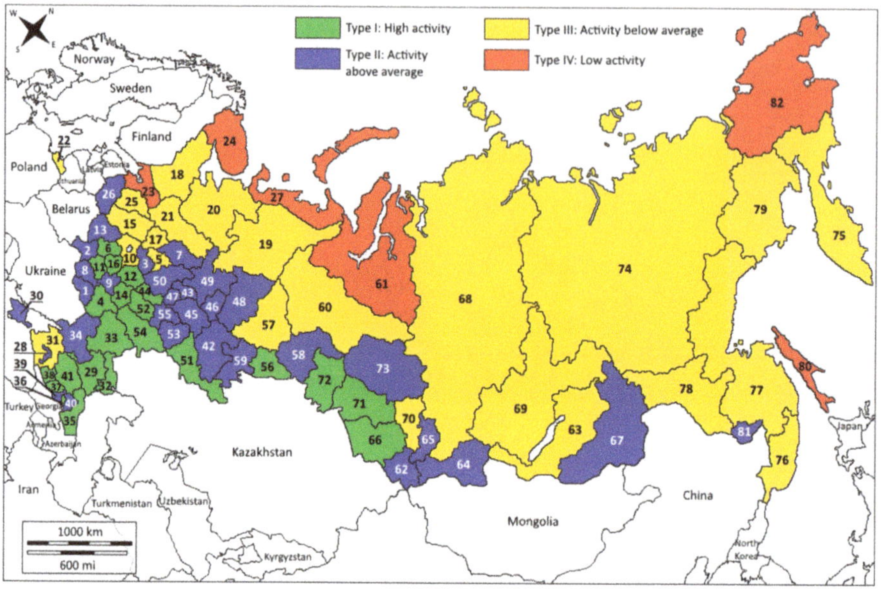

Figure 3. Russian territories: types of agricultural land activity. Note: 1 = Belgorod; 2 = Bryansk; 3 = Vladimir; 4 = Voronezh; 5 = Ivanovo; 6 = Kaluga; 7 = Kostroma; 8 = Kursk; 9 = Lipetsk; 10 = Moscow Oblast; 11 = Orel; 12 = Ryazan; 13 = Smolensk; 14 = Tambov; 15 = Tver; 16 = Tula; 17 = Yaroslavl; 18 = Karelia; 19 = Komi; 20 = Arkhangelsk; 21 = Vologda; 22 = Kaliningrad; 23 = Leningrad; 24 = Murmansk; 25 = Novgorod; 26 = Pskov; 27 = Nenets; 28 = Adygeya; 29 = Kalmykia; 30 = Crimea; 31 = Krasnodar; 32 = Astrakhan; 33 = Volgograd; 34 = Rostov; 35 = Dagestan; 36 = Ingushetia; 37 = Kabardino-Balkaria; 38 = Karachaevo-Cherkessia; 39 = North Osetia-Alania; 40 = Chechnya; 41 = Stavropol; 42 = Bashkortostan; 43 = Mari El; 44 = Mordovia; 45 = Tatarstan; 46 = Udmurtia; 47 = Chuvashia; 48 = Perm; 49 = Kirov; 50 = Nizhny Novgorod; 51 = Orenburg; 52 = Penza; 53 = Samara; 54 = Saratov; 55 = Ulyanovsk; 56 = Kurgan; 57 = Sverdlovsk; 58 = Tyumen; 59 = Chelyabinsk; 60 = Khanty-Mansi; 61 = Yamal-Nenets; 62 = Altay Republic; 63 = Buryatia; 64 = Tyva; 65 = Khakasia; 66 = Altay; 67 = Zabaikalsk; 68 = Krasnoyarsk; 69 = Irkutsk; 70 = Kemerovo; 71 = Novosibirsk; 72 = Omsk; 73 = Tomsk; 74 = Sakha Yakutia; 75 = Kamchatka; 76 = Primorye; 77 = Khabarovsk; 78 = Amur; 79 = Magadan; 80 = Sakhalin; 81 = Jewish AO; 82 = Chukotka. The Republic of Crimea was included in the study due to its current position as a territory under the de-facto control of Russia. This in no way reflects the authors' attitude to the international status of the area. Source: Authors' development.

3. Results

3.1. Composition of Land Funds

The analysis of land cadaster data across Russia's T_j territories (Appendix B, Tables A9–A16) allowed the discovery of a distinct regularity in the spatial distribution of agricultural lands. In southern and central parts of the country (green belt between 45° and 55° north latitude), croplands prevailed in the composition of the land funds (Figure 2). In the mountainous areas of North Caucasus, the blue belt comprised the territories where rangelands and other agricultural lands predominated. In most of the northern and eastern regions, the land funds were comprised of non-agricultural lands with a minor proportion of cropland.

3.2. Agricultural Land Activity

The ranking of Russia's territories on a parameter of agricultural land activity resulted in higher scores for the southern and central parts of the country than for Siberia and the Far East (Appendix C, Tables A17–A24). Concurrently, some less apparent findings were yielded (Appendix D, Table A25).

First, in the Southern Federal District, an agricultural granary for the country, the land fund composition was less agriculture-oriented compared to the Central and Volga districts and some territories of Siberia. Specifically, for Krasnodar and Rostov, two green belt territories with a considerable proportion of cropland in the structure of the land fund, the $\sum\overline{R_{ji}}$ values were well below the district average. In some territories in the south and center, high ranks of cropland and rangeland were negated by low ranks for barren lands, water reserve lands, residential, industrial, and infrastructure lands.

Second, in the Siberian Federal District, the $\sum\overline{R_{ji}}$ values nearly reached those values of the central and southern districts due to the high scores of hayfields in Omsk and Novosibirsk. The green belt by Altay was rated high for the proportion of cropland and other agricultural lands in the composition of the land fund.

Third, the yellow and red belts in the Far East feature the least agriculture-oriented macroregion in Russia. In Chukotka, Magadan, and Sakhalin, where woodlands and wetlands dominate the composition of the land fund, the agricultural land categories were ranked the lowest among the 82 territories examined here. However, in Primorye, Khabarovsk, Amur, and Jewish Autonomous Oblast, fallows, hayfields, and rangelands received high scores.

Following the obtained ranks, four R_j intervals were identified, each of which included T_j territories according to the degrees of agricultural land activity. The grouping reproduced the earlier revealed belt-like distribution of agricultural land, but with a modified configuration instead (Figure 3).

Generally, while the green belt shrank and shifted eastward, the blue one expanded and spread north of the 55° latitude mark. In some of the previously yellow belt territories of the Northwestern, Central, and Volga districts, perennial plantings and hayfields were ranked high enough to include those regions as type II regions. In Siberia, the green belt included Omsk and Novosibirsk due to the high rank of hayfields and the low rank of disturbed and barren lands. The blue belt stretched from Ural (Tyumen and Chelyabinsk) to Siberia (Tomsk, Khakasia, Tyva, and the Altay Republic) and farther to the Far East (Zabaikalsk and the Jewish Autonomous Oblast). In the south, the substantial activity of residential, industrial, transportation, communication, and disturbed lands downgraded Krasnodar to type III and Rostov and Crimea to type II. Ingushetia, Kabardino-Balkaria, Karachaevo-Cherkessia, and Dagestan, on the contrary, broke forth to the green belt due to high scores of perennial plantings and rangeland and low activity of wetlands, disturbed lands, and water reserve lands.

3.3. Correlation Analysis

In type I territories, the variations in the compositions of agricultural lands correlated with the changes in the acreage of non-agricultural land for infrastructure, primarily transportation and communication (the strongest correlation with cropland, perennial plantings, and hayfields) (Table 3). Strong correlations were also revealed between the proportions of croplands and fallows, on one side, and those of woodland and barren land on the other. The share of rangeland in the land fund was strongly correlated with that of barren land.

Table 3. Correlation matrix for type I territories.

Regressands	Regressors											
	ATR_{L1}	ATR_{L2}	ATR_{L3}	ATR_{L4}	ATR_{L5}	ATR_{L6}	ATR_{L7}	ATR_{L8}	ATR_{L9}	ATR_{L10}	ATR_{L11}	ATR_{L12}
ATR_{L2}	0.5317											
ATR_{L3}	0.6206	**0.7844**										
ATR_{L4}	0.4195	0.2973	**0.7109**									
ATR_{L5}	0.6619	0.6836	0.3088	0.6619								
ATR_{L6}	0.5854	**0.8127**	0.3917	0.1485	0.5193							
ATR_{L7}	**0.8863**	**0.7340**	0.6226	0.6993	0.4286	0.3275						
ATR_{L8}	0.4001	0.4872	0.4345	**0.7226**	0.6719	0.5015	0.4812					
ATR_{L9}	**0.8925**	0.5100	0.7001	0.6133	**0.7483**	0.4990	0.6291	**0.7255**				
ATR_{L10}	**0.9691**	0.5902	**0.9583**	**0.9121**	0.4128	0.3802	**0.8016**	0.3476	0.2196			
ATR_{L11}	**0.7779**	0.2014	0.4296	**0.7724**	0.1944	**0.8403**	0.2974	**0.7402**	0.3209	**0.7947**		
ATR_{L12}	0.1803	**0.7098**	0.5044	0.1725	0.4592	0.6274	0.3076	0.1577	**0.8182**	0.2619	0.5044	
ATR_{L13}	0.3956	**0.7317**	0.3712	0.4594	**0.8215**	0.6619	0.4764	0.1810	0.4464	0.3118	0.6275	0.2999

Note: ATR_{Li} = centered log-ratio-transformed data: ATR_{L1} = cropland; ATR_{L2} = fallow; ATR_{L3} = perennial plantings; ATR_{L4} = hayfields; ATR_{L5} = rangeland; ATR_{L6} = woodlands; ATR_{L7} = forest range; ATR_{L8} = water reserve lands; ATR_{L9} = residential and industrial lands; ATR_{L10} = lands under transportation and communication infrastructure; ATR_{L11} = wetlands; ATR_{L12} = disturbed lands; ATR_{L13} = barren; bold denotes a strong correlation, $C_{ATRli} > C_{cv}$ (0.7022 for type I territories). Source: Authors' development.

Similar to type I, in the type II group, a strong correlation was found between the shares of cropland and perennial plantings and those of lands for transportation and communication infrastructure (Table 4). Besides, since the blue belt predominantly was comprised of densely populated territories, there was a correlation between the shares of croplands and residential lands. In many type II territories, the contribution of woodlands and other forest ranges to the structure of the land fund was essential. This fact might explain the high correlation between the composition of agricultural lands and woodlands. In the south, where the climate and soil favor the development of horticulture and viniculture (i.e., in Crimea, Adygeya, and Rostov), C_{cv} emphasized a strong correlation between the proportions of perennial plantings and croplands within the agricultural land categories.

Table 4. Correlation matrix for type II territories.

Regressands	Regressors											
	ATR_{L1}	ATR_{L2}	ATR_{L3}	ATR_{L4}	ATR_{L5}	ATR_{L6}	ATR_{L7}	ATR_{L8}	ATR_{L9}	ATR_{L10}	ATR_{L11}	ATR_{L12}
ATR_{L2}	0.3291											
ATR_{L3}	**0.6719**	0.4417										
ATR_{L4}	0.5213	**0.8013**	0.5016									
ATR_{L5}	0.2706	0.3814	0.1928	0.4836								
ATR_{L6}	**0.8804**	**0.7512**	**0.7793**	0.3391	**0.8284**							
ATR_{L7}	0.4817	0.5788	0.4801	**0.6481**	0.2662	0.4571						
ATR_{L8}	0.2940	**0.6941**	0.5592	0.5702	0.1827	0.2719	**0.7027**					
ATR_{L9}	**0.7592**	0.4290	**0.7728**	0.4817	0.5011	0.0458	0.2664	0.5822				
ATR_{L10}	**0.8918**	0.2811	**0.9102**	0.1482	**0.7661**	0.3443	0.1988	0.5591	**0.6619**			
ATR_{L11}	0.1157	0.1792	0.2866	**0.7205**	**0.8003**	0.4509	0.4295	0.3619	**0.7268**	0.1384		
ATR_{L12}	**0.6834**	0.3810	0.3017	0.0133	0.4506	**0.7318**	**0.6040**	0.0744	**0.8112**	**0.6714**	0.2857	
ATR_{L13}	0.2375	0.027	**0.5993**	0.2915	**0.6266**	0.5011	0.1302	0.2599	0.2004	0.2777	0.5296	0.4018

Note: ATR_{Li} = centered log-ratio-transformed data: ATR_{L1} = cropland; ATR_{L2} = fallow; ATR_{L3} = perennial plantings; ATR_{L4} = hayfields; ATR_{L5} = rangeland; ATR_{L6} = woodlands; ATR_{L7} = forest range; ATR_{L8} = water reserve lands; ATR_{L9} = residential and industrial lands; ATR_{L10} = lands under transportation and communication infrastructure; ATR_{L11} = wetlands; ATR_{L12} = disturbed lands; ATR_{L13} = barren; bold denotes a strong correlation, $C_{ATRli} > C_{cv}$ (0.5904 for type II territories). Source: Authors' development.

The yellow belt included three types of territories, namely, northern territories, Siberia, and the Far East, occupying over half of the territory of Russia, but only representing 12.3% of its agricultural land, where the land use was primarily rangeland. The variations in the acreage of rangelands strongly correlated with those of woodlands, other forest ranges, and wetlands (Table 5). The northern locus included the territories of Russia's northwest, the Ural region, and central Russia (i.e., north of Moscow). In these highly industrialized but less populated territories, we revealed strong correlations between the proportions of croplands and barren land, as well as between those of perennial plantings and disturbed lands. In the south, the yellow belt included Krasnodar, the principal breadbasket territory of Russia. The share of cropland in the composition of Krasnodar's land fund was 52.8%. Krasnodar is also one of Russia's most densely populated regions and is the most popular resort area. The analysis demonstrated high correlations between the proportions of cropland and perennial plantings, on one side, and the shares of residential and industrial lands and lands under transportation and communication infrastructure on the other.

Table 5. Correlation matrix for type III territories.

Regressands	Regressors											
	ATR_{L1}	ATR_{L2}	ATR_{L3}	ATR_{L4}	ATR_{L5}	ATR_{L6}	ATR_{L7}	ATR_{L8}	ATR_{L9}	ATR_{L10}	ATR_{L11}	ATR_{L12}
ATR_{L2}	0.5638											
ATR_{L3}	**0.8819**	0.4291										
ATR_{L4}	**0.8025**	0.4010	**0.8211**									
ATR_{L5}	0.2811	0.6388	**0.9157**	0.5037								
ATR_{L6}	**0.9012**	0.5917	**0.8924**	0.4545	**0.7684**							
ATR_{L7}	0.4709	**0.7559**	0.6713	**0.7553**	**0.8315**	0.2819						
ATR_{L8}	0.6880	0.7000	0.5004	0.3819	**0.7700**	0.4196	0.3358					
ATR_{L9}	**0.8544**	0.3093	**0.8120**	0.6594	0.5428	**0.7920**	0.3902	0.4971				
ATR_{L10}	**0.7923**	0.4458	**0.7538**	0.2888	0.4111	**0.8328**	0.7010	0.6947	**0.7748**			
ATR_{L11}	0.7001	0.6219	0.4816	0.1329	**0.8148**	0.1887	0.6409	0.5068	**0.8591**	0.2509		
ATR_{L12}	0.5493	**0.7704**	0.3309	**0.8617**	0.1499	0.2796	**0.8419**	0.3991	0.4404	**0.7803**	0.3012	
ATR_{L13}	**0.8057**	0.1295	**0.7772**	0.6026	0.2891	0.4905	0.3948	0.4819	**0.9062**	**0.7696**	0.0180	**0.8016**

Note: ATR_{Li} = centered log-ratio-transformed data: ATR_{L1} = cropland; ATR_{L2} = fallow; ATR_{L3} = perennial plantings; ATR_{L4} = hayfields; ATR_{L5} = rangeland; ATR_{L6} = woodlands; ATR_{L7} = forest range; ATR_{L8} = water reserve lands; ATR_{L9} = residential and industrial lands; ATR_{L10} = lands under transportation and communication infrastructure; ATR_{L11} = wetlands; ATR_{L12} = disturbed lands; ATR_{L13} = barren; bold denotes a strong correlation, $C_{ATRli} > C_{cv}$ (0.7458 for type III territories). Source: Authors' development.

Type IV comprised the territories with the lowest activity of agricultural lands. The scarcity of agricultural lands represented intercategory variations in the composition of the agricultural land fund. The strongest correlations were identified between various categories of agricultural lands, specifically, cropland and hayfields, on one side, and perennial plantings and rangeland on the other (Table 6). The composition of the agricultural land fund was also affected by the proportions of barren land (in Chukotka and Nenets), woodlands (in Leningrad and Murmansk), wetlands (in Murmansk), and water reserve lands (in Yamal-Nenets).

Table 6. Correlation matrix for type IV territories.

Regressands	Regressors											
	ATR_{L1}	ATR_{L2}	ATR_{L3}	ATR_{L4}	ATR_{L5}	ATR_{L6}	ATR_{L7}	ATR_{L8}	ATR_{L9}	ATR_{L10}	ATR_{L11}	ATR_{L12}
ATR_{L2}	0.4018											
ATR_{L3}	**0.7301**	0.3884										
ATR_{L4}	0.3899	0.3892	**0.8496**									
ATR_{L5}	**0.6933**	0.2594	**0.7915**	**0.8101**								
ATR_{L6}	**0.6705**	0.3217	0.4024	0.1788	0.2894							
ATR_{L7}	**0.8111**	**0.7910**	0.3881	0.2519	0.3221	**0.7518**						
ATR_{L8}	0.3595	0.6159	0.2053	0.3706	0.1553	0.2995	0.4085					
ATR_{L9}	0.4276	**0.6757**	0.5829	0.4881	**0.7391**	0.2709	**0.7047**	**0.6586**				
ATR_{L10}	0.6083	0.4792	**0.7294**	0.3201	0.3899	0.3892	0.3999	0.5993	0.3788			
ATR_{L11}	0.4291	**0.7032**	0.5022	0.2718	0.0377	0.4920	0.4793	0.2819	0.3003	0.2709		
ATR_{L12}	0.1829	0.0377	0.4603	**0.6883**	**0.7418**	0.3207	**0.6991**	0.1842	**0.6309**	0.5346	0.1442	
ATR_{L13}	**0.6693**	0.4871	**0.7918**	0.5593	0.1294	**0.7622**	0.0412	0.3909	0.1899	**0.6511**	0.2895	0.1566

Note: ATR_{Li} = centered log-ratio-transformed data: ATR_{L1} = cropland; ATR_{L2} = fallow; ATR_{L3} = perennial plantings; ATR_{L4} = hayfields; ATR_{L5} = rangeland; ATR_{L6} = woodlands; ATR_{L7} = forest range; ATR_{L8} = water reserve lands; ATR_{L9} = residential and industrial lands; ATR_{L10} = lands under transportation and communication infrastructure; ATR_{L11} = wetlands; ATR_{L12} = disturbed lands; ATR_{L13} = barren; bold denotes a strong correlation, $C_{ATRli} > C_{cv}$ (0.6293 for type IV territories). Source: Authors' development.

4. Discussion

The results, as expected, demonstrated that the compositions of the land funds in Russia vary across territories. Echoing Bichler et al. [100], Chu [101], Smith et al. [102], and Bakker et al. [103], we found that the distribution of agricultural lands is largely affected by natural factors, while agricultural lands are spread unevenly across the country. At a regional scale, belt-type concentrations of cropland suggest an agriculture-focused land distribution pattern in the southern and central areas of Russia. This is consistent with the observations of Rounsevell et al. [104] and White and Engelen [105,106], who revealed that agricultural land use tends to become concentrated in locations, reflecting the influence of natural factors and neighboring land distribution patterns. Nevertheless, in particular territories, the proportion of agricultural lands in the land funds do not match the type of agricultural land activity.

Emulating earlier studies by Mazurkin and Mihailova [70], Shishkina et al. [74], Mazurkin [69], and Buckett [71], we revealed that the application of a land activity parameter could result in creating a picture of land distribution patterns that are different from that which might be expected from the knowledge of the proportions of individual land categories. Therefore, land distribution change maps are not sufficient to capture specific finer-scale variations in the compositions of land funds at a regional scale. In Russia, land tenure and demand for land have been the principal economic proxies to map agricultural land distribution. According to Shagaida [107], the demand for agricultural land varies significantly across Russia's territories, depending on the degree of land consolidation. In the course of land reform, the previously dominant state farms have transformed the organizational form of their land use but still have persisted as the backbone of the agricultural sector [34,108]. In the embryonic land market in the 1990–2000s, the establishment of new land tenure patterns had not involved immediate changes in the distribution of land from big ex-Soviet agricultural enterprises to individual owners [107]. Since land certificates do not specify land plots, most of the shareowners have not withdrawn their land property from joint use by former collective farms. Over 70% of land in Russia is still used by large enterprises for rent, 25% is contributed to the capital of large enterprises, and only 4% is retained by private owners [109]. In the breadbasket southern and central European territories of Russia, large agricultural holding companies have aggregated even more agricultural land property when compared to the Soviet period [110].

To a large extent, the existing demand-based distribution matches the land activity map (Figure 3), as the highest demand for land is identified in the central parts of the country close to Moscow. This demand primarily exists due to non-agricultural businesses. For type I and II territories, this correlates well with the finding of strong links between the proportions of agricultural land categories, on one side, and those of residential, industrial, and infrastructure lands on the other. In type III southern locus (Krasnodar), Lerman and Shagaida [20] reported high demand for land among corporate farms. In that classification, type I and II territories are considered as less developed areas in terms of agricultural production (sometimes even as "agriculturally depressed regions" ([20] p. 20)), where corporate farms tend to reduce their holdings and abandon land plots. Our results, on the contrary, demonstrated that in the south of European Russia, where the concentration of croplands is the highest, agricultural land activity is lower compared to many other territories of the country.

In the territories where a high proportion of croplands coexist with low agricultural land activity, many of the variations in the composition of a land fund could be explained by socio-economic factors. Van de Steeg et al. [111] and Gärtner et al. [112] confirmed that the distribution of agricultural land strongly correlates with the level of rural development, proximity to economic and market centers, urbanization, and the demand for agricultural land from non-agricultural industries. Our study revealed correlations between the proportions of agricultural and urban lands across type I–III territories, which could represent losing agricultural land due to urban development. In type II territories, the compositions of agricultural land funds are more affected by urban development than the compositions of type I and III. These results supported the findings of Daniels [113], Su et al. [114], Yeh and Huang [115], and Dredge [116], i.e., the proximity to urban development can be a powerful predictor of changes in agricultural land use. Many scholars, including Parsipour et al. [117], Li et al. [118], and Al-Kofahi et al. [119], among others, agree that the accelerating urbanization has been causing increasingly harmful effects on agricultural lands. In the case of Russia, we did not reveal the acceleration of agriculture land loss in urbanized type I–III territories. What was revealed, however, was the strengthening of the correlation between the variations in the compositions of agricultural land funds and residential, industrial, and infrastructure lands. As Zubair et al. [120] and Lucero and Tarlock [121] forecasted, such stronger associations would continue to put increasing pressure on agricultural lands and result in more fragmented agricultural land use in the future.

Along with urbanization, an orientation of a land fund composition towards agricultural production is determined by the population density [111,122]. In urbanized type I and II territories, agricultural land use is affected by the variations in the acreage of residential lands. In agriculture-oriented Krasnodar, Rostov, and Stavropol, the changes in agricultural land fund compositions are mainly linked with those of lands for transportation and communication. This result was consistent with what Ramadani and Bytyqi [123], Li et al. [118], and Al-Kofahi et al. [119] reported when assessing the effects of more significant concentrations of the population on the lower proportions of agricultural lands in a land fund.

Reversely, Meyfroidt et al. [124] and Nguyen et al. [125] revealed that in the industrialized areas in Russia, where the density of population is lower, the concentration of abandoned lands is higher. There is an array of studies that have reported a link between industrial growth, changes in agricultural land distribution, and the degradation of farming opportunities internationally. Explicitly, Oyebanji et al. [126] confirmed the existence of a positive long-term relationship between industrialization and land loss in Nigeria. Deng and Li [127] revealed that the soil sealing effect has resulted from industrial and infrastructure construction in China, while Müller and Sikor [128], Milanova et al. [33], and Müller et al. [129] linked changes in agricultural land distribution and agricultural abandonment in EU countries with unfavorable environmental conditions due to increasing industrialization. The expansion of urban and industrial infrastructure not only triggers agriculture-to-urban and agriculture-to-industry land transfers but also leads to the overexploitation and degradation of remaining agricultural lands [127]. Many areas in Russia may soon face a reduction in farming opportunities due to various kinds of environmental pollution. Many experts tend to explain the unprecedented increase of barren land in Russia (by four million ha during the past two decades) by the intensive exploitation of mineral resources and industrial construction [39,130]. Kashtanov [131] and Dobrovolski [132] associated the expansion of industrial infrastructure with long-term and irreversible losses of cropland in Russia. In support of the earlier findings of Sorokin et al. [130] and Solgerel et al. [133] concerning the close relationships between industrial development and arable acreage, strong correlations between the proportions of croplands, perennial plantings, and industrial lands are revealed in both urbanized type I and II territories and sparsely populated yellow belt areas.

Distinct from urbanization, industrialization may affect agricultural land use in remote areas. According to Sorokin et al. [130], most of the abandoned lands are located in the north of Russia. This agrees well with our finding of strong correlations between the variations in the acreage of croplands, disturbed lands, and barren lands in the north locus of the yellow belt. Prishchepov et al. [39] and MacDonald et al. [134] also reported abandoned agricultural land concentrated in remote and isolated industrialized areas in northern Russia. Nakvasina et al. [135] claimed that the proximity to urban areas might be used as a critical criterion to transfer disturbed and barren lands back into agricultural use. However, we did not identify strong correlations between the variations in agricultural land fund compositions and residential lands for type III territories.

In diverse land activity patterns across the Russian territories, changes in the compositions of agricultural and non-agricultural land funds depend on the degree of industrial development. As mentioned by Postek et al. [136] and Prishchepov et al. [39], agricultural land loss due to increasing industrialization causes the fragmentation of arable lands as smaller locations with lower productivity. However, according to Popov [137], fragmentation is not a problem in agriculture-oriented areas due to the excessive lease of agricultural land. The issue is particularly topical in territories where arable land is scarce, however [138,139]. Nefedova [140] reported that in northern and eastern parts of Russia, agricultural land distribution is extremely fragmented. Our results demonstrated that in the Russian North and Far East, low activity of cropland is coupled with the prevalence of hayfields in the composition of the agricultural land funds there. High intragroup correlations between the proportions of cropland, rangeland, hayfields, and perennial plantings in type IV territories confirm the observations of King and Burton [141], Tan et al. [142], and Dhakal and Khanal [143], i.e., the fragmentation results in the competition between the categories of agricultural lands.

We performed our analysis in the short-term, but it is commonly known that land transformations (particularly, for croplands and annual crops) can be rapid, whereas transformations are slower in grassland-livestock oriented areas and permanent crop areas. Nationally, the ongoing loss of croplands may not have an immediate effect on the agricultural output of Russia. Still, this represents enormous environmental, economic, and social costs that will be hard to absorb in terms of a long-term perspective [144]. Griewald et al. [34] and Hunt et al. [145] outlined five principal drivers of long-term change in agricultural land use in Russia, environmental drivers being one of them. Our findings would allow one to expect that the evolution of land-use change will be affected by the pressure exerted on ecosystems by various land management types [34]. While some authors, including Diputra and Baek [146] and Mahcene et al. [147], reported little evidence that industrialization causes a significant increase in disturbed land acreage, our results suggested that lower activity of agricultural land categories is correlated with a higher activity of barren land, disturbed land, and industrial land. Weaker, but still significant, correlations between the proportions of agricultural and industrial land categories are revealed in type I and II territories here. In type IV territories, the contributions of croplands and perennial plantings to regional land funds are also linked with variations in the acreage of barren lands.

Among the drivers of land-use change, in the long run, there are also economic, social, technological, and policy-related factors. Bukvareva et al. [148] stated that current land-use policies in Russia pay little attention to the environmental costs associated with the re-use of abandoned lands. In light of the economic recession that Russia has been experiencing since the mid-2010s, farmers tend to reinforce the exploitation of all available lands to ensure sufficient income inflow. Often, this is done regardless of whether some lands are of high environmental value or are socioeconomically marginal [29]. In the short-term perspective, we did not reveal an increase in the acreage of croplands due to the use of other categories of agricultural land. To some extent, however, the correlations between the proportions of agricultural land categories are identified in type III territories. In these yellow belt areas, land reclamation programs will require substantial investments for clearing forested land, liming, and other works. In the short-term, high reclamation costs along with poor soil quality may reduce expected economic returns [149]; however, in the long run, the incentives for reclamation may grow as both the availability and quality of croplands in type I and II territories degrade. Such a perspective highlights the need for a deeper investigation of the variations in land fund compositions within a sustainable agricultural land management approach as a component of the broader economic and environmental system [150,151].

5. Conclusions

In recent decades, there has been increasing concern for ensuring the effective utilization of agricultural land due to the limited area of highly productive arable land and the growing demand for food and farming products internationally. In Russia, an orientation of state policy towards the growth of agricultural production, along with a low level of environmental awareness among farmers, has impeded the prospects of sustainable land management as an integral aspect of land use planning. The degradation of agricultural lands due to irrational use has posed environmental, economic, and social threats to the national development objectives of land management in many territories of the country. As most studies in Russia have focused on land changes between the categories of agricultural land, the influence of agriculture-to-urban and agriculture-to-industry transfers has been downplayed.

We conducted this work, intending to study such variations by revealing the interdependencies between the proportions of agricultural land categories, on the one hand, and urban, industrial, and other types of land on the other. First, land distribution was mapped based on a share of agricultural lands in a composition of a land fund and, second, by a "land activity rating" of Russia's territories. Such a two-step approach to mapping allowed us to find that the proportions of agricultural lands in the composition of a land fund do not appropriately reveal the variations in the activities of agricultural land categories. In the territories, where agricultural lands dominated in the structure of a land fund, the agricultural land activity could be depressed by high proportions of non-agricultural lands. In urbanized and densely populated territories, the composition of the agricultural land fund was predominantly affected by the changes in the acreage of residential and industrial lands, as well as the lands for transportation and communication. In industrialized but underpopulated territories, the acreages of croplands and perennial plantings were strongly correlated with those of disturbed and barren lands. We also found that lower land activity tended to increase the variations within the agricultural land fund, which might indicate intercategory competition for more fertile, more productive, and better-located agricultural lands.

By establishing and testing the five-stage algorithm, we attempted to solve the scientific problem of low awareness in the causality between land-use processes and the composition of the land funds at regional scales. As distinguished from previous studies in the area, we investigated variations in the compositions of a land fund as interactions between the proportions of agricultural and non-agricultural lands. Practically, in territory-scale studies, such an approach might complement regionally adapted monitoring networks by targeting the mismatches between the cadaster-based mappings of agricultural land distributions and ranking-based activities of agricultural lands. Theoretically, such an algorithm allows one to capture the complex relationships of a variety of land categories and the resulting correlations between their proportions, therefore, being applicable for studying territorial land-use patterns, the simulation of agricultural land distribution systems, and the extrapolation of current trends into the future. Potentially, the algorithm is suitable for numerous locations. However, one of the limitations of the current study was that it used the Russian system of land statistics, which is built on thirteen land categories. Due to the different sources of land use data in different countries, an adjustment of the array of land categories to a national land reporting system is needed when implementing the method in a broader international context. Another limitation that could potentially challenge cross-country comparisons is the different sizes of territorial units. Russia's case demonstrates that this problem may arise even within one country, where territories substantially differ in size. In an attempt to overcome a data discrepancy obstacle, we conversed cadastral classification data into land-rating values. To address the diversity of territories, we used an agricultural land activity parameter. This allowed us to adjust agricultural and non-agricultural-oriented ranking systems to make them comparable. Nevertheless, further research is needed to assess to what extent the approach would be able to appropriately picture variations in agricultural land activity patterns in the conditions of information asymmetries among countries.

Author Contributions: V.E. conceptualized the research framework and wrote the paper; T.G. analyzed the data; A.I. performed the data collection. All authors have read and agreed to the published version of the manuscript.

Funding: This research and the APC were funded by the International and Regional Studies Fund of the Ministry of Education of the People's Republic of China (grant number 19GBQY117).

Conflicts of Interest: The authors declare no conflict of interest.

Appendix A

Table A1. Land acreage data of the Central Federal District in thousand hectares. Mean values for 2010–2018.

Territory	Total	L_1	L_2	L_3	L_4	L_5	L_6	L_7	L_8	L_9	L_{10}	L_{11}	L_{12}	L_{13}
T_1	2713.4	1645.2	0	34.0	55.8	399.3	241.9	90.5	25.1	73.1	57.9	22.5	6.5	61.6
T_2	3485.7	1174.9	121.4	26.0	205.5	346.5	1183.6	121.4	31.6	56.8	72.0	75.1	5.1	65.8
T_3	2908.4	605.7	46.6	20.0	163.9	159.1	1582.7	74.9	32.7	38.0	75.0	38.3	16.3	55.2
T_4	5221.6	3046.2	41.9	52.8	159.0	776.8	482.4	149.5	64.0	113.4	121.1	40.6	1.9	172.0
T_5	2143.7	565.9	9.8	9.0	124.1	112.5	1047.8	28.5	65.0	42.0	51.2	50.3	7.4	30.2
T_6	2977.7	956.1	36.1	21.0	131.2	232.2	1376.9	35.5	21.0	56.9	50.2	28.6	2.1	29.9
T_7	6021.1	655.0	31.2	5.6	154.5	148.3	4574.1	98.9	97.0	35.6	101.7	86.8	5.7	26.7
T_8	2999.7	1943.4	0.7	27.9	101.6	364.3	249.3	68.1	38.3	56.4	72.5	32.1	11.0	34.1
T_9	2404.7	1553.9	0.1	35.2	83.6	281.0	190.7	61.4	27.0	47.9	61.7	16.4	2.5	43.3
T_{10}	4579.9	1130.3	6.7	113.9	183.0	229.4	1998.3	35.2	90.1	303.1	158.8	50.6	34.7	98.8
T_{11}	2465.2	1570.0	55.7	25.3	58.6	341.5	203.1	74.2	14.4	21.9	72.8	3.8	0.7	23.2
T_{12}	3960.5	1535.2	26.1	24.6	202.6	722.4	1067.8	66.3	67.2	37.1	105.1	55.4	6.6	44.1
T_{13}	4977.9	1461.7	17.7	19.5	215.1	380.0	2167.6	357.6	53.7	55.7	86.5	115.3	18.0	29.5
T_{14}	3446.2	2127.5	9.6	32.4	166.0	388.8	371.7	97.9	42.8	55.1	60.8	43.9	1.7	48.0
T_{15}	8420.1	1504.3	19.4	14.7	379.1	501.0	4742.2	233.3	248.1	96.9	116.4	465.2	20.3	79.2
T_{16}	2567.9	1554.4	7.6	45.0	67.9	298.0	372.3	43.0	22.8	32.3	90.4	1.9	10.0	22.3
T_{17}	3617.7	793.3	0.3	14.6	123.7	196.1	1725.7	93.0	386.8	59.4	65.8	109.7	15.2	34.1

Note: T_1 = Belgorod; T_2 = Bryansk; T_3 = Vladimir; T_4 = Voronezh; T_5 = Ivanovo; T_6 = Kaluga; T_7 = Kostroma; T_8 = Kursk; T_9 = Lipetsk; T_{10} = Moscow Oblast; T_{11} = Orel; T_{12} = Ryazan; T_{13} = Smolensk; T_{14} = Tambov; T_{15} = Tver; T_{16} = Tula; T_{17} = Yaroslavl; $L_{(1-13)}$ = acreage of L_i category, thousand hectares: L_1 = cropland; L_2 = fallow; L_3 = perennial plantings; L_4 = hayfields; L_5 = rangeland; L_6 = woodlands; L_7 = forest range; L_8 = water reserve lands; L_9 = residential and industrial lands; L_{10} = lands under transportation and communication infrastructure; L_{11} = wetlands; L_{12} = disturbed lands; L_{13} = barren. Source: Authors' development.

Table A2. Land acreage data of Northwestern Federal District in thousand hectares. Mean values for 2010–2018.

Territory	Total	L_1	L_2	L_3	L_4	L_5	L_6	L_7	L_8	L_9	L_{10}	L_{11}	L_{12}	L_{13}
T_{18}	18,052.0	82.3	0.1	5.9	85.4	39.2	9850.2	22.1	4188.2	38.3	87.6	3543.6	13.4	95.7
T_{19}	41,677.4	102.4	0	6.5	239.6	69.6	31,093.5	135.6	641.5	48.2	144.8	4073.1	15.8	5106.8
T_{20}	41,310.3	302.5	1.8	9.1	304.1	109.8	22,948.6	126.3	811.5	93.3	131.3	5823.3	5.5	10,643.2
T_{21}	14,452.7	822.0	48.0	9.4	343.9	225.2	10,456.4	330.9	658.6	38.3	178.3	1271.8	22.2	47.7
T_{22}	1512.5	392.6	0	14.3	153.6	248.9	295.1	18.8	200.3	40.6	40.9	31.0	4.4	72.0
T_{23}	8390.8	434.1	0	44.4	194.6	125.4	5015.7	125.3	1266.8	58.7	112.7	830.0	23.0	160.1
T_{24}	14,490.2	19.4	0	3.1	2.8	0.3	5383.6	580.8	1191.5	37.1	31.3	5701.2	19.7	1519.4
T_{25}	5450.1	510.6	4.2	6.1	173.1	135.9	3580.9	138.6	174.8	25.5	69.8	548.5	10.4	71.7
T_{26}	5539.9	744.3	186.4	20.5	279.0	280.9	2249.0	785.3	375.3	34.8	71.9	476.2	8.9	27.4
T_{27}	17,681.0	0.2	0	0	19.8	5.7	1740.8	1439.2	1000.5	12.8	10.8	3381.8	2.5	10,066.9

Note: T_{18} = Karelia; T_{19} = Komi; T_{20} = Arkhangelsk; T_{21} = Vologda; T_{22} = Kaliningrad; T_{23} = Leningrad; T_{24} = Murmansk; T_{25} = Novgorod; T_{26} = Pskov; T_{27} = Nenets; $L_{(1-13)}$ = acreage of L_i category, thousand hectares: L_1 = cropland; L_2 = fallow; L_3 = perennial plantings; L_4 = hayfields; L_5 = rangeland; L_6 = woodlands; L_7 = forest range; L_8 = water reserve lands; L_9 = residential and industrial lands; L_{10} = lands under transportation and communication infrastructure; L_{11} = wetlands; L_{12} = disturbed lands; L_{13} = barren. Source: Authors' development.

Table A3. Land acreage data of the Southern Federal District in thousand hectares. Mean values for 2010–2018.

Territory	Total	L_1	L_2	L_3	L_4	L_5	L_6	L_7	L_8	L_9	L_{10}	L_{11}	L_{12}	L_{13}
T_{28}	779.2	259.6	0.3	9.3	4.9	85.7	288.8	7.7	53.5	22.1	18.8	4.0	0.3	24.2
T_{29}	7473.1	836.9	10.6	2.5	103.2	5363.6	32.6	42.3	175.6	32.2	65.1	123.5	4.0	681.0
T_{30}	2608.1	1271.6	10.6	75.8	1.9	433.6	266.2	35.0	211.7	118.8	43.4	5.2	1.5	132.8
T_{31}	7548.5	3985.4	0.2	125.2	63.1	531.1	1541.3	158.7	385.6	202.9	196.0	179.6	5.4	174.0
T_{32}	4902.4	352.0	6.7	9.8	404.8	2482.7	104.2	19.5	684.6	28.2	57.4	114.7	0.5	637.3
T_{33}	11,287.7	5854.0	4.7	42.8	206.9	2652.8	591.0	131.3	489.8	165.9	117.6	35.2	3.0	992.7
T_{34}	10,096.7	5907.3	0	58.2	88.4	2459.2	293.0	281.9	346.1	150.8	220.5	55.0	7.1	229.2

Note: T_{28} = Adygeya; T_{29} = Kalmykia; T_{30} = Crimea; T_{31} = Krasnodar; T_{32} = Astrakhan; T_{33} = Volgograd; T_{34} = Rostov; $L_{(1-13)}$ = acreage of L_i category, thousand hectares: L_1 = cropland; L_2 = fallow; L_3 = perennial plantings; L_4 = hayfields; L_5 = rangeland; L_6 = woodlands; L_7 = forest range; L_8 = water reserve lands; L_9 = residential and industrial lands; L_{10} = lands under transportation and communication infrastructure; L_{11} = wetlands; L_{12} = disturbed lands; L_{13} = barren. Source: Authors' development.

Table A4. Land acreage data of the North Caucasian Federal District in thousand hectares. Mean values for 2010–2018.

Territory	Total	L_1	L_2	L_3	L_4	L_5	L_6	L_7	L_8	L_9	L_{10}	L_{11}	L_{12}	L_{13}
T_{35}	5027.0	520.1	4.8	72.4	162.3	2588.6	585.0	57.2	176.9	34.5	63.0	20.6	2.5	739.1
T_{36}	362.8	111.0	0	4.7	9.7	96.6	101.0	2.3	1.7	4.5	5.5	0.1	0.1	25.6
T_{37}	1247.0	300.7	0	30.1	56.3	309.3	196.8	13.3	15.5	17.6	26.8	1.2	1.0	278.4
T_{38}	1427.7	161.1	3.8	4.9	140.9	353.2	431.2	9.7	22.5	13.9	14.1	1.3	0.8	270.3
T_{39}	798.7	202.4	0.4	5.1	23.2	169.7	205.9	9.7	11.5	19.1	12.0	0.5	0.3	138.9
T_{40}	1564.7	332.2	0.2	11.0	56.8	575.2	336.0	27.6	28.6	43.4	21.5	2.7	1.4	128.1
T_{41}	6616.0	3998.6	14.0	44.2	104.9	1625.8	110.2	144.1	127.0	107.5	147.9	28.8	3.4	159.6

Note: T_{35} = Dagestan; T_{36} = Ingushetia; T_{37} = Kabardino-Balkaria; T_{38} = Karachaevo-Cherkessia; T_{39} = North Osetia-Alania; T_{40} = Chechnya; T_{41} = Stavropol; $L_{(1-13)}$ = acreage of L_i category, thousand hectares: L_1 = cropland; L_2 = fallow; L_3 = perennial plantings; L_4 = hayfields; L_5 = rangeland; L_6 = woodlands; L_7 = forest range; L_8 = water reserve lands; L_9 = residential and industrial lands; L_{10} = lands under transportation and communication infrastructure; L_{11} = wetlands; L_{12} = disturbed lands; L_{13} = barren. Source: Authors' development.

Table A5. Land acreage data of the Volga Federal District in thousand hectares. Mean values for 2010–2018.

Territory	Total	L_1	L_2	L_3	L_4	L_5	L_6	L_7	L_8	L_9	L_{10}	L_{11}	L_{12}	L_{13}
T_{42}	14,294.7	3670.5	0	43.6	1266.7	2346.1	5765.6	227.9	149.9	132.1	260.1	50.8	17.2	364.2
T_{43}	2337.5	472.1	128.0	7.9	56.6	108.2	1340.6	18.9	85.0	26.2	39.5	33.1	1.4	20.0
T_{44}	2612.8	1084.8	56.8	14.5	62.3	437.2	726.1	64.8	20.8	33.5	53.0	15.9	1.5	41.6
T_{45}	6784.7	3420.6	0.7	41.1	144.2	932.8	1199.1	129.4	451.6	141.7	157.8	50.6	4.8	110.3
T_{46}	4206.1	1382.3	9.3	15.2	112.5	321.5	2019.1	102.0	53.8	36.2	99.5	16.7	5.3	32.7
T_{47}	1834.3	806.3	6.2	19.9	48.3	153.8	603.6	17.5	48.1	35.3	60.1	5.1	0.5	29.6
T_{48}	16,023.6	1980.7	67.8	25.4	388.8	376.5	11,749.2	145.5	399.6	124.1	209.1	369.8	8.5	178.6
T_{49}	12,037.4	2480.3	51.8	15.0	374.2	399.1	7949.0	150.6	118.0	48.7	148.4	133.3	12.9	156.1
T_{50}	7662.4	2035.8	180.0	33.8	218.6	642.5	3817.1	90.2	162.7	112.8	143.4	123.0	6.0	96.5
T_{51}	12,370.2	6115.3	0	23.0	698.0	3979.5	618.6	199.3	111.3	158.7	184.7	15.3	13.0	253.5
T_{52}	4335.2	2263.6	153.4	22.5	71.4	528.1	975.7	77.2	42.2	59.7	89.7	13.5	0.9	37.3
T_{53}	5356.5	2937.5	103.5	42.3	67.0	847.5	685.6	104.5	226.0	103.0	123.7	42.0	3.9	70.0
T_{54}	10,124.0	5981.1	0	39.9	122.2	2400.5	614.2	121.2	357.9	113.3	149.4	19.2	2.4	202.7
T_{55}	3718.1	1655.7	105.8	17.7	37.8	390.3	1035.2	55.0	228.5	34.8	85.6	10.7	1.4	59.6

Note: T_{42} = Bashkortostan; T_{43} = Mari El; T_{44} = Mordovia; T_{45} = Tatarstan; T_{46} = Udmurtia; T_{47} = Chuvashia; T_{48} = Perm; T_{49} = Kirov; T_{50} = Nizhny Novgorod; T_{51} = Orenburg; T_{52} = Penza; T_{53} = Samara; T_{54} = Saratov; T_{55} = Ulyanovsk; $L_{(1-13)}$ = acreage of L_i category, thousand hectares: L_1 = cropland; L_2 = fallow; L_3 = perennial plantings; L_4 = hayfields; L_5 = rangeland; L_6 = woodlands; L_7 = forest range; L_8 = water reserve lands; L_9 = residential and industrial lands; L_{10} = lands under transportation and communication infrastructure; L_{11} = wetlands; L_{12} = disturbed lands; L_{13} = barren. Source: Authors' development.

Table A6. Land acreage data of the Ural Federal District in thousand hectares. Mean values for 2010–2018.

Territory	Total	L_1	L_2	L_3	L_4	L_5	L_6	L_7	L_8	L_9	L_{10}	L_{11}	L_{12}	L_{13}
T_{56}	7148.8	2402.6	459.3	12.4	559.0	1024.8	1759.5	37.2	318.7	49.1	86.3	383.9	1.1	54.9
T_{57}	19,430.7	1470.4	99.5	32.4	624.3	351.1	13,631.8	230.7	262.3	162.4	228.5	2046.2	61.8	229.3
T_{58}	16,012.2	1353.0	364.7	11.7	895.8	756.7	7112.8	144.9	508.5	80.0	96.1	4609.1	4.6	74.3
T_{59}	8852.9	3058.8	55.0	38.3	591.1	1352.0	2707.3	75.2	275.9	137.8	145.5	192.7	31.8	191.5
T_{60}	53,480.1	13.1	3.0	10.5	343.8	259.7	28,693.6	156.5	3185.4	141.6	170.7	19,913.4	55.7	533.1
T_{61}	76,925.0	0.9	0	0.2	165.3	57.3	18,763.5	4380.3	13,319.9	120.5	170.7	14,798.8	103.7	25,043.9

Note: T_{56} = Kurgan; T_{57} = Sverdlovsk; T_{58} = Tyumen; T_{59} = Chelyabinsk; T_{60} = Khanty-Mansi; T_{61} = Yamal-Nenets; $L_{(1-13)}$ = acreage of L_i category, thousand hectares: L_1 = cropland; L_2 = fallow; L_3 = perennial plantings; L_4 = hayfields; L_5 = rangeland; L_6 = woodlands; L_7 = forest range; L_8 = water reserve lands; L_9 = residential and industrial lands; L_{10} = lands under transportation and communication infrastructure; L_{11} = wetlands; L_{12} = disturbed lands; L_{13} = barren. Source: Authors' development.

Table A7. Land acreage data of the Siberian Federal District in thousand hectares. Mean values for 2010–2018.

Territory	Total	L_1	L_2	L_3	L_4	L_5	L_6	L_7	L_8	L_9	L_{10}	L_{11}	L_{12}	L_{13}
T_{62}	9290.3	143.5	2.2	1.7	120.9	1522.8	4357.7	190.0	86.3	10.9	23.1	73.3	0.4	2757.5
T_{63}	35,133.4	829.6	61.6	8.2	389.6	1856.8	23,660.6	220.7	2409.0	73.2	86.3	487.3	7.8	5042.7
T_{64}	16,860.4	191.3	147.9	0.9	76.5	3416.6	8667.2	450.1	228.1	21.7	29.3	1026.4	5.5	2598.9
T_{65}	6156.9	685.0	40.0	7.3	160.4	1022.5	3288.9	23.1	112.2	30.0	39.3	32.1	12.7	703.4
T_{66}	16,799.6	6654.4	298.9	27.8	1235.6	2789.7	4029.3	205.8	442.6	131.9	195.5	374.7	3.6	409.8
T_{67}	43,189.2	484.1	951.5	5.7	1722.6	4481.7	30,782.9	497.5	318.7	152.1	114.3	1076.9	24.2	2577.0
T_{68}	236,679.7	3120.1	136.4	37.4	781.8	1334.1	120,936.8	3185.0	9221.5	175.3	182.5	22,690.2	17.3	74,861.3
T_{69}	77,484.6	1734.5	3.3	30.0	390.1	640.8	66,080.5	235.1	2639.0	165.1	260.9	1709.4	26.3	3569.6
T_{70}	9572.5	1539.4	0.1	27.1	471.3	582.5	6074.7	163.2	91.7	107.5	174.5	90.5	83.4	166.6
T_{71}	17,775.6	3772.1	81.0	33.6	2197.9	2315.0	4799.2	280.3	766.5	102.4	166.8	3059.6	1.7	199.5
T_{72}	14,114.0	4156.6	175.9	26.5	1096.2	1265.5	4667.7	89.4	289.8	93.9	150.7	2026.8	5.0	70.0
T_{73}	31,439.1	675.9	1.3	9.4	479.9	204.5	19,939.9	88.1	608.3	42.5	87.9	9173.9	7.1	120.4

Note: T_{62} = Altay Republic; T_{63} = Buryatia; T_{64} = Tyva; T_{65} = Khakasia; T_{66} = Altay; T_{67} = Zabaikalsk; T_{68} = Krasnoyarsk; T_{69} = Irkutsk; T_{70} = Kemerovo; T_{71} = Novosibirsk; T_{72} = Omsk; T_{73} = Tomsk; $L_{(1-13)}$ = acreage of L_i category, thousand hectares: L_1 = cropland; L_2 = fallow; L_3 = perennial plantings; L_4 = hayfields; L_5 = rangeland; L_6 = woodlands; L_7 = forest range; L_8 = water reserve lands; L_9 = residential and industrial lands; L_{10} = lands under transportation and communication infrastructure; L_{11} = wetlands; L_{12} = disturbed lands; L_{13} = barren. Source: Authors' development.

Table A8. Land acreage data of the Far Eastern Federal District in thousand hectares. Mean values for 2010–2018.

Territory	Total	L_1	L_2	L_3	L_4	L_5	L_6	L_7	L_8	L_9	L_{10}	L_{11}	L_{12}	L_{13}
T_{74}	308,352.3	105.3	19.0	1.0	719.5	795.4	164,862.0	1837.7	13,087.5	82.6	129.1	19,783.6	30.9	106,898.7
T_{75}	46,427.5	64.3	1.0	5.3	97.3	307.7	26,810.0	305.8	844.5	16.3	17.0	2523.3	2.9	15,432.1
T_{76}	16,467.3	755.0	60.8	25.9	361.8	445.9	13,023.3	407.6	424.6	111.1	101.3	466.9	16.8	266.3
T_{77}	78,763.3	98.4	25.1	16.8	401.9	123.4	59,571.6	231.8	1476.3	79.3	95.7	5605.9	6.1	11,031.0
T_{78}	36,190.8	1577.2	244.0	11.9	418.0	482.5	26,136.8	268.4	1151.0	54.1	136.3	4794.1	12.7	903.8
T_{79}	46,246.4	23.8	3.5	0.1	51.5	42.6	28,467.1	340.8	477.3	9.5	14.5	4815.4	77.4	11,922.9
T_{80}	8710.1	51.2	0	7.6	63.6	60.0	6607.9	347.5	233.2	34.0	33.1	642.0	10.5	619.5
T_{81}	3627.1	94.6	70.3	3.1	119.2	250.0	1783.2	139.1	35.3	12.1	20.7	914.5	1.5	183.5
T_{82}	72,148.1	0.1	0	0	8.2	0.3	13,015.1	3878.3	2442.7	4.5	22.2	2833.0	47.5	49,896.2

Note: T_{74} = Sakha Yakutia; T_{75} = Kamchatka; T_{76} = Primorye; T_{77} = Khabarovsk; T_{78} = Amur; T_{79} = Magadan; T_{80} = Sakhalin; T_{81} = Jewish AO; T_{82} = Chukotka; $L_{(1-13)}$ = acreage of L_i category, thousand hectares: L_1 = cropland; L_2 = fallow; L_3 = perennial plantings; L_4 = hayfields; L_5 = rangeland; L_6 = woodlands; L_7 = forest range; L_8 = water reserve lands; L_9 = residential and industrial lands; L_{10} = lands under transportation and communication infrastructure; L_{11} = wetlands; L_{12} = disturbed lands; L_{13} = barren. Source: Authors' development.

Appendix B

Table A9. Activity per land category in Russia, Central Federal District. Mean values for 2010–2018.

Parameter	T_1	T_2	T_3	T_4	T_5	T_6	T_7	T_8	T_9	T_{10}	T_{11}	T_{12}	T_{13}	T_{14}	T_{15}	T_{16}	T_{17}	$\overline{A_{T(1-17)Li}}$
L_1	0.606	0.337	0.208	0.583	0.264	0.321	0.109	0.648	0.646	0.255	0.637	0.388	0.294	0.617	0.179	0.605	0.219	0.367
V_{L1}	−0.002	+0.006	-	−0.003	−0.003	-	−0.002	−0.001	-	−0.008	-	-	-	−0.001	-	−0.001	-	
L_2	0	0.035	0.016	0.008	0.005	0.012	0.005	0	0	0.002	0.023	0.007	0.004	0.003	0.002	0.003	0	0.007
V_{L2}	-	−0.007	-	-	+0.001	-	-	-	-	+0.002	-	-	-	−0.002	-	-	-	
L_3	0.013	0.007	0.007	0.010	0.004	0.007	0.001	0.009	0.015	0.026	0.010	0.006	0.004	0.009	0.002	0.018	0.004	0.008
V_{L3}	-	-	-	-	-	-	-	-	-	+0.001	-	-	-	-	-	-	-	
L_4	0.021	0.059	0.056	0.030	0.058	0.044	0.026	0.034	0.035	0.041	0.024	0.051	0.043	0.048	0.045	0.026	0.034	0.040
V_{L4}	-	+0.001	-	-	-	-	-	-	-	−0.001	-	-	-	+0.006	-	−0.001	−0	
L_5	0.147	0.099	0.055	0.149	0.052	0.078	0.025	0.121	0.117	0.052	0.139	0.182	0.076	0.113	0.060	0.116	0.054	0.090
V_{L5}	-	-	-	+0.002	-	-	-	-	-	−0.003	-	-	-	+0.010	-	−0.001	-	
L_6	0.089	0.340	0.544	0.092	0.489	0.462	0.760	0.083	0.079	0.451	0.082	0.270	0.435	0.108	0.563	0.145	0.477	0.363
V_{L6}	-	-	-	+0.006	-	-	-	-	-	+0.001	-	+0.001	-	-	+0.002	-	+0.001	
L_7	0.033	0.035	0.026	0.029	0.013	0.012	0.016	0.023	0.026	0.008	0.030	0.017	0.072	0.028	0.028	0.017	0.026	0.027
V_{L7}	-	-	-	−0.006	-	-	+0.002	-	-	−0.001	-	-	−0.001	+0.006	−0.002	-	-	
L_8	0.009	0.009	0.011	0.012	0.030	0.007	0.016	0.013	0.011	0.020	0.006	0.017	0.011	0.012	0.029	0.009	0.107	0.020
V_{L8}	-	-	-	-	-	-	-	-	-	-	-	-	-	-	-	-	-	
L_9	0.027	0.016	0.013	0.022	0.020	0.019	0.006	0.019	0.020	0.068	0.009	0.009	0.011	0.016	0.012	0.013	0.016	0.019
V_{L9}	+0.001	-	-	+0.001	+0.001	-	-	+0.001	-	+0.005	-	-	-	-	-	+0.003	+0.001	
L_{10}	0.021	0.021	0.026	0.023	0.024	0.017	0.017	0.024	0.026	0.036	0.030	0.027	0.017	0.018	0.014	0.035	0.018	0.022
V_{L10}	+0.001	-	-	-	-	-	-	-	-	-	-	-	-	-	-	-	-	
L_{11}	0.008	0.022	0.013	0.008	0.023	0.010	0.014	0.011	0.007	0.011	0.002	0.014	0.023	0.013	0.065	0.001	0.030	0.019
V_{L11}	-	-	-	-	-	-	-	-	-	-	-	-	-	-	-	-	-	
L_{12}	0.002	0.001	0.006	0	0.003	0.001	0.001	0.004	0.001	0.008	0	0.002	0.004	0	0.002	0.004	0.004	0.003
V_{L12}	-	-	-	-	-	-	-	-	-	-	-	-	-	-	-	-	-	
L_{13}	0.023	0.019	0.018	0.033	0.014	0.010	0.004	0.011	0.018	0.022	0.010	0.011	0.006	0.013	0.009	0.009	0.009	0.015
V_{L13}	-	−0.005	−0.006	−0.007	−0.003	−0.001	−0.002	−0.004	−0.004	−0.003	−0.005	−0.005	−0.002	−0.004	−0.001	−0.003	−0.004	

Note: T_1 = Belgorod; T_2 = Bryansk; T_3 = Vladimir; T_4 = Voronezh; T_5 = Ivanovo; T_6 = Kaluga; T_7 = Kostroma; T_8 = Kursk; T_9 = Lipetsk; T_{10} = Moscow Oblast; T_{11} = Orel; T_{12} = Ryazan; T_{13} = Smolensk; T_{14} = Tambov; T_{15} = Tver; T_{16} = Tula; T_{17} = Yaroslavl; $L_{(1-13)}$ = portion of L_i category in a composition of the land fund in T_j territory, percentage: L_1 = cropland; L_2 = fallow; L_3 = perennial plantings; L_4 = hayfields; L_5 = rangeland; L_6 = woodlands; L_7 = forest range; L_8 = water reserve lands; L_9 = residential and industrial lands; L_{10} = lands under transportation and communication infrastructure; L_{11} = wetlands; L_{12} = disturbed lands; L_{13} = barren; $V_{L(1-13)}$ = variability of $L_{(1-13)}$, i.e., change in 2018 compared to 2010; "-" = no change or insignificant change. Source: Authors' development.

Table A10. Activity per land category in Russia, Northwestern Federal District. Mean values for 2010–2018.

Parameter	T_{18}	T_{19}	T_{20}	T_{21}	T_{22}	T_{23}	T_{24}	T_{25}	T_{26}	T_{27}	$\overline{A_{T(18–27)Li}}$
L_1	0.005	0.002	0.007	0.057	0.260	0.052	0.001	0.094	0.134	0	0.020
V_{L1}	-	-	-	+0.004	−0.005	+0.002	-	−0.002	+0.003	-	
L_2	0	0	0	0.003	0	0	0	0.001	0.034	0	0.001
V_{L2}	-	-	-	-	-	-	-	-	−0.003	-	
L_3	0	0	0	0.001	0.009	0.005	0	0.001	0.004	0	0.001
V_{L3}	-	-	-	-	+0.001	−0.001	-	-	-	-	
L_4	0.005	0.006	0.007	0.024	0.102	0.023	0	0.032	0.050	0.001	0.011
V_{L4}	+0.001	−0.001	+0.001	−0.003	+0.004	−0.002	-	+0.001	+0.003	-	
L_5	0.002	0.002	0.003	0.016	0.165	0.015	0	0.025	0.051	0	0.007
V_{L5}	-	-	-	+0.002	−0.005	+0.001	-	+0.001	−0.003	-	
L_6	0.546	0.746	0.556	0.723	0.195	0.598	0.372	0.657	0.406	0.098	0.549
V_{L6}	+0.012	−0.009	+0.004	−0.033	+0.017	−0.005	+0.023	−0.008	+0.014	+0.002	
L_7	0.001	0.003	0.003	0.023	0.012	0.015	0.040	0.025	0.142	0.081	0.022
V_{L7}	-	-	-	+0.002	-	-	−0.001	−0.001	−0.011	+0.002	
L_8	0.232	0.015	0.020	0.046	0.132	0.151	0.082	0.032	0.068	0.057	0.062
V_{L8}	+0.006	−0.001	-	+0.002	+0.004	−0.003	-	-	+0.001	+0.002	
L_9	0.002	0.001	0.002	0.003	0.027	0.007	0.003	0.005	0.006	0.001	0.003
V_{L9}	-	-	-	-	−0.003	-	-	-	+0.001	-	
L_{10}	0.005	0.003	0.003	0.012	0.027	0.013	0.002	0.013	0.013	0.001	0.005
V_{L10}	−0.001	-	-	+0.001	−0.002	-	-	-	−0.001	-	
L_{11}	0.196	0.098	0.141	0.088	0.020	0.099	0.393	0.101	0.086	0.191	0.152
V_{L11}	+0.004	−0.003	+0.010	+0.003	-	−0.002	+0.007	−0.013	−0.002	+0.004	
L_{12}	0.001	0	0	0.002	0.003	0.003	0.001	0.002	0.002	0	0.001
V_{L12}	-	-	-	-	-	-	-	-	-	-	
L_{13}	0.005	0.123	0.258	0.003	0.048	0.019	0.105	0.013	0.005	0.569	0.165
V_{L13}	−0.004	−0.023	+0.011	-	+0.014	+0.008	+0.025	+0.005	−0.004	+0.054	

Note: T_{18} = Karelia; T_{19} = Komi; T_{20} = Arkhangelsk; T_{21} = Vologda; T_{22} = Kaliningrad; T_{23} = Leningrad; T_{24} = Murmansk; T_{25} = Novgorod; T_{26} = Pskov; T_{27} = Nenets; $L_{(1–13)}$ = portion of L_i category in a composition of the land fund in T_j territory, percentage: L_1 = cropland; L_2 = fallow; L_3 = perennial plantings; L_4 = hayfields; L_5 = rangeland; L_6 = woodlands; L_7 = forest range; L_8 = water reserve lands; L_9 = residential and industrial lands; L_{10} = lands under transportation and communication infrastructure; L_{11} = wetlands; L_{12} = disturbed lands; L_{13} = barren; $V_{L(1–13)}$ = variability of $L_{(1–13)}$, i.e., change in 2018 compared to 2010; "-" = no change or insignificant change. Source: Authors' development.

Table A11. Activity per land category in Russia, Southern Federal District. Mean values for 2010–2018.

Parameter	T_{28}	T_{29}	T_{30}	T_{31}	T_{32}	T_{33}	T_{34}	$\overline{A_{T(28–34)Li}}$
L_1	0.333	0.112	0.488	0.528	0.072	0.519	0.585	0.413
V_{L1}	−0.002	−0.007	-	−0.001	+0.003	−0.007	+0.004	
L_2	0	0.001	0.004	0	0.001	0	0	0.001
V_{L2}	-	-	+0.001	-	-	-	-	
L_3	0.012	0	0.029	0.017	0.002	0.004	0.006	0.007
V_{L3}	+0.002	-	−0.003	+0.002	-	-	+0.001	
L_4	0.006	0.014	0.001	0.008	0.083	0.018	0.009	0.020
V_{L4}	-	−0.001	-	-	+0.004	+0.002	−0.005	
L_5	0.110	0.718	0.166	0.070	0.506	0.235	0.244	0.313
V_{L5}	−0.006	+0.005	−0.003	−0.004	−0.010	+0.009	+0.003	
L_6	0.371	0.004	0.102	0.204	0.021	0.052	0.029	0.070
V_{L6}	+0.004	-	−0.002	+0.005	−0.001	−0.002	−0.003	
L_7	0.010	0.006	0.013	0.021	0.004	0.012	0.028	0.015
V_{L7}	-	-	+0.001	−0.002	-	-	+0.002	
L_8	0.069	0.023	0.081	0.051	0.140	0.043	0.034	0.052
V_{L8}	−0.006	+0.001	−0.003	+0.002	−0.005	-	-	
L_9	0.028	0.004	0.046	0.027	0.006	0.015	0.015	0.016
V_{L9}	+0.002	-	−0.001	+0.001	-	-	+0.001	
L_{10}	0.024	0.009	0.017	0.026	0.012	0.010	0.022	0.016

Table A11. *Cont.*

Parameter	T_{28}	T_{29}	T_{30}	T_{31}	T_{32}	T_{33}	T_{34}	$A_{T(28–34)Li}$
V_{L10}	−0.001	-	+0..002	-	-	-	−0.001	
L_{11}	0.005	0.017	0.002	0.024	0.023	0.003	0.005	0.012
V_{L11}	-	+0.001	-	-	+0.001	-	-	
L_{12}	0	0.001	0.001	0.001	0	0	0.001	0
V_{L12}	-	-	-	-	-	-	-	
L_{13}	0.031	0.091	0.051	0.023	0.130	0.088	0.023	0.064
V_{L13}	+0.013	+0.014	-	+0.004	+0.033	+0.021	+0.004	

Note: T_{28} = Adygeya; T_{29} = Kalmykia; T_{30} = Crimea; T_{31} = Krasnodar; T_{32} = Astrakhan; T_{33} = Volgograd; T_{34} = Rostov; $L_{(1-13)}$ = portion of L_i category in a composition of the land fund in T_j territory, percentage: L_1 = cropland; L_2 = fallow; L_3 = perennial plantings; L_4 = hayfields; L_5 = rangeland; L_6 = woodlands; L_7 = forest range; L_8 = water reserve lands; L_9 = residential and industrial lands; L_{10} = lands under transportation and communication infrastructure; L_{11} = wetlands; L_{12} = disturbed lands; L_{13} = barren; $V_{L(1-13)}$ = variability of $L_{(1-13)}$, i.e., change in 2018 compared to 2010; "-" = no change or insignificant change. Source: Authors' development.

Table A12. Activity per land category in Russia, North Caucasian Federal District. Mean values for 2010–2018.

Parameter	T_{35}	T_{36}	T_{37}	T_{38}	T_{39}	T_{40}	T_{41}	$A_{T(35–41)Li}$
L_1	0.103	0.306	0.241	0.113	0.253	0.212	0.604	0.330
V_{L1}	+0.002	−0.003	+0.006	+0.002	−0.003	+0.004	+0.011	
L_2	0.001	0	0	0.003	0.001	0	0.002	0.001
V_{L2}	-	-	-	-	-	-	-	
L_3	0.014	0.013	0.024	0.003	0.006	0.007	0.007	0.010
V_{L3}	−0.001	+0.002	−0.005	-	-	+0.001	−0.001	
L_4	0.032	0.027	0.045	0.099	0.029	0.036	0.016	0.033
V_{L4}	+0.003	+0.004	-	+0.002	+0.002	−0.001	+0.002	
L_5	0.515	0.266	0.248	0.247	0.212	0.368	0.246	0.336
V_{L5}	−0.006	−0.004	−0.007	+0.002	+0.011	+0.006	−0.004	
L_6	0.116	0.278	0.158	0.302	0.258	0.215	0.017	0.115
V_{L6}	−0.003	+0.006	+0.012	−0.008	−0.003	−0.005	+0.002	
L_7	0.011	0.006	0.011	0.007	0.012	0.018	0.022	0.015
V_{L7}	-	-	−0.001	-	+0.003	−0.002	+0.004	
L_8	0.035	0.005	0.012	0.016	0.014	0.018	0.019	0.023
V_{L8}	+0.004	-	−0.001	−0.002	+0.003	+0.001	−0.003	
L_9	0.007	0.012	0.014	0.010	0.024	0.028	0.016	0.014
V_{L9}	-	+0.001	+0.002	-	−0.001	+0.003	+0.001	
L_{10}	0.013	0.015	0.021	0.010	0.015	0.014	0.022	0.017
V_{L10}	−0.002	-	-	-	+0.001	−0.002	−0.003	
L_{11}	0.004	0	0.001	0.001	0.001	0.002	0.004	0.003
V_{L11}	-	-	-	-	-	-	−0.001	
L_{12}	0	0	0.001	0.001	0	0.001	0.001	0.001
V_{L12}	-	-	-	-	-	-	-	
L_{13}	0.147	0.071	0.223	0.189	0.174	0.082	0.024	0.102
V_{L13}	+0.010	+0.009	+0.021	+0.021	+0.044	+0.005	+0.005	

Note: T_{35} = Dagestan; T_{36} = Ingushetia; T_{37} = Kabardino-Balkaria; T_{38} = Karachaevo-Cherkessia; T_{39} = North Osetia-Alania; T_{40} = Chechnya; T_{41} = Stavropol; $L_{(1-13)}$ = portion of L_i category in a composition of the land fund in T_j territory, percentage: L_1 = cropland; L_2 = fallow; L_3 = perennial plantings; L_4 = hayfields; L_5 = rangeland; L_6 = woodlands; L_7 = forest range; L_8 = water reserve lands; L_9 = residential and industrial lands; L_{10} = lands under transportation and communication infrastructure; L_{11} = wetlands; L_{12} = disturbed lands; L_{13} = barren; $V_{L(1-13)}$ = variability of $L_{(1-13)}$, i.e., change in 2018 compared to 2010; "-" = no change or insignificant change. Source: Authors' development.

Table A13. Activity per land category in Russia, Volga Federal District. Mean values for 2010–2018.

Parameter	T_{42}	T_{43}	T_{44}	T_{45}	T_{46}	T_{47}	T_{48}	T_{49}	T_{50}	T_{51}	T_{52}	T_{53}	T_{54}	T_{55}	$\overline{A_{T(42-55)Li}}$
L_1	0.257	0.202	0.415	0.504	0.329	0.440	0.124	0.206	0.266	0.494	0.522	0.548	0.591	0.445	0.350
V_{L1}	+0.003	+0.011	+0.010	+0.009	+0.003	+0.003	+0.007	+0.009	+0.002	+0.018	+0.012	+0.004	+0.008	+0.007	
L_2	0	0.055	0.022	0	0.002	0.003	0.004	0.004	0.023	0	0.035	0.019	0	0.028	0.008
V_{L2}	-	−0.003	−0.001	-	-	-	+0.001	-	−0.002	-	−0.004	−0.002	-	−0.003	
L_3	0.003	0.003	0.006	0.006	0.004	0.011	0.002	0.001	0.004	0.002	0.005	0.008	0.004	0.005	0.003
V_{L3}	-	-	+0.001	+0.001	-	−0.002	-	-	+0.001	-	-	+0.001	-	-	
L_4	0.089	0.024	0.024	0.021	0.027	0.026	0.024	0.031	0.029	0.056	0.016	0.013	0.012	0.010	0.035
V_{L4}	−0.004	−0.003	−0.005	−0.002	−0.003	−0.002	−0.004	−0.003	−0.005	−0.004	−0.001	−0.002	−0.001	−0.002	
L_5	0.164	0.046	0.167	0.137	0.076	0.084	0.023	0.033	0.084	0.322	0.122	0.158	0.237	0.105	0.134
V_{L5}	+0.001	+0.002	+0.004	+0.001	+0.002	+0.004	−0.001	−0.001	+0.003	+0.003	+0.005	+0.006	+0.007	−0.003	
L_6	0.403	0.574	0.278	0.177	0.480	0.329	0.733	0.660	0.498	0.050	0.225	0.128	0.061	0.278	0.377
V_{L6}	+0.003	+0.005	+0.003	−0.002	−0.012	−0.007	+0.013	+0.017	+0.012	−0.001	+0.004	−0.003	-	+0.003	
L_7	0.016	0.008	0.025	0.019	0.024	0.010	0.009	0.013	0.012	0.016	0.018	0.020	0.012	0.015	0.015
V_{L7}	−0.002	-	−0.002	+0.003	+0.001	−0.001	-	-	+0.001	+0.002	−0.002	−0.001	+0.002	+0.002	
L_8	0.010	0.036	0.008	0.067	0.013	0.026	0.025	0.010	0.021	0.009	0.010	0.042	0.035	0.061	0.024
V_{L8}	+0.003	+0.006	+0.001	+0.008	−0.001	−0.001	+0.003	+0.001	−0.002	-	-	+0.003	−0.001	+0.003	
L_9	0.009	0.011	0.013	0.021	0.009	0.019	0.008	0.004	0.015	0.013	0.014	0.019	0.011	0.009	0.011
V_{L9}	-	-	+0.001	−0.002	-	+0.002	-	-	−0.002	+0.004	-	−0.004	-	-	
L_{10}	0.018	0.017	0.020	0.023	0.024	0.033	0.013	0.012	0.019	0.015	0.021	0.023	0.015	0.023	0.017
V_{L10}	+0.003	+0.002	+0.001	+0.002	+0.001	−0.001	−0.001	−0.001	−0.003	+0.001	+0.002	+0.002	+0.001	+0.001	
L_{11}	0.004	0.014	0.006	0.007	0.004	0.003	0.023	0.011	0.016	0.001	0.003	0.008	0.002	0.003	0.009
V_{L11}	−0.001	−0.001	−0.002	+0.001	-	-	−0.004	−0.002	−0.005	-	-	+0.001	-	-	
L_{12}	0.001	0.001	0.001	0.001	0.001	0	0.001	0.001	0.001	0.001	0	0.001	0	0	0.001
V_{L12}	-	-	-	-	-	-	-	-	-	-	-	-	-	-	
L_{13}	0.025	0.009	0.016	0.016	0.008	0.016	0.011	0.013	0.013	0.020	0.009	0.013	0.020	0.016	0.016
V_{L13}	+0.002	-	-	-	-	−0.003	−0.003	−0.002	-	−0.002	-	-	-	-	

Note: T_{42} = Bashkortostan; T_{43} = Mari El; T_{44} = Mordovia; T_{45} = Tatarstan; T_{46} = Udmurtia; T_{47} = Chuvashia; T_{48} = Perm; T_{49} = Kirov; T_{50} = Nizhny Novgorod; T_{51} = Orenburg; T_{52} = Penza; T_{53} = Samara; T_{54} = Saratov; T_{55} = Ulyanovsk; $L_{(1-13)}$ = portion of L_i category in a composition of the land fund in T_j territory, percentage: L_1 = cropland; L_2 = fallow; L_3 = perennial plantings; L_4 = hayfields; L_5 = rangeland; L_6 = woodlands; L_7 = forest range; L_8 = water reserve lands; L_9 = residential and industrial lands; L_{10} = lands under transportation and communication infrastructure; L_{11} = wetlands; L_{12} = disturbed lands; L_{13} = barren; $V_{L(1-13)}$ = variability of $L_{(1-13)}$, i.e., change in 2018 compared to 2010; "-" = no change or insignificant change. Source: Authors' development.

Table A14. Activity per land category in Russia, Ural Federal District. Mean values for 2010–2018.

Parameter	T_{56}	T_{57}	T_{58}	T_{59}	T_{60}	T_{61}	$\overline{A_{T(56\text{–}61)Li}}$
L_1	0.336	0.076	0.084	0.346	0	0	0.046
V_{L1}	+0.012	−0.002	−0.001	+0.003	-	-	
L_2	0.064	0.005	0.023	0.006	0	0	0.005
V_{L2}	+0.002	-	−0.002	−0.001	-	-	
L_3	0.002	0.002	0.001	0.004	0	0	0.001
V_{L3}	+0.001	-	-	+0.001	-	-	
L_4	0.078	0.032	0.056	0.067	0.006	0.002	0.017
V_{L4}	+0.004	−0.002	+0.003	+0.004	+0.001	-	
L_5	0.143	0.018	0.047	0.153	0.005	0.001	0.021
V_{L5}	−0.005	−0.003	−0.001	−0.005	+0.002	-	
L_6	0.246	0.702	0.444	0.306	0.537	0.244	0.400
V_{L6}	−0.003	−0.004	+0.002	−0.003	−0.015	−0.004	
L_7	0.005	0.012	0.009	0.008	0.003	0.057	0.028
V_{L7}	-	+0.002	+0.003	+0.001	-	+0.005	
L_8	0.045	0.013	0.032	0.031	0.060	0.173	0.098
V_{L8}	−0.013	−0.011	−0.002	−0.003	−0.004	−0.003	
L_9	0.007	0.008	0.005	0.016	0.003	0.002	0.004
V_{L9}	+0.001	+0.002	-	+0.003	-	-	
L_{10}	0.012	0.012	0.006	0.016	0.003	0.002	0.005
V_{L10}	+0.002	+0.003	+0.002	+0.004	+0.001	-	
L_{11}	0.054	0.105	0.288	0.022	0.372	0.192	0.231
V_{L11}	+0.001	−0.002	−0.004	−0.002	−0.003	−0.004	
L_{12}	0	0.003	0	0.004	0.001	0.001	0.001
V_{L12}	-	+0.001	-	+0.002	-	-	
L_{13}	0.008	0.012	0.005	0.022	0.010	0.326	0.144
V_{L13}	+0.004	+0.002	-	+0.004	+0.002	+0.004	

Note: T_{56} = Kurgan; T_{57} = Sverdlovsk; T_{58} = Tyumen; T_{59} = Chelyabinsk; T_{60} = Khanty-Mansi; T_{61} = Yamal-Nenets; $L_{(1\text{–}13)}$ = portion of L_i category in a composition of the land fund in T_j territory, percentage: L_1 = cropland; L_2 = fallow; L_3 = perennial plantings; L_4 = hayfields; L_5 = rangeland; L_6 = woodlands; L_7 = forest range; L_8 = water reserve lands; L_9 = residential and industrial lands; L_{10} = lands under transportation and communication infrastructure; L_{11} = wetlands; L_{12} = disturbed lands; L_{13} = barren; $V_{L(1\text{–}13)}$ = variability of $L_{(1\text{–}13)}$, i.e., change in 2018 compared to 2010; "-" = no change or insignificant change. Source: Authors' development.

Table A15. Activity per land category in Russia, Siberian Federal District. Mean values for 2010–2018.

Parameter	T_{62}	T_{63}	T_{64}	T_{65}	T_{66}	T_{67}	T_{68}	T_{69}	T_{70}	T_{71}	T_{72}	T_{73}	$\overline{A_{T(62\text{–}73)Li}}$
L_1	0.015	0.024	0.011	0.111	0.396	0.011	0.013	0.022	0.161	0.212	0.295	0.021	0.047
V_{L1}	+0.002	+0.001	−0.001	+0.003	+0.004	−0.003	−0.004	+0.003	−0.005	−0.004	+0.006	+0.001	
L_2	0	0.002	0.009	0.006	0.018	0.022	0.001	0	0	0.005	0.012	0	0.004
V_{L2}	-	-	+0.001	−0.002	−0.002	+0.001	-	-	-	+0.001	−0.003	-	
L_3	0	0	0	0.001	0.002	0	0	0	0.003	0.002	0.002	0	0
V_{L3}	-	-	-	-	-	-	-	-	+0.001	-	-	-	
L_4	0.013	0.011	0.005	0.026	0.074	0.040	0.003	0.005	0.049	0.124	0.078	0.015	0.018
V_{L4}	−0.002	−0.001	−0.001	−0.003	−0.004	−0.002	-	-	−0.004	−0.006	−0.001	-	
L_5	0.164	0.053	0.203	0.166	0.166	0.104	0.006	0.008	0.061	0.130	0.090	0.007	0.042
V_{L5}	+0.002	+0.001	+0.004	+0.003	+0.001	+0.003	+0.001	+0.001	−0.001	+0.003	+0.005	+0.001	
L_6	0.469	0.673	0.514	0.534	0.240	0.713	0.511	0.853	0.635	0.270	0.331	0.634	0.578
V_{L6}	−0.002	−0.005	−0.003	−0.006	−0.003	−0.012	−0.014	−0.018	−0.013	−0.003	−0.005	−0.008	
L_7	0.020	0.006	0.027	0.004	0.012	0.012	0.013	0.003	0.017	0.016	0.006	0.003	0.011
V_{L7}	−0.003	−0.001	−0.003	-	−0.002	−0.001	−0.002	-	−0.004	−0.003	-	-	
L_8	0.009	0.069	0.014	0.018	0.026	0.007	0.039	0.034	0.010	0.043	0.021	0.019	0.033
V_{L8}	-	−0.001	-	-	−0.001	-	+0.002	+0.001	-	−0.002	+0.001	+0.003	
L_9	0.001	0.002	0.001	0.005	0.008	0.004	0.001	0.002	0.011	0.006	0.007	0.001	0.002
V_{L9}	-	-	-	+0.001	+0.002	+0.001	-	-	+0.002	+0.001	+0.001	-	

Table A15. *Cont.*

Parameter	T_{62}	T_{63}	T_{64}	T_{65}	T_{66}	T_{67}	T_{68}	T_{69}	T_{70}	T_{71}	T_{72}	T_{73}	$\overline{A_{T(62-73)Li}}$
L_{10}	0.002	0.002	0.002	0.006	0.012	0.003	0.001	0.003	0.018	0.009	0.011	0.003	0.003
V_{L10}	-	-	-	+0.002	+0.003	-	-	+0.001	+0.001	+0.001	+0.002	-	
L_{11}	0.008	0.014	0.061	0.005	0.022	0.025	0.096	0.022	0.009	0.172	0.144	0.292	0.081
V_{L11}	+0.001	−0.001	−0.002	-	−0.001	−0.001	−0.002	−0.002	−0.001	+0.005	+0.003	−0.003	
L_{12}	0	0	0	0.002	0	0.001	0	0	0.009	0	0	0	0
V_{L12}	-	-	-	+0.001	-	-	-	-	+0.001	-	-	-	
L_{13}	0.297	0.144	0.154	0.114	0.024	0.060	0.316	0.046	0.017	0.011	0.005	0.004	0.181
V_{L13}	+0.004	+0.003	+0.005	+0.004	-	+0.004	+0.005	+0.002	-	-	-	-	

Note: T_{62} = Altay Republic; T_{63} = Buryatia; T_{64} = Tyva; T_{65} = Khakasia; T_{66} = Altay; T_{67} = Zabaikalsk; T_{68} = Krasnoyarsk; T_{69} = Irkutsk; T_{70} = Kemerovo; T_{71} = Novosibirsk; T_{72} = Omsk; T_{73} = Tomsk;$L_{(1-13)}$ = portion of L_i category in a composition of the land fund in T_j territory, percentage: L_1 = cropland; L_2 = fallow; L_3 = perennial plantings; L_4 = hayfields; L_5 = rangeland; L_6 = woodlands; L_7 = forest range; L_8 = water reserve lands; L_9 = residential and industrial lands; L_{10} = lands under transportation and communication infrastructure; L_{11} = wetlands; L_{12} = disturbed lands; L_{13} = barren; $V_{L(1-13)}$ = variability of $L_{(1-13)}$, i.e., change in 2018 compared to 2010; "-" = no change or insignificant change. Source: Authors' development.

Table A16. Activity per land category in Russia, Far Eastern Federal District. Mean values for 2010–2018.

Parameter	T_{74}	T_{75}	T_{76}	T_{77}	T_{78}	T_{79}	T_{80}	T_{81}	T_{82}	$\overline{A_{T(74-82)Li}}$
L_1	0	0.001	0.046	0.001	0.044	0.001	0.006	0.026	0	0.004
V_{L1}	-	-	+0.001	-	+0.001	-	-	−0.001	-	
L_2	0	0	0.004	0	0.007	0	0	0.019	0	0.001
V_{L2}	-	-	−0.001	-	−0.002	-	-	+0.002	-	
L_3	0	0	0.002	0	0	0	0.001	0.001	0	0
V_{L3}	-	-	-	-	-	-	-	-	-	
L_4	0.002	0.002	0.022	0.005	0.012	0.001	0.007	0.033	0	0.004
V_{L4}	-	-	+0.003	+0.001	+0.001	-	−0.001	+0.002	-	
L_5	0.003	0.007	0.027	0.002	0.013	0.001	0.007	0.069	0	0.004
V_{L5}	+0.001	+0.001	+0.002	-	+0.003	-	+0.001	+0.004	-	
L_6	0.535	0.577	0.791	0.756	0.722	0.616	0.759	0.492	0.180	0.552
V_{L6}	−0.002	−0.004	−0.012	−0.011	−0.015	−0.012	−0.021	−0.020	−0.017	
L_7	0.006	0.007	0.025	0.003	0.007	0.007	0.040	0.038	0.054	0.013
V_{L7}	−0.001	−0.001	−0.002	-	-	−0.001	−0.003	−0.004	−0.011	
L_8	0.042	0.018	0.026	0.019	0.032	0.010	0.027	0.010	0.034	0.033
V_{L8}	−0.002	−0.001	+0.001	+0.001	−0.003	-	-	+0.001	−0.002	
L_9	0	0	0.007	0.001	0.001	0	0.004	0.003	0	0.001
V_{L9}	-	-	+0.001	-	-	-	+0.001	+0.001	-	
L_{10}	0	0	0.006	0.001	0.004	0	0.004	0.006	0	0.001
V_{L10}	-	-	+0.001	-	+0.001	-	+0.001	+0.001	-	
L_{11}	0.064	0.054	0.028	0.071	0.132	0.104	0.074	0.252	0.039	0.069
V_{L11}	−0.002	−0.002	−0.001	−0.005	−0.004	−0.006	−0.001	−0.009	−0.003	
L_{12}	0	0	0.001	0	0	0.002	0.001	0	0.001	0
V_{L12}	-	-	-	-	-	-	-	-	-	
L_{13}	0.347	0.332	0.016	0.140	0.025	0.258	0.071	0.051	0.692	0.320
V_{L13}	+0.019	+0.024	+0.003	+0.005	-	+0.031	+0.004	+0.005	+0.040	

Note: T_{74} = Sakha Yakutia; T_{75} = Kamchatka; T_{76} = Primorye; T_{77} = Khabarovsk; T_{78} = Amur; T_{79} = Magadan; T_{80} = Sakhalin; T_{81} = Jewish AO; T_{82} = Chukotka; $L_{(1-13)}$ = portion of L_i category in a composition of the land fund in T_j territory, percentage: L_1 = cropland; L_2 = fallow; L_3 = perennial plantings; L_4 = hayfields; L_5 = rangeland; L_6 = woodlands; L_7 = forest range; L_8 = water reserve lands; L_9 = residential and industrial lands; L_{10} = lands under transportation and communication infrastructure; L_{11} = wetlands; L_{12} = disturbed lands; L_{13} = barren; $V_{L(1-13)}$ = variability of $L_{(1-13)}$, i.e., change in 2018 compared to 2010; "-" = no change or insignificant change. Source: Authors' development.

Appendix C

Table A17. Ranking of T_j territories on land activity, Central Federal District.

Parameter	T_1	T_2	T_3	T_4	T_5	T_6	T_7	T_8	T_9	T_{10}	T_{11}	T_{12}	T_{13}	T_{14}	T_{15}	T_{16}	T_{17}	$\overline{R_{(1-17)i}}$
R_1	77	58	40	72	49	54	30	81	80	46	79	60	51	78	37	76	43	59
R_2	0	64	53	49	42	51	44	15	6	26	59	47	36	32	30	33	12	35
R_3	72	63	60	68	48	62	24	65	75	79	69	57	45	66	33	77	47	59
R_4	31	72	69	48	71	61	40	54	56	59	35	67	60	64	62	43	55	56
R_5	56	43	32	57	29	39	21	50	49	28	54	67	37	47	33	48	31	42
R_6	71	43	23	70	31	35	2	72	74	36	73	52	38	67	20	64	33	47
R_7	9	8	16	11	43	49	35	24	18	62	10	34	2	12	14	33	17	23
R_8	73	74	63	62	33	79	52	59	64	43	80	51	65	61	34	76	5	57
R_9	4	17	26	8	11	14	51	15	10	0	40	37	35	19	31	29	16	21
R_{10}	20	22	7	14	11	32	31	9	8	0	3	5	29	28	39	1	27	17
R_{11}	54	39	47	57	32	52	43	51	59	49	75	45	34	48	24	79	28	48
R_{12}	13	20	2	57	8	39	31	5	29	1	63	17	6	51	12	4	3	21
R_{13}	27	40	39	29	43	59	79	62	45	46	65	61	74	51	69	66	67	54
$\sum R_{j(1-13)}$	507	563	477	602	451	626	483	562	573	475	705	600	512	624	438	629	384	542

Note: T_1 = Belgorod; T_2 = Bryansk; T_3 = Vladimir; T_4 = Voronezh; T_5 = Ivanovo; T_6 = Kaluga; T_7 = Kostroma; T_8 = Kursk; T_9 = Lipetsk; T_{10} = Moscow Oblast; T_{11} = Orel; T_{12} = Ryazan; T_{13} = Smolensk; T_{14} = Tambov; T_{15} = Tver; T_{16} = Tula; T_{17} = Yaroslavl; $R_{(1-13)}$ = ranks of land activity per land categories: R_1 = cropland; R_2 = fallow; R_3 = perennial plantings; R_4 = hayfields; R_5 = rangeland; R_6 = woodlands; R_7 = forest range; R_8 = water reserve lands; R_9 = residential and industrial lands; R_{10} = lands under transportation and communication infrastructure; R_{11} = wetlands; R_{12} = disturbed lands; R_{13} = barren. Source: Authors' development.

Table A18. Ranking of T_j territories on land activity, Northwestern Federal District.

Parameter	T_{18}	T_{19}	T_{20}	T_{21}	T_{22}	T_{23}	T_{24}	T_{25}	T_{26}	T_{27}	$\overline{R_{(18-27)i}}$
R_1	10	9	12	24	48	23	7	28	35	1	20
R_2	1	0	8	34	0	0	0	22	63	0	13
R_3	16	7	13	22	67	53	12	25	43	0	26
R_4	10	13	17	36	80	34	1	50	66	4	31
R_5	7	6	9	18	62	17	1	22	27	2	17
R_6	22	5	21	7	60	17	41	13	39	69	29
R_7	81	75	76	23	45	39	5	19	0	1	36
R_8	0	54	44	17	4	2	6	29	10	14	18
R_9	67	74	65	62	6	45	64	56	50	77	57
R_{10}	61	64	67	46	4	41	73	44	43	77	52
R_{11}	5	16	10	18	40	15	0	14	19	7	14
R_{12}	35	54	74	19	10	11	21	15	18	73	33
R_{13}	71	16	8	80	28	60	18	55	76	1	41
$\sum R_{j(1-13)}$	386	393	424	406	454	357	249	392	489	326	388

Note: T_{18} = Karelia; T_{19} = Komi; T_{20} = Arkhangelsk; T_{21} = Vologda; T_{22} = Kaliningrad; T_{23} = Leningrad; T_{24} = Murmansk; T_{25} = Novgorod; T_{26} = Pskov; T_{27} = Nenets; $R_{(1-13)}$ = ranks of land activity per land categories: R_1 = cropland; R_2 = fallow; R_3 = perennial plantings; R_4 = hayfields; R_5 = rangeland; R_6 = woodlands; R_7 = forest range; R_8 = water reserve lands; R_9 = residential and industrial lands; R_{10} = lands under transportation and communication infrastructure; R_{11} = wetlands; R_{12} = disturbed lands; R_{13} = barren. Source: Authors' development.

Table A19. Ranking of T_j territories on land activity, Southern Federal District.

Parameter	T_{28}	T_{29}	T_{30}	T_{31}	T_{32}	T_{33}	T_{34}	$\overline{R_{(28-34)i}}$
R_1	56	32	65	70	25	68	73	56
R_2	18	25	38	4	24	19	0	18
R_3	71	18	80	76	37	44	55	54
R_4	14	26	2	18	77	30	19	27
R_5	46	81	65	36	79	70	72	64
R_6	42	81	68	59	79	76	78	69
R_7	56	71	42	26	73	52	13	48
R_8	8	40	7	15	3	18	26	17
R_9	2	57	1	5	53	23	21	23
R_{10}	10	56	33	6	50	53	18	32
R_{11}	63	41	72	31	33	68	61	53
R_{12}	53	47	43	37	75	66	40	52
R_{13}	53	21	72	75	25	31	33	44
$\sum R_{j(1-13)}$	492	596	588	458	633	618	509	556

Note: T_{28} = Adygeya; T_{29} = Kalmykia; T_{30} = Crimea; T_{31} = Krasnodar; T_{32} = Astrakhan; T_{33} = Volgograd; T_{34} = Rostov; $R_{(1-13)}$ = ranks of land activity per land categories: R_1 = cropland; R_2 = fallow; R_3 = perennial plantings; R_4 = hayfields; R_5 = rangeland; R_6 = woodlands; R_7 = forest range; R_8 = water reserve lands; R_9 = residential and industrial lands; R_{10} = lands under transportation and communication infrastructure; R_{11} = wetlands; R_{12} = disturbed lands; R_{13} = barren. Source: Authors' development.

Table A20. Ranking of T_j territories on land activity, North Caucasian Federal District.

Parameter	T_{35}	T_{36}	T_{37}	T_{38}	T_{39}	T_{40}	T_{41}	$\overline{R_{(35-41)i}}$
R_1	29	53	44	33	45	42	75	46
R_2	23	0	0	31	20	14	28	17
R_3	74	73	78	41	58	61	59	63
R_4	52	44	63	79	47	57	28	53
R_5	80	76	75	74	69	78	73	75
R_6	66	49	63	47	53	58	80	59
R_7	54	67	55	65	47	31	25	49
R_8	25	81	60	53	55	48	46	53
R_9	47	30	24	36	7	3	18	24
R_{10}	45	35	19	54	36	40	17	35
R_{11}	65	81	77	78	80	74	64	74
R_{12}	50	64	33	45	56	32	49	47
R_{13}	14	24	10	12	7	20	50	20
$\sum R_{j(1-13)}$	624	677	601	648	580	558	612	614

Note: T_{35} = Dagestan; T_{36} = Ingushetia; T_{37} = Kabardino-Balkaria; T_{38} = Karachaevo-Cherkessia; T_{39} = North Osetia-Alania; T_{40} = Chechnya; T_{41} = Stavropol; $R_{(1-13)}$ = ranks of land activity per land categories: R_1 = cropland; R_2 = fallow; R_3 = perennial plantings; R_4 = hayfields; R_5 = rangeland; R_6 = woodlands; R_7 = forest range; R_8 = water reserve lands; R_9 = residential and industrial lands; R_{10} = lands under transportation and communication infrastructure; R_{11} = wetlands; R_{12} = disturbed lands; R_{13} = barren. Source: Authors' development.

Table A21. Ranking of T_j territories on land activity, Volga Federal District.

Parameter	T_{42}	T_{43}	T_{44}	T_{45}	T_{46}	T_{47}	T_{48}	T_{49}	T_{50}	T_{51}	T_{52}	T_{53}	T_{54}	T_{55}	$\overline{R_{(42-55)i}}$
R_1	47	38	62	67	55	63	34	39	50	66	69	71	74	64	57
R_2	0	66	57	13	29	35	39	40	61	0	65	55	0	62	37
R_3	39	40	54	56	42	70	29	27	50	34	52	64	46	51	47
R_4	78	38	37	32	45	42	39	49	46	70	29	24	23	20	41
R_5	61	25	66	53	38	40	20	24	41	77	51	59	71	45	48
R_6	40	19	50	62	32	45	6	12	29	77	57	65	75	48	44
R_7	37	61	20	29	22	57	58	44	51	36	30	28	48	40	40
R_8	66	23	77	11	58	37	39	68	41	75	69	21	24	12	44
R_9	39	33	28	9	41	12	44	58	22	27	25	13	34	38	30
R_{10}	26	30	23	13	12	2	42	47	24	37	21	15	38	16	25
R_{11}	67	44	60	58	66	71	35	50	42	76	69	56	73	70	60
R_{12}	25	42	44	38	23	65	48	26	34	27	71	36	67	55	43
R_{13}	36	81	56	58	64	44	49	37	57	35	70	54	48	47	53
$\sum R_{j(1-13)}$	561	540	634	499	527	583	482	521	548	637	678	561	621	568	569

Note: T_{42} = Bashkortostan; T_{43} = Mari El; T_{44} = Mordovia; T_{45} = Tatarstan; T_{46} = Udmurtia; T_{47} = Chuvashia; T_{48} = Perm; T_{49} = Kirov; T_{50} = Nizhny Novgorod; T_{51} = Orenburg; T_{52} = Penza; T_{53} = Samara; T_{54} = Saratov; T_{55} = Ulyanovsk; $R_{(1-13)}$ = ranks of land activity per land categories: R_1 = cropland; R_2 = fallow; R_3 = perennial plantings; R_4 = hayfields; R_5 = rangeland; R_6 = woodlands; R_7 = forest range; R_8 = water reserve lands; R_9 = residential and industrial lands; R_{10} = lands under transportation and communication infrastructure; R_{11} = wetlands; R_{12} = disturbed lands; R_{13} = barren. Source: Authors' development.

Table A22. Ranking of T_j territories on land activity, Ural Federal District.

Parameter	T_{56}	T_{57}	T_{58}	T_{59}	T_{60}	T_{61}	$\overline{R_{(56-61)i}}$
R_1	57	26	27	59	3	2	29
R_2	67	43	60	45	9	0	37
R_3	32	31	21	49	10	2	24
R_4	76	51	68	73	15	6	48
R_5	55	19	26	58	10	3	29
R_6	54	10	37	46	24	55	38
R_7	72	50	59	60	79	3	54
R_8	16	57	31	32	13	1	25
R_9	46	42	54	20	63	69	49
R_{10}	48	49	59	34	66	72	55
R_{11}	26	12	3	38	1	6	14
R_{12}	72	9	62	7	28	22	33
R_{13}	73	52	68	32	41	6	45
$\sum R_{j(1-13)}$	694	451	575	553	362	247	480

Note: T_{56} = Kurgan; T_{57} = Sverdlovsk; T_{58} = Tyumen; T_{59} = Chelyabinsk; T_{60} = Khanty-Mansi; T_{61} = Yamal-Nenets; $R_{(1-13)}$ = ranks of land activity per land categories: R_1 = cropland; R_2 = fallow; R_3 = perennial plantings; R_4 = hayfields; R_5 = rangeland; R_6 = woodlands; R_7 = forest range; R_8 = water reserve lands; R_9 = residential and industrial lands; R_{10} = lands under transportation and communication infrastructure; R_{11} = wetlands; R_{12} = disturbed lands; R_{13} = barren. Source: Authors' development.

Table A23. Ranking of T_j territories on land activity, Siberian Federal District.

Parameter	T_{62}	T_{63}	T_{64}	T_{65}	T_{66}	T_{67}	T_{68}	T_{69}	T_{70}	T_{71}	T_{72}	T_{73}	$\overline{R_{(62-73)i}}$
R_1	16	19	14	31	61	13	15	18	36	41	52	17	28
R_2	16	27	50	46	54	58	21	7	2	41	52	5	32
R_3	9	14	4	26	30	6	8	19	38	36	35	15	20
R_4	25	21	9	41	74	58	8	11	65	81	75	27	41
R_5	60	30	68	64	63	44	11	15	34	52	42	12	41
R_6	34	11	27	26	56	9	28	0	14	51	44	15	26
R_7	27	69	15	74	46	53	41	77	32	38	68	80	52
R_8	72	9	56	49	36	78	22	27	71	19	42	45	44
R_9	73	68	72	55	43	60	76	66	32	52	49	71	60
R_{10}	70	71	74	57	51	69	76	65	25	55	52	68	61
R_{11}	55	46	23	62	36	30	17	37	53	8	9	2	32
R_{12}	81	69	61	14	70	46	79	60	0	77	58	68	57
R_{13}	9	15	13	19	34	17	5	23	42	63	78	77	33
$\sum R_{j(1-13)}$	547	469	486	564	654	541	407	425	444	614	656	502	526

Note: T_{62} = Altay Republic; T_{63} = Buryatia; T_{64} = Tyva; T_{65} = Khakasia; T_{66} = Altay; T_{67} = Zabaikalsk; T_{68} = Krasnoyarsk; T_{69} = Irkutsk; T_{70} = Kemerovo; T_{71} = Novosibirsk; T_{72} = Omsk; T_{73} = Tomsk; $R_{(1-13)}$ = ranks of land activity per land categories: R_1 = cropland; R_2 = fallow; R_3 = perennial plantings; R_4 = hayfields; R_5 = rangeland; R_6 = woodlands; R_7 = forest range; R_8 = water reserve lands; R_9 = residential and industrial lands; R_{10} = lands under transportation and communication infrastructure; R_{11} = wetlands; R_{12} = disturbed lands; R_{13} = barren. Source: Authors' development.

Table A24. Ranking of T_j territories on land activity, Far Eastern Federal District.

Parameter	T_{74}	T_{75}	T_{76}	T_{77}	T_{78}	T_{79}	T_{80}	T_{81}	T_{82}	$\overline{R_{(74-82)i}}$
R_1	4	8	22	6	21	5	11	20	0	11
R_2	10	3	37	17	48	11	0	56	0	20
R_3	3	5	28	11	17	1	23	22	0	12
R_4	7	5	33	12	22	3	16	53	0	17
R_5	8	13	23	5	16	4	14	35	0	13
R_6	25	18	1	4	8	16	3	30	61	18
R_7	70	66	21	78	63	64	6	7	4	42
R_8	20	50	38	47	30	67	35	70	28	43
R_9	79	78	48	75	70	80	59	61	81	70
R_{10}	78	79	58	75	63	80	62	60	81	71
R_{11}	22	25	29	21	11	13	20	4	27	19
R_{12}	76	80	30	78	59	16	24	52	41	51
R_{13}	2	4	38	11	30	3	22	26	0	15
$\sum R_{j(1-13)}$	404	434	406	440	458	363	295	496	323	402

Note: T_{74} = Sakha Yakutia; T_{75} = Kamchatka; T_{76} = Primorye; T_{77} = Khabarovsk; T_{78} = Amur; T_{79} = Magadan; T_{80} = Sakhalin; T_{81} = Jewish AO; T_{82}= Chukotka; $R_{(1-13)}$ = ranks of land activity per land categories: R_1 = cropland; R_2 = fallow; R_3 = perennial plantings; R_4 = hayfields; R_5 = rangeland; R_6 = woodlands; R_7 = forest range; R_8 = water reserve lands; R_9 = residential and industrial lands; R_{10} = lands under transportation and communication infrastructure; R_{11} = wetlands; R_{12} = disturbed lands; R_{13} = barren. Source: Authors' development.

Appendix D

Table A25. Ranking of T_j territories on a parameter of agricultural land activity, federal districts grouping.

Land Category	Parameter	Central	Northwestern	Southern	North Caucasian	Volga	Ural	Siberian	Far Eastern
L_1	$\overline{A_{jL1}}$	0.367	0.020	0.413	0.330	0.350	0.046	0.047	0.004
	$\overline{R_{j1}}$	59	20	56	46	57	29	28	11
	$\sum R_{j1}$	1011	197	389	321	799	174	333	97
L_2	$\overline{A_{jL2}}$	0.007	0.001	0.001	0.001	0.008	0.005	0.004	0.001
	$\overline{R_{j2}}$	35	13	18	17	37	37	32	20
	$\sum R_{j2}$	599	128	128	116	522	224	379	182
L_3	$\overline{A_{jL3}}$	0.008	0.001	0.007	0.010	0.003	0.001	0.000	0
	$\overline{R_{j3}}$	59	26	54	63	47	24	20	12
	$\sum R_{j3}$	1010	258	381	444	654	145	240	110
L_4	$\overline{A_{jL4}}$	0.040	0.011	0.020	0.033	0.035	0.017	0.018	0.004
	$\overline{R_{j4}}$	56	31	27	53	41	48	41	17
	$\sum R_{j4}$	947	311	186	370	572	289	495	151
L_5	$\overline{A_{jL5}}$	0.090	0.007	0.313	0.336	0.134	0.021	0.042	0.004
	$\overline{R_{j5}}$	42	17	64	75	48	29	41	13
	$\sum R_{j5}$	721	171	449	525	671	171	495	118
L_6	$\overline{A_{jL6}}$	0.363	0.549	0.070	0.115	0.377	0.400	0.578	0.552
	$\overline{R_{j6}}$	47	29	69	59	44	38	26	18
	$\sum R_{j6}$	804	294	483	416	617	226	315	166
L_7	$\overline{A_{jL7}}$	0.027	0.022	0.015	0.015	0.015	0.028	0.011	0.013
	$\overline{R_{j7}}$	23	36	48	49	40	54	52	42
	$\sum R_{j7}$	397	364	333	344	561	323	620	379
L_8	$\overline{A_{jL8}}$	0.020	0.062	0.052	0.023	0.024	0.098	0.033	0.033
	$\overline{R_{j8}}$	57	18	17	53	44	25	44	43
	$\sum R_{j8}$	974	180	117	368	621	150	526	385
L_9	$\overline{A_{jL9}}$	0.019	0.003	0.016	0.014	0.011	0.004	0.002	0.001
	$\overline{R_{j9}}$	21	57	23	24	30	49	60	70
	$\sum R_{j9}$	363	566	162	165	423	294	717	631

Table A25. *Cont.*

Land Category	Parameter	Central	Northwestern	Southern	North Caucasian	Volga	Ural	Siberian	Far Eastern
L_{10}	$\overline{A_{jL10}}$	0.022	0.005	0.016	0.017	0.017	0.005	0.003	0.001
	$\overline{R_{j10}}$	17	52	32	35	25	55	61	71
	$\sum R_{j10}$	286	520	226	246	346	328	733	636
L_{11}	$\overline{A_{jL11}}$	0.019	0.152	0.012	0.003	0.009	0.231	0.081	0.069
	$\overline{R_{j11}}$	48	14	53	74	60	14	32	19
	$\sum R_{j11}$	816	144	369	519	837	86	378	172
L_{12}	$\overline{A_{jL12}}$	0.003	0.001	0	0.001	0.001	0.001	0	0
	$\overline{R_{j12}}$	21	33	52	47	43	33	57	51
	$\sum R_{j12}$	361	330	361	329	601	200	683	456
L_{13}	$\overline{A_{jL13}}$	0.004	0.085	0.011	0.042	0.005	0.071	0.092	0.235
	$\overline{R_{j13}}$	54	41	44	20	53	45	33	15
	$\sum R_{j13}$	922	413	310	137	736	272	395	136
	$\sum \overline{R_{ji}}$ per district	542	388	556	614	569	480	526	402
	$\sum R_{ji}$ per district	9211	3876	3894	4300	7960	2882	6309	3619

Note: L_1 = cropland; L_2 = fallow; L_3 = perennial plantings; L_4 = hayfields; L_5 = rangeland; L_6 = woodlands; L_7 = forest range; L_8 = water reserve lands; L_9 = residential and industrial lands; L_{10} = lands under transportation and communication infrastructure; L_{11} = wetlands; L_{12} = disturbed lands; L_{13} = barren; $\overline{A_{jLi}}$ is averaged in respect to individual values of A_{jLi} in T_j territories per districts; $\overline{R_{ji}}$ is averaged in respect to individual rankings R_{ji} in T_j territories per districts; $\sum R_{ji}$ = sum of land activity rankings per districts. Source: Authors' development.

References

1. Klein Goldewijk, K.; Beusen, A.; Van Drecht, G.; De Vos, M. The HYDE 3.1 Spatially explicit database of human-induced global land-use change over the past 12,000 Years. *Glob. Ecol. Biogeogr.* **2011**, *20*, 73–86. [CrossRef]
2. Lambin, E.; Turner, B.; Geist, H.; Agbola, S.; Angelsen, A.; Bruce, J.; Coomes, O.; Dirzo, R.; Fischer, G.; Folke, C.; et al. The causes of land-use and land-cover change: Moving beyond the myths. *Glob. Environ. Chang.* **2001**, *11*, 261–269. [CrossRef]
3. Zhang, H.; Xu, E.; Zhu, H. Ecological-living-productive land classification system in China. *J. Resour. Ecol.* **2017**, *8*, 121–128.
4. Diogo, V. *Agricultural Land Systems: Explaining and Simulating Agricultural Land-Use Patterns*; Vrije Universiteit: Amsterdam, The Netherlands, 2018.
5. Bruinsma, J. *World Agriculture: Towards 2015/2030, an FAO Perspective*; Earthscan Publications: London, UK, 2003.
6. Gibbs, H.K.; Ruesch, A.S.; Achard, F.; Clayton, M.K.; Holmgren, P.; Ramankutty, N.; Foley, J.A. Tropical forests were the primary sources of new agricultural land in the 1980s and 1990s. *Proc. Natl. Acad. Sci. USA* **2010**, *107*, 16732–16737. [CrossRef]
7. Smith, P. Delivering food security without increasing pressure on land. *Glob. Food Secur.* **2013**, *2*, 18–23. [CrossRef]
8. Smith, P.; Gregory, P.; van Vuuren, D.; Obersteiner, M.; Rounsevell, M.; Woods, J.; Havlik, P.; Stehfest, E.; Bellarby, J. Competition for land. *Philos. Trans. R. Soc. B* **2010**, *365*, 2941–2957. [CrossRef]
9. Erokhin, V.; Gao, T. *Handbook of Research on Globalized Agricultural Trade and New Challenges for Food Security*; IGI Global: Hershey, PA, USA, 2020.
10. DeFries, R.S.; Rudel, T.; Uriarte, M.; Hansen, M. Deforestation driven by urban population growth and agricultural trade in the twenty-first century. *Nat. Geosci.* **2010**, *3*, 178–181. [CrossRef]
11. Ajani, J. The global wood market, wood resource productivity and price trends: An examination with special attention to China. *Environ. Conserv.* **2011**, *38*, 53–63. [CrossRef]
12. Lambin, E. Global land availability: Malthus versus Ricardo. *Glob. Food Secur.* **2012**, *1*, 83–87. [CrossRef]
13. Platt, R. The Loss of farmland: Evolution of public response. *Geogr. Rev.* **1977**, *67*, 93–101. [CrossRef]
14. Briggs, D.; Yurman, E. Disappearing farmland: A national concern. *Soil Conserv.* **1980**, *45*, 4–7.
15. Vining, D.; Plaut, T.; Beri, K. Urban encroachment on prime agricultural land in the United States. *Int. Reg. Sci. Rev.* **1977**, *2*, 143–156. [CrossRef]
16. Sioen, G.B.; Terada, T.; Sekiyama, M.; Yokohari, M. Resilience with mixed agricultural and urban land uses in Tokyo, Japan. *Sustainability* **2018**, *10*, 435. [CrossRef]
17. Ramadani, I.; Gashi, G.; Ejupi, A.; Bytyqi, V. The environment impacts of power plants in Kosovo and sustainable development. *J. Int. Environ. Appl. Sci.* **2011**, *6*, 332–338.
18. O'Meara, M. *Reinventing Cities for People and the Planet*; Worldwatch Institute: Washington, DC, USA, 1999.
19. Mather, A.S.; Needle, C.L. The forest transition: A theoretical basis. *Area* **1998**, *30*, 117–124. [CrossRef]
20. Lerman, Z.; Shagaida, N. Land policies and agricultural land markets in Russia. *Land Use Policy* **2007**, *24*, 14–23. [CrossRef]
21. Fadeeva, O.; Soliev, I. Institutional analysis of land tenure system in post-socialist Russia: Actors, Rules and Land Use. In *KULUNDA: Climate Smart Agriculture. Innovations in Landscape Research*; Frühauf, M., Guggenberger, G., Meinel, T., Theesfeld, I., Lentz, S., Eds.; Springer: Cham, Switzerland, 2020; pp. 259–273.
22. Pismennaya, E.; Loshakov, A.; Shevchenko, D.; Odincov, S.; Kipa, L. Comprehensive approach for evaluating the potential of the Stavropol agricultural territory. *Int. J. Econ. Financ. Issues* **2015**, *5*, 113–120.
23. Trukhachev, V.; Ivolga, A.; Lescheva, M. Enhancement of land tenure relations as a factor of sustainable agricultural development: Case of Stavropol Krai, Russia. *Sustainability* **2015**, *7*, 164–179. [CrossRef]
24. Rozhkov, V. The Analysis of the Factors that influence on the land market of the Russian Federation. *Russ. Entrep.* **2016**, *21*, 2939–2952.
25. Visser, O.; Spoor, M.; Mamonova, N. Is Russia the emerging global "breadbasket"? re-cultivation, agroholdings and grain production. *Eur. -Asia Stud.* **2014**, *10*, 1589–1610. [CrossRef]
26. Smirnov, V.; Mulendeeva, A. A Structural analysis of land use in Russia. *Natl. Interests Priorities Secur.* **2019**, *15*, 1057–1074. [CrossRef]

27. Prishchepov, A.; Müller, D.; Baumann, M.; Kuemmerle, T.; Alcantara, C.; Radeloff, V.C. Underlying Drivers and Spatial Determinants of Post-Soviet Agricultural Land Abandonment in Temperate Eastern Europe. In *Land-Cover and Land-Use Changes in Eastern Europe after the Collapse of the Soviet Union in 1991*; Gutman, G., Radeloff, V., Eds.; Springer: Cham, Switzerland, 2017; pp. 91–117.
28. Milanova, E. Land use/cover Change in Russia within the Context of Global Challenges. *Rom. J. Geogr.* **2012**, *56*, 105–116.
29. Prishchepov, A.; Schierhorn, F.; Dronin, N.; Ponkina, E.; Müller, D. 800 Years of Agricultural Land-Use Change in Asian (Eastern) Russia. In *KULUNDA: Climate Smart Agriculture. Innovations in Landscape Research*; Frühauf, M., Guggenberger, G., Meinel, T., Theesfeld, I., Lentz, S., Eds.; Springer: Cham, Switzerland, 2020; pp. 67–87.
30. Grishchenko, M. *Land Use Changes in Russia and Their Impact on Migrating Geese*; Wageningen University & Research: Wageningen, The Netherlands, 2018.
31. Pismennaya, E.; Stukalo, V.; Kipa, L. Development of a project of land tenure in the Stavropol territory. *Sib. J. Life Sci. Agric.* **2016**, *11*, 74–88. [CrossRef]
32. Stanfield, D. *Creation of Land Markets in Transition Countries: Implications for the Institutions of Land Administration*; University of Wisconsin-Madison: Madison, WI, USA, 1999.
33. Milanova, E.; Lioubimtseva, E.; Tcherkashin, P.; Yanvareva, L. Land use/cover change in Russia: Mapping and GIS. *Land Use Policy* **1999**, *16*, 153–159. [CrossRef]
34. Griewald, Y.; Clemens, G.; Kamp, J.; Gladun, E.; Hölzel, N.; von Dressler, H. Developing land use scenarios for stakeholder participation in Russia. *Land Use Policy* **2017**, *68*, 264–276. [CrossRef]
35. United Nations Convention to Combat Decertification. Sustainable Land Management (SLM). Available online: https://knowledge.unccd.int/topics/sustainable-land-management-slm (accessed on 30 May 2020).
36. Small, C.; Pozzi, F.; Elvidge, C.D. Spatial Analysis of Global Urban Extents from the DMSP-OLS Night-Lights. *Remote Sens. Environ.* **2005**, *96*, 277–291. [CrossRef]
37. Elnaggar, A. Spatial and temporal changes in agricultural lands eastern Nile-delta, Egypt. *J. Soil Sci. Agric. Eng.* **2013**, *4*, 187–201. [CrossRef]
38. Baker, J.M.; Everett, Y.; Liegel, L.; Van Kirk, R. Patterns of irrigated agricultural land conversion in a western U.S. watershed: Implications for landscape-level water management and land-use planning. *Soc. Nat. Resour.* **2014**, *27*, 1145–1160. [CrossRef]
39. Prishchepov, A.; Müller, D.; Dubinin, M.; Baumann, M.; Radeloff, V. Determinants of agricultural land abandonment in post-soviet European Russia. *Land Use Policy* **2013**, *30*, 873–884. [CrossRef]
40. Brueckner, J.K. Urban sprawl: Diagnosis and remedies. *Int. Reg. Sci. Rev.* **2000**, *23*, 160–171. [CrossRef]
41. Brown, D.G.; Johnson, K.M.; Loveland, T.R.; Theobald, D.M. Rural land-use trends in the conterminous United States, 1950–2000. *Ecol. Appl.* **2005**, *15*, 1851–1863. [CrossRef]
42. Erma, B.; Zhang, Y.; Yu, M.; Shu, J. Study of the Russian land market. *Nat. Inn. Asia* **2018**, *6*, 77–85.
43. Verburg, P.H.; Schot, P.P.; Dijst, M.J.; Veldkamp, A. Land use change modelling: Current practice and research priorities. *GeoJournal* **2004**, *61*, 309–324. [CrossRef]
44. Van Doorn, A.; Bakker, M. The Destination of Arable Land in a Marginal Agricultural Landscape in South Portugal: An Exploration of Land Use Change Determinants. *Landsc. Ecol.* **2007**, *22*, 1073–1087. [CrossRef]
45. Nainggolan, D.; de Vente, J.; Boix-Fayos, C.; Termansen, M.; Hubacek, K.; Reed, M. Afforestation, agricultural abandonment and intensification: Competing trajectories in semi-arid mediterranean agro-ecosystems. *Agric. Ecosyst. Environ.* **2012**, *159*, 90–104. [CrossRef]
46. Diogo, V.; Koomen, E. Land-use change in Portugal 1990–2006: Main processes and underlying factors. *Cartographica* **2012**, *47*, 237–248. [CrossRef]
47. Stacherzak, A.; Heldak, M.; Hajek, L.; Przybyla, K. State interventionism in agricultural land turnover in Poland. *Sustainability* **2019**, *11*, 1534. [CrossRef]
48. Lovell, S.T. Multifunctional Urban Agriculture for Sustainable Land Use Planning in the United States. *Sustainability* **2010**, *2*, 2499–2522. [CrossRef]
49. Moss, M. Land processes and land classification. *J. Environ. Manag.* **1985**, *20*, 295–319.
50. Rowe, S. Soil, Site and Land Classification. *For. Chron.* **1962**, *38*, 420–432. [CrossRef]
51. Shagaida, N. On criticism of agricultural land classification. *Russ. Agric. Sci.* **2007**, *33*, 59–61. [CrossRef]
52. Nosov, S. Fundamentals of productive land classification for suitability for agricultural use. *Econ. Labor Manag. Agric.* **2018**, *43*, 49–54. [CrossRef]

53. Macht, V.; Makenova, S.; Karpova, O. The analysis of the existing method of land classification. *Vestn. Voronezh State Agrar. Univ.* **2017**, *52*, 253–258. [CrossRef]
54. Loshakov, A. Methods and results of zoning of agricultural landscapes on the susceptibility to degradation processes and suitability for agricultural land use in the Stavropol territory. *Mosc. Econ. J.* **2019**, *11*, 48–57.
55. DeMers, M. Land classification research: A retrospective and agenda. *Int. J. Appl. Geospat. Res.* **2014**, *5*, 82–92. [CrossRef]
56. Federal State Statistics Service. Regions of Russia. Social and Economic Indicators. Available online: https://www.gks.ru/folder/210/document/13204 (accessed on 11 February 2020).
57. Federal State Statistics Service. All-Russian Agricultural Census. 2016. Available online: https://www.gks.ru/519 (accessed on 11 February 2020).
58. Nikonova, G.; Trafimov, A. All-russian agricultural census as a source of information about the development of agricultural sector. *Bull. St. -Petersburg State Agrar. Univ.* **2017**, *49*, 207–212.
59. Federal Service for State Registration, Cadastre and Cartography. Availability and Allocation of Lands in the Russian Federation. Available online: https://rosreestr.ru/site/activity/sostoyame-zemerrossii/gosudarstvennyy-natsionalnyy-doklad-o-sostoyanii-i-ispolzovanii-zemel-v-rossiyskoy-federatsii/ (accessed on 11 February 2020).
60. Wright, L.E.; Zitzmann, W.; Young, K.; Googins, R. LESA—agricultural land evaluation and site assessment. *J. Soil Water Conserv.* **1983**, *38*, 82–86.
61. Cocks, K.D.; Ive, J.R.; Davis, J.R.; Baird, I.A. SIRO-PLAN and LUPLAN: An australian approach to land-use planning. 1. The SIRO-PLAN land-use planning method. *Environ. Plan. B Plan. Des.* **1983**, *10*, 331–345. [CrossRef]
62. Burrough, P.A.; Frank, A.U. *Geographic Objects with Indeterminate Boundaries*; Taylor and Francis: London, UK, 1996.
63. Munroe, D.K.; Croissant, C.; York, A. Land use policy and landscape fragmentation in an urbanizing region: Assessing the impact of zoning. *Appl. Geogr.* **2005**, *25*, 121–141. [CrossRef]
64. Grčman, H.; Vozel, S.; Zupanc, V. Soil Characteristics and agricultural land evaluation. *Geod. Vestn.* **2017**, *61*, 13–22. [CrossRef]
65. Odak, I.; Tomić, H.; Mastelić Ivić, S. Valuation of agricultural land fragmentation. *Geod. List* **2017**, *71*, 215–232.
66. Boitt, M.K.; Langat, F.C.; Kapoi, J.K. Geospatial agro-climatic characterization for assessment of potential agricultural areas in Somalia, Africa. *J. Agric. Inform.* **2018**, *9*, 18–35. [CrossRef]
67. Lata, S. *Irrigation Water Management for Agricultural Development in Uttar Pradesh, India*; Springer: Cham, Switzerland, 2019.
68. Espinosa, J.; Moreno, J. Agricultural Land Use. In *The Soils of Ecuador*; Espinosa, J., Moreno, J., Bernal, G., Eds.; Springer: Cham, Switzerland, 2018; pp. 151–162.
69. Mazurkin, P. Activity of land Resources in Russia Districts. In Proceedings of the International Forum "Actual Problems of Contemporary Land Management", Moscow, Russia, 25 December 2014.
70. Mazurkin, P.; Mihailova, S. Factor Analysis Categories of Land for the District Russia. In Proceedings of the International Forum "Actual Problems of Contemporary Land Management", Moscow, Russia, 25 December 2014.
71. Buckett, M. *An Introduction to Farm Organization and Management*; Elsevier: New York, NY, USA, 1988.
72. Artamonova, I.; Baturina, I.; Mikhajluk, O.; Poverinova, E. Improving methodologies of assessing the efficiency of agricultural land use. *Adv. Soc. Sci. Educ. Humanit. Res.* **2020**, *392*, 121–124.
73. Stupen, R.; Stupen, M.; Ryzhok, Z.; Stupen, O. Modeling of the effective functioning of the agricultural lands market in Ukraine. *Geod. Cartogr.* **2019**, *45*, 96–101. [CrossRef]
74. Shishkina, N.; Yushkova, V.; Mamistova, E. Assessment of the use of the potential of agricultural lands in Voronezh region. *Kne Life Sci.* **2019**, *14*, 467–477. [CrossRef]
75. Yerseitova, A.; Issakova, S.; Jakisheva, L.; Nauryzbekova, A.; Moldasheva, A. Efficiency of using agricultural land in Kazakhstan. *Entrep. Sustain. Issues* **2018**, *6*, 558–576. [CrossRef]
76. Kotykova, O.; Kuzmenko, O.; Semenchuk, I. Sustainable agricultural land use in the post-socialist camp countries: Monitoring and evaluation. *Balt. J. Econ. Stud.* **2019**, *5*, 101–111. [CrossRef]
77. Zhildikbaeva, A.; Sabirova, A.; Pentaev, T.; Omarbekova, A. Improving the agricultural land use system in the Republic of Kazakhstan. *J. Environ. Manag. Tour.* **2018**, *9*, 1585–1592. [CrossRef]

78. Koomen, E.; Rietveld, P.; De Nijs, T. Modelling land-use change for spatial planning support. *Ann. Reg. Sci.* **2008**, *42*, 1–10. [CrossRef]
79. Alcamo, J.; Kok, K.; Busch, G.; Priess, J. Searching for the future of land: Scenarios from the local to global scale. In *Land-Use and Land-Cover Change: Local Processes and Global Impacts*; Lambin, E., Geist, H., Eds.; Springer-Verlag: Berlin, Germany, 2006; pp. 71–116.
80. Lavalle, C.; Baranzelli, C.; Mubareka, S.; Rocha Gomes, C.; Hiederer, R.; Batista e Silva, F.; Estreguil, C. *Implementation of the CAP Policy Options with the Land Use Modelling Platform: A First Indicator-Based Analysis*; Publications Office of the European Union: Luxembourg, 2011.
81. Bakker, M.M.; Govers, G.; Kosmas, C.; Venacker, V.; Van Oost, K.; Rounsevell, M. Soil erosion as a driver of land-use change. *Agric. Ecosyst. Environ.* **2005**, *105*, 467–481. [CrossRef]
82. Hatna, E.; Bakker, M. Abandonment and expansion of arable land in europe. *Ecosystems* **2011**, *14*, 720–731. [CrossRef]
83. Millington, J.; Perry, G.; Romero-Calcerrada, R. Regression Techniques for examining land use/cover change: A case study of a Mediterranean landscape. *Ecosystems* **2007**, *10*, 562–578. [CrossRef]
84. Feranec, J.; Jaffrain, G.; Soukup, T.; Hazeu, G. Determining changes and flows in European landscapes 1990–2000 using CORINE Land cover data. *Appl. Geogr.* **2010**, *30*, 19–35. [CrossRef]
85. Filzmoser, P.; Hron, K. Correlation analysis of compositional data. *Math. Geosci.* **2009**, *41*, 905–919. [CrossRef]
86. Aitchison, J. A Concise Guide to Compositional Data Analysis. Available online: http://ima.udg.edu/activitats/codawork05/A_concise_guide_to_compositional_data_analysis.pdf (accessed on 8 May 2020).
87. Pawlowsky-Glahn, V.; Egozcue, J.J.; Tolosana-Delgado, R. Lecture Notes on Compositional Data Analysis. Available online: https://www.semanticscholar.org/paper/Lecture-Notes-on-Compositional-Data-Analysis-Pawlowsky-Glahn-Egozcue/89dac9d573264a0fa6a550f9dd8ee8f6d13f2178 (accessed on 7 May 2020).
88. Greenacre, M. *Compositional Data Analysis in Practice*; Chapman and Hall/CRC: London, UK, 2018.
89. Aitchison, J. *The Statistical Analysis of Compositional Data*; Chapman and Hall: London, UK, 1986.
90. Pawlowsky-Glahn, V.; Egozcue, J.J.; Tolosana-Delgado, R. *Modeling and Analysis of Compositional Data*; Wiley: Hoboken, NJ, USA, 2015.
91. Aitchison, J. The statistical analysis of compositional data. *J. R. Stat. Soc. Ser. B (Methodol.)* **1982**, *44*, 139–177. [CrossRef]
92. Long, W.; Wang, Q. Two methods of correlation coefficient on compositional data. *Procedia Comput. Sci.* **2013**, *18*, 1757–1763. [CrossRef]
93. Van den Boogaart, G.; Tolosana-Delgado, R. *Analyzing Compositional Data with R*; Springer-Verlag: Berlin/Heidelberg, Germany, 2013.
94. Egozcue, J.J.; Pawlowsky-Glahn, V.; Mateu-Figueraz, G.; Barceló-Vidal, C. Isometric Logratio Transformations for compositional data analysis. *Math. Geol.* **2003**, *35*, 279–300. [CrossRef]
95. Egozcue, J.J.; Pawlowsky-Glahn, V. Groups of parts and their balances in compositional data analysis. *Math. Geol.* **2005**, *37*, 795–828. [CrossRef]
96. Korhonová, M.; Hron, K.; Klimcíková, D.; Müller, L.; Bednář, P.; Barták, P. Coffe aroma—Statistical analysis of compositional data. *Talanta* **2009**, *80*, 710–715. [CrossRef]
97. Thió-Henestrosa, S.; Martín-Fernández, J.A. Dealing with Compositional Data: The Freeware CoDaPack. *Math. Geol.* **2005**, *37*, 773–793. [CrossRef]
98. Egozcue, J.J.; Pawlowsky-Glahn, V. Compositional data: The sample space and its structure. *Test* **2019**, *28*, 599–638. [CrossRef]
99. Muriithi, F.K. Centered Log-Ratio (clr) Transformation and robust principal component analysis of long-term NDVI data reveal vegetation activity linked to climate processes. *Climate* **2015**, *3*, 135–149.
100. Bichler, B.; Haering, A.M.; Dabbert, S. The determinants of the spatial distribution of organic farming in Germany. *Ber. Uber Landwirtsch. -Hambg.* **2005**, *83*, 50–75.
101. Chu, D. Spatial Distribution of Land-Use Types. In *Remote Sensing of Land Use and Land Cover in Mountain Region*; Chu, D., Ed.; Springer: Singapore, 2020; pp. 67–80.
102. Smith, O.; Cohen, A.; Reganold, J.; Jones, M.; Orpet, R.; Taylor, J.; Thurman, J.; Cornell, K.; Olsson, R.; Ge, Y.; et al. Landscape context affects the sustainability of organic farming systems. *Proc. Natl. Acad. Sci. USA* **2020**, *117*, 2870–2878. [CrossRef] [PubMed]
103. Bakker, M.M.; Hatna, E.; Kuhlman, T.; Mücher, C. Changing environmental characteristics of European cropland. *Agric. Syst.* **2011**, *104*, 522–532. [CrossRef]

104. Rounsevell, M.D.A.; Annetts, J.E.; Audsley, E.; Mayr, T.; Reginster, I. Modelling the spatial distribution of agricultural land use at the regional level. *Agric. Ecosyst. Environ.* **2003**, *95*, 465–479. [CrossRef]
105. White, R.; Engelen, G. Cellular dynamics and GIS: Modelling spatial complexity. *Geogr. Syst.* **1994**, *1*, 237–253.
106. White, R.; Engelen, G. Cellular automata as the basis of integrated dynamic regional modelling. *Environ. Plan. B* **1997**, *24*, 235–246. [CrossRef]
107. Shagaida, N. Agricultural land market in Russia: Living with constraints. *Comp. Econ. Stud.* **2005**, *47*, 127–140. [CrossRef]
108. Franks, J.R.; Davydova, I. Reforming the farming sector in Russia: New options for old problems. *Outlook Agric.* **2005**, *34*, 97–103. [CrossRef]
109. Patsiorkovsky, V.; O'Brien, D.; Wergen, S.K. Land reform and land relations in rural Russia. *East. Eur. Countrys.* **2005**, *11*, 5–18.
110. Ioffe, G.; Nefedova, T.; de Beurs, K. Agrarian transformation in the Russian breadbasket: Contemporary trends as manifest in Stavropol. *Post-Sov. Aff.* **2014**, *30*, 441–463. [CrossRef]
111. Van de Steeg, J.A.; Verburg, P.H.; Baltenweck, I.; Staal, S.J. Characterization of the spatial distribution of farming systems in the Kenyan highlands. *Appl. Geogr.* **2010**, *30*, 239–253. [CrossRef]
112. Gärtner, D.; Keller, A.; Schulin, R. A simple regional downscaling approach for spatial distributing land use types for agricultural land. *Agric. Syst.* **2013**, *120*, 10–19. [CrossRef]
113. Daniels, T. *When City and Country Collide: Managing Growth in the Metropolitan Fringe*; Island Press: Washington, DC, USA, 1999.
114. Su, S.; Jiang, Z.; Zhang, Q.; Zhang, Y. Transformation of agricultural landscapes under rapid urbanization: A threat to sustainability in Hang-Jia-Hu Region, China. *Appl. Geogr.* **2011**, *31*, 439–449. [CrossRef]
115. Yeh, C.-T.; Huang, S.-L. Investigating spatiotemporal patterns of landscape diversity in response to urbanization. *Landsc. Urban Plan.* **2009**, *93*, 151–162. [CrossRef]
116. Dredge, D. Sustainable rapid urban expansion: The case of Xalapa, Mexico. *Habitat Int.* **1995**, *19*, 317–329. [CrossRef]
117. Parsipour, H.; Popović, S.; Behzadfar, M.; Skataric, G.; Spalevic, V. Cities expansion and land use changes of agricultural and garden lands in peri-urban villages (case study: Bojnurd). *Agric. For.* **2019**, *65*, 173–187. [CrossRef]
118. Li, E.; Endter-Wada, J.; Li, S. Dynamics of Utah's agricultural landscapes in response to urbanization: A comparison between irrigated and non-irrigated agricultural lands. *Appl. Geogr.* **2019**, *105*, 58–72. [CrossRef]
119. Al-Kofahi, S.; Nammouri, N.; Sawalhah, M.; Al-Hammouri, A.; Aukour, F. Assessment of the urban sprawl on agriculture lands of two major municipalities in Jordan using supervised classification techniques. *Arab. J. Geosci.* **2018**, *11*, 45. [CrossRef]
120. Zubair, O.; Ji, W.; Festus, O. Urban expansion and the loss of prairie and agricultural lands: A satellite remote-sensing-based analysis at a sub-watershed scale. *Sustainability* **2019**, *11*, 4673. [CrossRef]
121. Lucero, L.; Tarlock, A.D. Water supply and urban growth in New Mexico: Same old, same old or a new era. *Nat. Resour. J.* **2003**, *43*, 803–835.
122. Staal, S.J.; Baltenweck, I.; Waithaka, M.M.; de Wolff, T.; Njoroge, L. Location and uptake: Integrated household and GIS analysis of technology adoption and land use, with application to smallholder dairy farms in Kenya. *Agric. Econ.* **2002**, *27*, 295–315. [CrossRef]
123. Ramadani, I.; Bytyqi, V. Processes affecting sustainable use of agricultural land in Kosovo. *Quaest. Geogr.* **2018**, *37*, 53–66. [CrossRef]
124. Meyfroidt, P.; Schierhorn, F.; Prishchepov, A.; Müller, D.; Kuemmerle, T. Drivers, constraints and trade-offs associated with recultivating abandoned cropland in Russia, Ukraine and Kazakhstan. *Glob. Environ. Chang.* **2016**, *37*, 1–15. [CrossRef]
125. Nguyen, H.; Hölzel, N.; Völker, A.; Kamp, J. Patterns and determinants of post-soviet cropland abandonment in the Western Siberian Grain Belt. *Remote Sens.* **2018**, *10*, 1973. [CrossRef]
126. Oyebanji, I.J.; Adeniyi, B.; Khobai, H.; Le Roux, P. Green Growth and Environmental Sustainability in Nigeria. *Int. J. Energy Econ. Policy* **2017**, *7*, 216–223.
127. Deng, X.; Li, Z. Economics of Land Degradation in China. In *Economics of Land Degradation and Improvement—A Global Assessment for Sustainable Development*; Nkonya, E., Mirzabaev, A., von Braun, J., Eds.; Springer: Cham, Switzerland, 2016; pp. 385–399.
128. Müller, D.; Sikor, T. Effects of postsocialist reforms on land cover and land use in south-eastern Albania. *Appl. Geogr.* **2006**, *26*, 175–191. [CrossRef]

129. Müller, D.; Kuemmerle, T.; Rusu, M.; Griffiths, P. Lost in transition: Determinants of post-socialist cropland abandonment in Romania. *J. Land Use Sci.* **2009**, *4*, 109–129. [CrossRef]
130. Sorokin, A.; Bryzzhev, A.; Strokov, A.; Mirzabaev, A.; Johnson, T.; Kiselev, S. The economics of land degradation in Russia. In *Economics of Land Degradation and Improvement—A Global Assessment for Sustainable Development*; Nkonya, E., Mirzabaev, A., von Braun, J., Eds.; Springer: Cham, Switzerland, 2016; pp. 541–576.
131. Kashtanov, A. Concept of sustainable development of agriculture of Russia in XXI century. *Eurasian Soil Sci.* **2001**, *3*, 263–265.
132. Dobrovolski, G. *Soil Degradation and Preservation*; Moscow State University: Moscow, Russia, 2002.
133. Solgerel, P.; Narantuya, A.; Amarmend, C.; Erdenechimeg, A.; Purvee, L. Assessment of vulnerability for land degradation. *J. Adv. Nat. Sci.* **2018**, *5*, 438–443.
134. MacDonald, D.; Crabtree, J.R.; Wiesinger, G.; Dax, T.; Stamou, N.; Fleury, P.; Lazpita, J.G.; Gibon, A. Agricultural abandonment in mountain areas of Europe: Environmental consequences and policy response. *J. Environ. Manag.* **2000**, *59*, 47–69. [CrossRef]
135. Nakvasina, E.; Parinova, T.; Romanov, E.; Volkov, A.; Golubeva, L.; Popova, A. Monitoring of agricultural lands in Arkhangelsk Region. *Iop Conf. Ser. Earth Environ. Sci.* **2019**, *263*, 012025. [CrossRef]
136. Postek, P.; Leń, P.; Stręk, Ż. The proposed indicator for fragmentation of agricultural land. *Ecol. Indic.* **2019**, *103*, 581–588. [CrossRef]
137. Popov, A. Assessment of land fragmentation of agricultural enterprises in Ukraine. *Econ. Ann. -XXI* **2017**, *164*, 56–60. [CrossRef]
138. Merot, A.; Belhouchette, H. Hierarchical patch dynamics perspective in farming system design. *Agronomy* **2019**, *9*, 604. [CrossRef]
139. Ripoche, A.; Rellier, J.P.; Martin-Clouaire, R.; Paré, N.; Biarnès, A.; Gary, C. Modelling adaptive management of intercropping in vineyards to satisfy agronomic and environmental performances under Mediterranean climate. *Environ. Model. Softw.* **2011**, *26*, 1467–1480. [CrossRef]
140. Nefedova, T. *Ten Topical Issues about Rural Russia: A Geographer's Viewpoint*; LENAND: Moscow, Russia, 2013.
141. King, R.; Burton, S. Land fragmentation: Notes on a fundamental rural spatial problem. *Prog. Hum. Geogr.* **1982**, *6*, 475–494. [CrossRef]
142. Tan, S.; Heerink, N.; Qu, F. Land fragmentation and its driving forces in China. *Land Use Policy* **2006**, *23*, 272–285. [CrossRef]
143. Dhakal, B.N.; Khanal, N.R. Causes and consequences of fragmentation of agricultural land: A case of Nawalparasi district, Nepal. *Geogr. J. Nepal* **2018**, *11*, 95–112. [CrossRef]
144. Gill, A.R.; Viswanathan, K.K.; Hassan, S. Is environmental Kuznets curve (EKC) still relevant? *Int. J. Energy Econ. Policy* **2017**, *7*, 156–165.
145. Hunt, D.V.L.; Lombardi, D.R.; Atkinson, S.; Barber, A.R.G.; Barnes, M.; Boyko, C.T.; Brown, J.; Bryson, J.; Butler, D.; Caputo, S.; et al. Scenario archetypes: Converging rather than diverging themes. *Sustainability* **2012**, *4*, 740–772. [CrossRef]
146. Diputra, E.M.; Baek, J. Is growth good or bad for the environment in Indonesia? *Int. J. Energy Econ. Policy* **2018**, *8*, 1–4.
147. Mahcene, Z.; Abdellah, M.; Zergoune, M.; Lacheheb, M. Land degradation and economic development in Algeria. *Int. J. Energy Econ. Policy* **2019**, *9*, 137–142.
148. Bukvareva, E.; Grunewald, K.; Bobylev, S.; Zamolodchikov, D.; Zimenko, A.; Bastian, O. The current state of knowledge of ecosystems and ecosystem services in Russia: A Status Report. *Ambio* **2015**, *44*, 491–507. [CrossRef] [PubMed]
149. Tchebakova, N.; Parfenova, E.; Lysanova, G.; Soja, A. Agroclimatic potential across central Siberia in an altered twenty-first century. *Environ. Res. Lett.* **2011**, *6*, 045207. [CrossRef]
150. Foley, J.A.; DeFries, R.; Asner, G.P.; Barford, C.; Bonan, G.; Carpenter, S.R.; Chapin, F.S.; Coe, M.T.; Daily, G.C.; Gibbs, H.K.; et al. Global consequences of land use. *Science* **2005**, *309*, 570–574. [CrossRef]
151. Janssen, M.A.; Anderies, J.M.; Ostrom, E. Robustness of social-ecological systems to spatial and temporal variability. *Soc. Nat. Resour.* **2007**, *20*, 307–322. [CrossRef]

MDPI
St. Alban-Anlage 66
4052 Basel
Switzerland
Tel. +41 61 683 77 34
Fax +41 61 302 89 18
www.mdpi.com

Land Editorial Office
E-mail: land@mdpi.com
www.mdpi.com/journal/land

www.ingramcontent.com/pod-product-compliance
Lightning Source LLC
Chambersburg PA
CBHW040752250926
51850CB00007BB/1480
* 9 7 8 3 0 3 6 5 4 0 4 9 8 *